SO-AXV-396

*Advances in*
# ORGANOMETALLIC CHEMISTRY

## VOLUME 49

# Advances in Organometallic Chemistry

EDITED BY

ROBERT WEST

DEPARTMENT OF CHEMISTRY
UNIVERSITY OF WISCONSIN
MADISON, WISCONSIN

ANTHONY F. HILL

AUSTRALIAN NATIONAL UNIVERSITY
RESEARCH SCHOOL OF CHEMISTRY
INSTITUTE OF ADVANCED STUDIES
CANBERRA, ACT, AUSTRALIA

FOUNDING EDITOR

F. GORDON A. STONE

VOLUME 49

ACADEMIC PRESS

An imprint of Elsevier Science

Amsterdam   Boston   Heidelberg   London   New York   Oxford
Paris   San Diego   San Francisco   Singapore   Sydney   Tokyo

Academic Press
*An imprint of Elsevier Science*
525 B Street, Suite 1900
San Diego, California 92101-4495
USA

© 2003 Elsevier Science (USA) All rights reserved.

This work is protected under copyright by Elsevier Science, and the following terms and conditions apply to its use:

Photocopying
Single photocopies of single chapters may be made for personal use as allowed by national copyright laws. Permission of the Publisher and payment of a fee is required for all other photocopying, including multiple or systematic copying, copying for advertising or promotional purposes, resale, and all forms of document delivery. Special rates are available for educational institutions that wish to make photocopies for non-profit educational classroom use.

Permissions may be sought directly from Elsevier's Science & Technology Rights Department in Oxford, UK: phone: (+44) 1865 843830, fax: (+44) 1865 853333, e-mail: permissions@ elsevier.com. You may also complete your request on-line via the Elsevier Science homepage (http:// www.elsevier.com), by selecting 'Customer Support' and then 'Obtaining Permissions'.

In the USA, users may clear permissions and make payments through the Copyright Clearance Center, Inc., 222 Rosewood Drive, Danvers, MA 01923, USA; phone: (+1) (978) 7508400, fax: (+1) (978) 7504744, and in the UK through the Copyright Licensing Agency Rapid Clearance Service (CLARCS), 90 Tottenham Court Road, London W1P 0LP, UK; phone: (+44) 207 631 5555; fax: (+44) 207 631 5500. Other countries may have a local reprographic rights agency for payments.

Derivative Works
Tables of contents may be reproduced for internal circulation, but permission of Elsevier Science is required for external resale or distribution of such material.
Permission of the Publisher is required for all other derivative works, including compilations and translations.

Electronic Storage or Usage
Permission of the Publisher is required to store or use electronically any material contained in this work, including any chapter or part of a chapter.

Except as outlined above, no part of this work may be reproduced, stored in a retrieval system or transmitted in any form or by any means, electronic, mechanical, photocopying, recording or otherwise, without prior written permission of the Publisher.
Address permissions requests to: Elsevier's Science & Technology Rights Department, at the phone, fax and e-mail addresses noted above.

Notice
No responsibility is assumed by the Publisher for any injury and/or damage to persons or property as a matter of products liability, negligence or otherwise, or from any use or operation of any methods, products, instructions or ideas contained in the material herein. Because of rapid advances in the medical sciences, in particular, independent verification of diagnoses and drug dosages should be made.

First edition 2003

ISBN: 0-12-031149-6
ISSN: 0065-3055

⊛ The paper used in this publication meets the requirements of ANSI/NISO Z39.48-1992 (Permanence of Paper). Printed in The Netherlands.

# Contents

## Decamethylsilicocene: Synthesis, Structure, Bonding and Chemistry

### THORSTEN KÜHLER and PETER JUTZI

## Organometallic Phosphorous and Arsenic Betaines

### NIKOLAI N. ZEMLYANSKY, IRINA V. BORISOVA and YURI A. USTYNYUK

## Coordination Compounds with Organoantimony and Sb_n Ligands

### HANS JOACHIM BREUNIG and IOAN GHESNER

# Ladder Polysilanes

## SOICHIRO KYUSHIN and HIDEYUKI MATSUMOTO

# Structure–Reactivity Relationships in the Cyclo-Oligomerization of 1,3-Butadiene Catalyzed by Zerovalent Nickel Complexes

## SVEN TOBISCH

# Group 13/15 Organometallic Compounds—Synthesis, Structure, Reactivity and Potential Applications

## STEPHAN SCHULZ

# Contents

# Contributors

*Numbers in parentheses indicate the pages on which the authors' contributors begin.*

THORSTEN KÜHLER and PETER JUTZI (1), Faculty of Chemistry, University of Bielefeld, Universitätsstraße 25, D-33615 Bielefeld, Germany

NIKOLAI N. ZEMLYANSKY and IRINA V. BORISOVA (35), A. V. Topchiev Institute of Petrochemical Synthesis, Russian Academy of Sciences (TIPS RAS), 29, Leninsky prospect, 119991-GSP-1, Moscow, Russian Federation

YURI A. USTYNYUK (35), Department of Chemistry, M. V. Lomonosov Moscow State University, Vorob'evy Gory, 119899 Moscow, Russian Federation

HANS JOACHIM BREUNIG and IOAN GHESNER (95), Institut für Anorganische und Physikalische Chemie (Fb 2), Universität Bremen, D-28334 Bremen, Germany

SOICHIRO KYUSHIN and HIDEYUKI MATSUMOTO (133), Department of Applied Chemistry, Faculty of Engineering, Gunma University, Kiryu, Gunma 376-8515, Japan

SVEN TOBISCH (167), Institut für Anorganische Chemie der Martin-Luther-Universität Halle-Wittenberg, Fachbereich Chemie, Kurt-Mothes-Straße 2, D-06120 Halle, Germany

STEPHAN SCHULZ (225), Institut für Anorganische Chemie der Universität Bonn, Gerhard-Domagk-Str. 1, D-53121 Bonn, Germany

# Decamethylsilicocene: Synthesis, Structure, Bonding and Chemistry

## THORSTEN KÜHLER and PETER JUTZI

*Faculty of Chemistry, University of Bielefeld,*
*Universitätsstraße 25, D-33615 Bielefeld, Germany*

## I

## INTRODUCTION

The metallocene era started with the synthesis of ferrocene and with the right description of the structure and bonding of this compound.[1] Only a few years after this epochal discovery, the first metallocenes of the group 14 elements tin and lead were synthesized, namely stannocene, $(H_5C_5)_2Sn$,[2] and plumbocene $(H_5C_5)_2Pb$;[3] the analogous germanium compound germanocene, $(H_5C_5)_2Ge$, was prepared nearly two decades later.[4]

Some years later, the introduction of the pentamethylcyclopentadienyl ligand into the π-complex chemistry of the group 14 elements has allowed the synthesis of the deca-methylmetallocenes of germanium, $(Me_5C_5)_2Ge$,[5] tin, $(Me_5C_5)_2Sn$[5] and lead, $(Me_5C_5)_2Pb$:[6] the enhanced stability but still high reactivity of these permethylated complexes has led to extensive reactivity studies and to the isolation and characterization of many derivatives.[7]

The decamethylmetallocenes of germanium, tin and lead as well as the parent metallocenes were prepared by reaction of divalent inorganic

1

©2003 Elsevier Science (USA)
All rights reserved.

substrates (usually halides) with the corresponding cyclopentadienyl transfer agents (usually alkaline metal derivatives). Later on, we could show that the decamethylmetallocenes of germanium and tin can also be prepared by reduction of the tetravalent bis(pentamethylcyclopentadienyl) dihalides of these elements under properly chosen reducing conditions to avoid over reduction and the elimination of the pentamethylcyclopentadienide ligands.[8] This interesting observation was the starting point for experiments with the aim to synthesize decamethylsilicocene, $(Me_5C_5)_2Si$ (**1**), in an analogous manner. At that time, no monomeric and under ordinary conditions stable inorganic or organometallic species with divalent silicon was known. Such species existed only as reactive and transient intermediates; some of them had been isolated in low-temperature matrices.[9] To change this situation, π-complexation[7] seemed to us to be a very promising tool. In this article, we summarize the results concerning synthesis, structure, bonding and chemistry of decamethylsilicocene (**1**), which was the first divalent silicon compound and at the same time also the first stable π-complex with silicon as the central atom. Since the synthesis of **1** in 1986,[10] the field of stable silicon(II) compounds has shown an interesting development. Meanwhile, at least four other classes of compounds have been described in the literature: one containing a hyper-coordinated silicon center(class II),[11] and three containing dicoordinated silicon centers (class IIIa,[12,13] IIIb,[13,14], IIIc,[13,15] class IV[13,16] and class V,[17] see Fig. 1). With regard to other π-complexes with silicon as the central atom, silicon(IV) compounds of

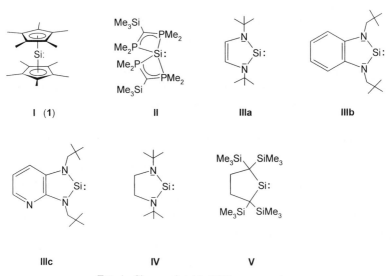

FIG. 1. Classes of stable Si(II) compounds.

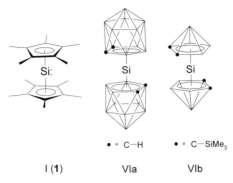

I (1)            VIa            VIb

FIG. 2. π-Complexes of silicon.

type VIa[18] and VIb[19] containing carbollide ligands were prepared shortly after the synthesis of **1** (see Fig. 2).

## II

## SYNTHESIS AND STRUCTURE OF DECAMETHYLSILICOCENE

As already mentioned, the tetravalent bis(pentamethylcyclopentadienyl)-silicon dihalides $(Me_5C_5)_2Si(Hal)_2$ (Hal = F (**2**), Cl (**3**), Br (**4**)) were used as precursors for the synthesis of **1**. Their preparation was straightforward only in the case of the difluoro compound **2**; due to the steric effects exerted by two pentamethylcyclopentadienyl substituents, the dichloro (**3**) and the dibromo compound **4** could not be synthesized by a simple metathesis route. They were prepared following a strategy where the fifth methyl group at the cyclopentadienyl ring had to be introduced in the last step of the procedure.[20] The difluoro compound **2** reacted with alkali metals or with alkali metal naphthalenides to give mainly the corresponding alkali metal cyclopentadienides and a Si–F-containing polymer, thus indicating that in **2** the pentamethylcyclopentadienyl unit is a better leaving group than the fluoro substituent. In the reaction of the dichloro compound **3** with lithium-, sodium-, or potassium-naphthalenide, a grey-black suspension was formed. After filtration, elemental silicon remained as residue. From the colorless solution, **1** could be isolated after removal of the solvent and fractional sublimation. The highest yield of **1** was obtained by performing the reaction at −50 °C in dimethoxyethane as the solvent and using a 50% excess of the sodium compound.[10] The undesired formation of elemental silicon was avoided by using the dibromo compound **4** as the precursor and potassium anthracenide as the reducing agent.[10] A clean reduction to

**1** without any formation of silicon was observed in the reaction of **3** with decamethylsamarocene.[21] It was found only recently that the reduction of 1,2-bis(pentamethylcyclopentadienyl)tetrachlorodisilane (**5**) with alkali metal naphthalenides leads to **1** and to elemental silicon in stoichiometric amounts.[22] The different strategies for the preparation of **1** are collected in Scheme 1.

Crystallization from *n*-pentane gives colorless crystals of **1**, which are soluble in all common aprotic organic solvents. Compound **1** is monomeric in benzene solution, sensitive towards hydrolysis, but stable in air for short periods of exposure. It melts at 171 °C without decomposition, but decomposes under MOCVD-conditions to elemental silicon at about 600 °C. At room temperature **1** is regarded to be indefinitely persistent in the solid state and in solution; in toluene solution it survives unchanged after heating to 110° for several days.

Cyclic voltammetry measurements in dichloromethane as solvent have shown that **1** cannot be reduced in the region available (up to $-1.7$ V versus SCE). An irreversible oxidation takes place at $+0.4$ V; presumably the $(Me_5C_5)_2Si^+$ radical cation (**1**$^+$) is formed which is unstable due to the easy loss of the pentamethylcyclopentadienyl radical. The fate of the remaining $Me_5C_5Si^+$ cation is uncertain. Further irreversible oxidation processes are observed in the region from $+0.8$ to $+1.5$ V. In the mass spectrum of **1** (EI and CI), the molecular ion $(Me_5C_5)_2Si^+$ (**1**$^+$) is not observed; the fragment with the highest mass ($m/z = 163$) corresponds to the $Me_5C_5Si^+$ cation. These observations complete those of the CV studies and demonstrate that **1**$^+$ is rather unstable not only in solution, but also in the gas phase.[10]

The results of an X-ray crystal structure analysis[10] of **1** are presented in Figs. 3 and 4. Surprisingly, two geometrical isomers, **1a** and **1b**, in the ratio 1:2 are present in the unit cell (space group $C2/c$; $Z = 12$). Isomer **1a** is isostructural with decamethylferrocene, and the silicon lone-pair is not

$(Me_5C_5)_2SiCl_2$ **3** + 2 $MC_{10}H_8$ $\xrightarrow[\substack{-2\ MCl \\ -2\ C_{10}H_8}]{}$ $(Me_5C_5)_2Si$ **1** (+ Si + $MC_5H_5$)

$(Me_5C_5)_2SiBr_2$ **4** + 2 $KC_{14}H_{10}$ $\xrightarrow[\substack{-2\ KBr \\ -2\ C_{14}H_{10}}]{}$ $(Me_5C_5)_2Si$ **1**

$(Me_5C_5)_2SiCl_2$ **3** + 2 $(Me_5C_5)_2Sm$ $\xrightarrow{\hspace{2cm}}$ 2 $(Me_5C_5)_2SmCl$ + $(Me_5C_5)_2Si$ **1**

$(Me_5C_5)_2Si_2Cl_4$ **5** + 4 $MC_{10}H_8$ $\xrightarrow[\substack{-4\ MCl \\ -4\ C_{10}H_8}]{}$ $(Me_5C_5)_2Si$ **1** + Si

SCHEME 1. Syntheses of decamethylsilicocene (**1**).

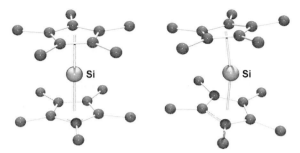

FIG. 3. Molecular structure of the linear (**1a**) ($D_{5d}$) and of the bent (**1b**) ($C_{2v}$) isomer of decamethylsilicocene in the solid state.

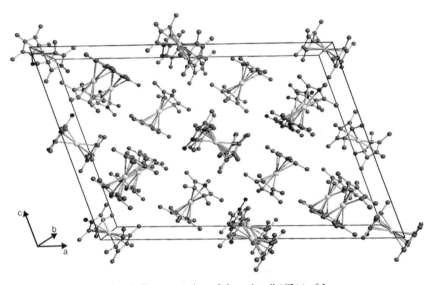

FIG. 4. Representation of the unit cell ($C2/c$) of **1**.

stereochemically active. Isomer **1b** is of the bent-metallocene type with an interplane angle of 25.3° and with pentamethylcyclopentadienyl rings asymmetrically bonded in a staggered conformation; the silicon lone-pair is regarded to be stereochemically active. The Si–C separations are equidistant in **1a** [2.42(1) Å], but different in **1b** [ranging from 2.323(7) to 2.541(7) Å]. The distance between the silicon atom and the cyclopentadienyl ring centroids is 2.11 Å in **1a** and 2.12 Å in **1b**. Space-filling models clearly demonstrate the interplane angle in **1b** to be of the largest possible value. According to GED studies, **1** has a bent-metallocene type structure in the gas phase with an interplane angle of 22.3°.

TABLE I

X-RAY CRYSTAL STRUCTURE AND GED DATA OF THE
DECAMETHYLMETALLOCENES OF SILICON, GERMANIUM, TIN, AND
LEAD

| Compound | d $[\text{Å}]^a$ | $r_{El-C}$ $[\text{Å}]^b$ | $\alpha$ $[°]^c$ |
|---|---|---|---|
| $(Me_5C_5)_2Si$ | 2.11 | 2.42(1) | 0 |
|  | 2.12 | 2.42(6) | 25.3 |
| $(Me_5C_5)_2Ge^d$ | 2.21 | 2.52 | 23 |
| $(Me_5C_5)_2Sn$ | 2.39 | 2.68 | 36 |
| $(Me_5C_5)_2Pb$ | 2.48 | 2.79 | 43 |

$^a$Distance from central atom to ring centroid.
$^b$Averaged El–C distances.

TABLE II

NMR DATA OF THE DECAMETHYLMETALLOCENES OF SILICON, GERMANIUM, TIN, AND LEAD

| Compound | $El_{(solid)}{}^a$ | $El_{(solv)}$ | $^1H$ | $^{13}C$ |
|---|---|---|---|---|
| $(Me_5C_5)_2Si$ | −403.2, −423.4 | −398.0 | 1.89 | 10.0, 119.1 |
| $(Me_5C_5)_2Ge$ | – | – | 1.99 | 9.8, 118.1 |
| $(Me_5C_5)_2Sn$ | −2136.6, −2140.2 | −2146 | 2.06 | 10.5, 117.0 |
| $(Me_5C_5)_2Pb$ | −4474 | −4390 | 2.18 | 10.1, 117.4 |

$^a$ $^{29}Si$-, $^{119}Sn$-, $^{207}Pb$- CP-MAS-NMR of powder samples.

A comparison of some important structural data of the decamethylmetallocenes of silicon, germanium, tin, and lead is given in Table I. As expected, the distance from the respective central atom to the cyclopentadienyl ring centroid or to the ring carbon atoms rises on going to the heavier homologues. In the same direction (with the exception of **1b**), a widening of the angle between the cyclopentadienyl ring planes is observed (see discussion of the bonding situation). The implications of these structural features on bonding and reactivity are discussed elsewhere.[7]

The solid-state CP-MAS $^{29}Si$ NMR spectrum of **1** reflects the gross structural features known from the X-ray analysis.[23] The nuclear shielding in **1a** ($\delta = -423.4$ ppm) is found to be higher ($\Delta\delta = 20.2$ ppm) than that in **1b** ($\delta = -403.2$ ppm) (see also Table II). Interestingly, the measured chemical shift values for **1** are at the high-field end of the $^{29}Si$ NMR scale! The observed small shielding anisotropies $\Delta\sigma$ can only be explained if one assumes that the silicon lone-pair in **1a** or **1b** provides a fairly homogeneous source of electron density for the silicon nucleus.

Information concerning the structure of **1** in solution stems from $^1H$, $^{13}C$, and $^{29}Si$ NMR spectra.[10] They show that only one isomer is present and that this isomer is highly dynamic. A resonance at $\delta = -398$ ppm in the $^{29}Si$ NMR spectrum indicates the presence of the bent structure **1b**, and averaged signals in the $^1$H and in the $^{13}$C NMR spectrum for the ring carbons and for the methyl groups correspond to very rapidly rotating π-pentamethylcyclopentadienyl ligands. The dynamic behavior of **1** is similar to that of the heavier homologues. The NMR data of all the group 14 decamethylmetallocenes are collected in Table II; they are in accord with the general observation for π-complexes of main-group elements, that heteronuclear resonances for the central atoms (here for Si, Sn, and Pb) appear at very high field and that fluxionality within the element-cyclopentadienyl unit leads to averaged $^1$H and $^{13}$C NMR chemical shifts.[7,24,25]

## III

## BONDING IN DECAMETHYLSILICOCENE

Information about the bonding stems from calculations for the parent but still experimentally unknown silicocene molecule ($Si(C_5H_5)_2$) and for the permethylated derivative **1** as well as from the He(I) spectrum of the latter compound. Several calculations have been performed for silicocene, some of them probably at insufficient levels of theory. It is evident that the silicocene potential surface is rather flat with respect to the interconversion between several conformers and to the easy rotation of the cyclopentadienyl rings. At the SCF DZP level, the molecule is predicted to adopt a bent $C_s$ structure. However, the energy difference between the low symmetry conformers ($C_s$, $C_2$, $C_{2v}$) is only 2.4 kcal mol$^{-1}$; the $D_{5d}$ conformer is 8.8 kcal mol$^{-1}$ higher in energy. The total electron population at the silicon center is calculated to be 13.5; d-orbitals at silicon do not play an important role in π-bonding. The Si-(πCp) dissociation energy has been calculated to be in the range of 55 kcal mol$^{-1}$.[26] Quantum chemical calculations for **1** have been performed at the DFT level without any geometrical constraints in the energy optimization.[27] The computed structure is in perfect agreement with the solid state structure of the bent isomer. An NBO analysis of silicocene ($D_{5d}$) assigns a +0.852 charge to the silicon atom; for comparison, a +0.117 charge is calculated for the iron atom in the ferrocene molecule.

On a more qualitative level, the bonding in the more stable isomer **1b** can be explained on the basis of the general molecular orbital scheme for bent ($C_{2v}$) metallocenes containing 14 valence electrons, as shown in Fig. 5. The localization of three electron pairs in bonding orbitals ($1_{a1}$, $2_{b1}$, $2_{b2}$) is primarily responsible for the Si–Cp interaction; the absence of a silicon orbital of $a_2$ symmetry imposes the presence of a ligand-based non-bonding orbital. Structural adjustment from $D_{5d}$ (ferrocene type) to $C_{2v}$

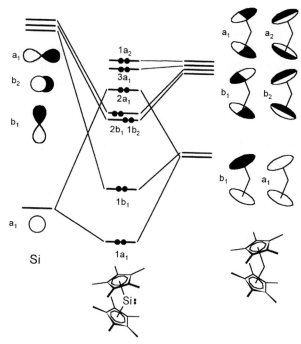

FIG. 5. Schematic MO diagram of decamethylsilicocene (**1**) (only occupied MO's are shown).

TABLE III

PES DATA OF THE DECAMETHYLMETALLOCENES OF SILICON, GERMANIUM, TIN, AND LEAD

| Orbital | $(Me_5C_5)_2Si$ | $(Me_5C_5)_2Ge$ | $(Me_5C_5)_2Sn$ | $(Me_5C_5)_2Pb$ |
|---------|-----------------|-----------------|-----------------|-----------------|
| $1_{a2}$ | 6.70 | 6.60 | 6.60 | 6.33 |
| $3_{a1}$ | 6.96 | 6.75 | 6.60 | 6.88 |
| $2_{b1}$ | 8.06 | 7.91 | 7.64 | 7.38 |
| $1_{b2}$ | 8.30 | 8.05 | 7.64 | 7.38 |
| $2_{a1}$ | 7.50 | 8.36 | 8.40 | 8.93 |

symmetry gives rise to a stabilization of the $2_{a1}$ orbital as a consequence of the second-order Jahn-Teller effect (incorporation of the $a_1$ p orbital at silicon). In more simplified terms, the resultant $2_{a1}$ orbital may be regarded as a non-bonding electron pair, the so-called "lone pair".

Experimental support for the MO sequence in Fig. 5 is provided by the photoelectron spectroscopic (PE) data of **1**,[10] which are presented in Table III together with the corresponding data for the heavier homologues. The following conclusions can be drawn from these data: (1) the HOMO's in all metallocenes are non-bonding with respect to element–Cp interactions; (2) the $2_{b1}$ and $1_{b2}$ type orbitals provide the strongest contributions to element–Cp bonding, the respective ionization energies depending strongly on the element involved; (3) the "lone-pair" orbital is rather low in energy for the germanium, tin, and lead compounds, but is higher in energy for the silicon compound. The observed differences in energy have strong implications on the chemical reactivity. The "lone pair" in **1**, which is located in the region of the frontier orbitals, is involved in nearly all reactions performed so far. This is not the case in the chemistry of the heavier homologues.[7]

Summarizing the available bonding information, decamethylsilicocene (**1**) is regarded as an electron-rich silicon(II) compound containing a hypercoordinated silicon atom which is sandwiched between two rather weakly π-bonded pentamethylcyclopentadienyl ligands and thus is effectively shielded; the "lone-pair" orbital at silicon is part of the frontier orbitals of the molecule.

## IV

## CHEMISTRY OF DECAMETHYLSILICOCENE (1)

Experiments performed during the last fifteen years have shown that the chemistry of **1** is mainly determined by the nucleophilicity of the silicon "lone pair" and by the weakness of the π-bonds between silicon and the pentamethylcyclopentadienyl ligands. The "lone-pair" activity is responsible for the donor qualities of **1**. Easy haptotropic shifts ($\eta^5$–$\eta^1$) of the cyclopentadienyl ligands cause a vacant orbital at silicon and thus create a silylene-like reaction center; this bonding situation enables oxidative addition and cycloaddition processes. The inherent weakness of the Si–σCp bonds in the primary reaction products might cause migration, rearrangement or elimination reactions; as a result, rather complicated and surprising reaction pathways are observed in some cases. In the following, the chemistry of **1** is described in more detail.

### A. Attempts to Prepare the Radical Cation and the Radical Anion of 1

Whereas decamethylferrocene—here regarded as the prototype of a metallocene—can be easily transferred to the corresponding radical cation or anion, this is not the case for decamethylsilicocene (**1**). CV measurements have already shown that the cation **1**$^+$ is only a very short-lived transient species.[10] Consequently, chemical experiments to prepare salts containing **1**$^+$ have also failed so far. Similar observations have been made concerning

$(Me_5C_5)_2Si$ **1** + 2 $MC_{10}H_8$ $\longrightarrow$ Si + 2 $MC_5Me_5$ + 2 $C_{10}H_8$

$(Me_5C_5)_2Si$ **1** + 2 $MC_{14}H_{10}$ $\xrightarrow{\quad//\quad}$

SCHEME 2. Reduction of **1**.

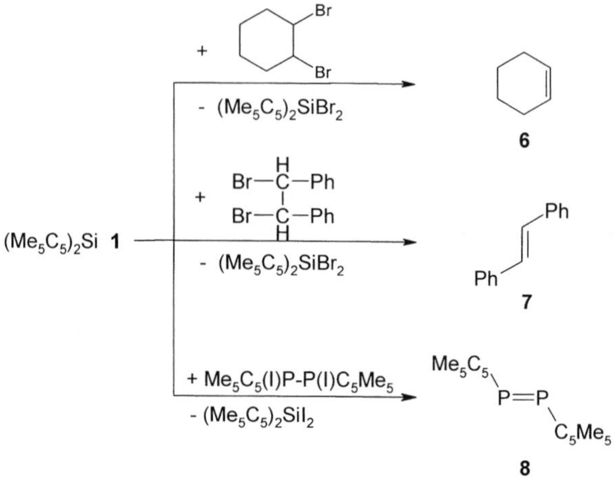

SCHEME 3. **1** as reducing agent for vicinal dihalogeno compounds.

the reduction of **1**. It was not possible to stabilize or even to identify the anion $1^-$, not even in experiments with tunable reducing systems. Thus, reaction with stoichiometric amounts of alkali metal naphthalenides resulted in the quantitative formation of elemental silicon and of the respective alkali metal pentamethylcyclopentadienide, whereas alkali metal anthracenides did not react at all[10] (see Scheme 2). The different reactivity of decamethylferrocene and of **1** can be understood with the help of their MO schemes. Whereas the HOMO in the ferrocene is metal-centered, the HOMO in **1** is centered at the ligands. As a result, single-electron transfer reagents attack compound **1** preferentially at the pentamethylcyclopenta-dienyl ligands and lead to decomposition. This behavior of **1** is characteristic of cyclopentadienyl compounds of p-block elements.[28]

## B. Reduction of Geminal and Vicinal Dihalogeno Compounds

In the reaction of **1** with vicinal and geminal dihalogeno compounds, dehalogenation under formation of the corresponding bis(pentamethylcy-clopentadienyl)dihalogenosilane takes place under mild reaction conditions. In Scheme 3, examples for the reduction of vicinal dibromo- or

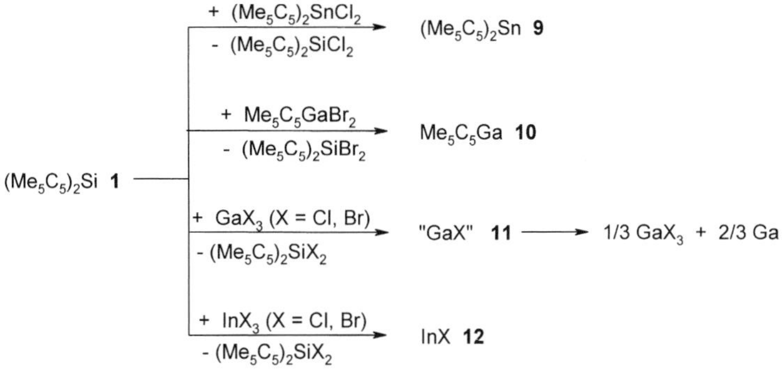

$$(Me_5C_5)_2Si \;\; \mathbf{1} \longrightarrow$$

$$\xrightarrow[- (Me_5C_5)_2SiCl_2]{+ \; (Me_5C_5)_2SnCl_2} \;\; (Me_5C_5)_2Sn \;\; \mathbf{9}$$

$$\xrightarrow[- (Me_5C_5)_2SiBr_2]{+ \; Me_5C_5GaBr_2} \;\; Me_5C_5Ga \;\; \mathbf{10}$$

$$\xrightarrow[- (Me_5C_5)_2SiX_2]{+ \; GaX_3 \; (X = Cl, \, Br)} \;\; \text{"GaX"} \;\; \mathbf{11} \longrightarrow 1/3 \; GaX_3 + 2/3 \; Ga$$

$$\xrightarrow[- (Me_5C_5)_2SiX_2]{+ \; InX_3 \; (X = Cl, \, Br)} \;\; InX \;\; \mathbf{12}$$

SCHEME 4. **1** as reducing agent for geminal dihalogeno compounds.

diiodo- compounds are presented, whereby species with element–element double bonds are generated. In organic chemistry, trans-1,2-dibromocyclo-hexane was converted into cyclohexene (**6**), and meso-1,2-dibromo-1,2-diphenylethane was transformed into trans-stilbene (**7**); in phosphorus chemistry, 1,2-diiodo-1,2-bis(pentamethylcyclopentadienyl)diphosphane was selectively reduced to bis(pentamethylcyclopentadienyl)diphosphene (**8**).[29]

In Scheme 4, examples for the dehalogenation of geminal dihalogeno compounds are collected. Thus, decamethylstannocene (**9**) was formed in the reaction of **1** with bis(pentamethylcyclopentadienyl)dichlorostannane.[29] Similarly, pentamethylcyclopentadienylgallium (**10**) was generated from the reaction of **1** with dibromo(pentamethylcyclopentadienyl)gallane.[30] The trichlorides and –bromides of gallium and indium were reduced to the corresponding monohalides **11** and **12**.[30] Of these, the gallium compounds **11** are metastable species,[31] which disproportionate to elemental gallium and to the corresponding gallium trihalide.

The examples from Schemes 3 and 4 demonstrate that **1** can be used in synthetic chemistry as an effective dehalogenating agent, especially when mild reaction conditions are required.

## C. Insertion into Non-polar and into Polar Bonds (X–X, H–X, El–Hal, El–C)

The insertion of **1** into element–element bonds is a crucial step in an extensive series of remarkable transformations. From the perspective of the particular substrate, such reactions are oxidative additions to a silylene-like center. The primary insertion products all possess reactive pentamethylcy-clopentadienyl-silicon $\sigma$-bonds. In this chapter only those insertion reactions are described, which are not followed by rather complicated rearrangement or elimination processes; the latter reactions are illustrated in Chapter IV.D. Typical examples for simple reactions or reaction sequences are collected in Schemes 5–9.

$(Me_5C_5)_2Si$ **1** $\xrightarrow{+ I_2}$ $(Me_5C_5)_2SiI_2$ **13**

$(Me_5C_5)_2Si$ **1** $\xrightarrow{+ RSSR}$ $(Me_5C_5)_2Si(SR)_2$ **14**

SCHEME 5. Insertion of **1** into non-polar bonds.

$(Me_5C_5)_2Si$ **1** + HX ⟶ $(Me_5C_5)_2Si(H)X$ **15 - 24**

| X | F | Cl | Br | $MeCO_2$ | $CF_3CO_2$ | $CF_3SO_3$ |
|---|---|---|---|---|---|---|
| | **15** | **16** | **17** | **18** | **19** | **20** |

| X | $C_6H_5O$ | p-$MeC_6H_4S$ | Me(H)C=NO | tBu(H)N |
|---|---|---|---|---|
| | **21** | **22** | **23** | **24** |

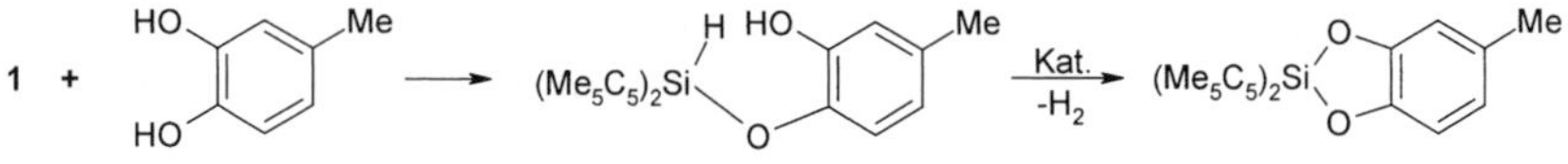


**25** **26**

$(Me_5C_5)_2Si$ **1** + 2 $HMn(CO)_5$ ⟶ $Me_5C_5Si(H)[Mn(CO)_5]_2$ **27** + $Me_5C_5H$

SCHEME 6. Insertion reactions of **1** with protic substrates.

$(Me_5C_5)_2Si$ **1** + $PX_3$ ⟶ $Me_5C_5(X)_2Si\text{-}P(X)C_5Me_5$

X = Cl: **28;** X = Br: **29**

SCHEME 7. Reaction of **1** with phosphorous trihalides.

$(Me_5C_5)_2Si$ **1** + $HgX_2$ ⟶ $(Me_5C_5)_2(X)SiHgX$ [X = Cl: **30**; X = Br: **31**]

2 $(Me_5C_5)_2Si$ **1** + $HgX_2$ ⟶ $[(Me_5C_5)_2(X)Si]_2Hg$ **32** [X = Cl, Br]

$(Me_5C_5)_2Si$ **1** + $H_5C_5FeC_5H_4HgCl$ ⟶ $(Me_5C_5)_2(Cl)SiHg(H_4C_5FeC_5H_5)$ **33**

$(Me_5C_5)_2Si$ **1** + $(H_5C_5)Ni(PPh_3)Cl$ ⟶ $(Me_5C_5)_2(Cl)SiNiC_5H_5$ **34** + $PPh_3$

$(Me_5C_5)_2Si$ **1** + LAuCl ⟶ $(Me_5C_5)_2(Cl)SiAuL$

L = $PPh_3$: **35**; CNR: **36**; $NC_5H_5$: **37**; $SC_4H_8$ **38**

SCHEME 8. Reaction of **1** with metal halides.

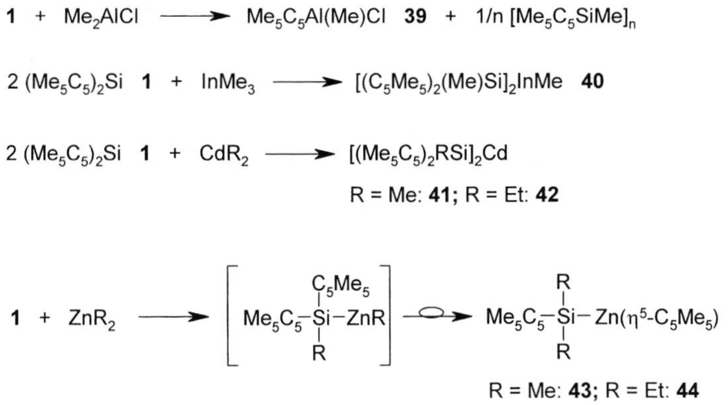

$\mathbf{1}$ + Me$_2$AlCl $\longrightarrow$ Me$_5$C$_5$Al(Me)Cl $\mathbf{39}$ + 1/n [Me$_5$C$_5$SiMe]$_n$

2 (Me$_5$C$_5$)$_2$Si $\mathbf{1}$ + InMe$_3$ $\longrightarrow$ [(C$_5$Me$_5$)$_2$(Me)Si]$_2$InMe $\mathbf{40}$

2 (Me$_5$C$_5$)$_2$Si $\mathbf{1}$ + CdR$_2$ $\longrightarrow$ [(Me$_5$C$_5$)$_2$RSi]$_2$Cd

R = Me: $\mathbf{41}$; R = Et: $\mathbf{42}$

$$\mathbf{1} \; + \; ZnR_2 \; \longrightarrow \; \left[ Me_5C_5 \overset{C_5Me_5}{\underset{R}{-Si-ZnR}} \right] \; \longrightarrow \; Me_5C_5 \overset{R}{\underset{R}{-Si-}} Zn(\eta^5\text{-}C_5Me_5)$$

R = Me: $\mathbf{43}$; R = Et: $\mathbf{44}$

Scheme 9. Reaction of $\mathbf{1}$ with metal alkyls.

## 1. Insertion into X–X Bonds

Rather weak element–element bonds are required for the insertion of $\mathbf{1}$, so that this type of reaction seems to be only of marginal importance. Two examples[29] are presented in Scheme 5. Compound $\mathbf{1}$ reacted with iodine to give bis(pentamethylcyclopentadienyl)diiodosilane ($\mathbf{13}$) in good yields, although this compound could not be prepared following the classical metathesis route. It is worth mentioning that the reaction of $\mathbf{1}$ with bromine led under Me$_5$C$_5$–Si bond splitting to the formation of Me$_5$C$_5$SiBr$_3$, thus demonstrating the leaving-group character of the pentamethylcyclopenta-dienyl group. The bis(pentamethylcyclopentadienyl)diarylthiolatosilane $\mathbf{14}$ was prepared in high yields from the reaction of $\mathbf{1}$ with the corresponding diaryldisulfide.

## 2. Insertion into H–X Bonds

The insertion of $\mathbf{1}$ into H–X bonds of protic substrates is of much greater importance, as documented by the many examples, which are collected in Scheme 6. Insertion reactions led in high yields and under mild conditions to products of the type $(\sigma\text{-Me}_5C_5)_2$Si(H)X, in which three different types of leaving groups are present (H, X, Me$_5$C$_5$).[32] In the reaction with the protic species HF, HCl, HBr, EtCO$_2$H, F$_3$CCO$_2$H, and F$_3$CSO$_3$H, the compounds $\mathbf{15–20}$ were formed in high yields; the pyridine adduct H$_5$C$_5$N · HF was used as a source of HF. Phenol and $p$-methyl-phenylthiol added to $\mathbf{1}$ to give the insertion products $\mathbf{21}$ and $\mathbf{22}$, respectively, and with ethanaloxime compound $\mathbf{23}$ was formed. Interestingly, insertion was also observed into the less protic N–H bond of tert-butylamine to give compound $\mathbf{24}$, but only under more vigorous conditions.[33] The reaction with 4-methyl-catechol led to the phenoxysilane **25**, which contains a protic OH and a hydridic SiH group; elimination of dihydrogen and formation of the *o*-phenylenedioxysilane **26** was observed in a catalyzed process.[32] An unexpected compound was obtained from the reaction of **1** with HMn(CO)$_5$. Complex **27** containing two Si–Mn bonds and pentamethylcyclopentadienyl was formed independently from the stoichiometry of the reactants.[33] The molecular structure of **27** is presented in Fig. 7.

### 3. Insertion into El–Hal Bonds

Many compounds of p- and d-block elements containing El–Cl or El–Br bonds react with **1** under insertion into these bonds. In some cases, the formed insertion products are stable under ordinary conditions. In other cases, the insertion products undergo further sometimes rather complicated rearrangement and elimination reactions; such situations are described in Chapter IV.D.

Insertion into a P–Hal bond was observed in the reaction of **1** with PCl$_3$ or PBr$_3$ which resulted in the formation of the silylphophanes **28** and **29**, as described in Scheme 7.[33] In contrast, reaction with the corresponding arsenic and antimony trihalides led under pentamethylcyclopentadienyl group transfer to the formation of the pentamethylcyclopentadienylelement dihalides Me$_5$C$_5$ElHal$_2$ and of polymeric pentamethylcyclopentadienyl (halogeno)silicon compounds with undefined structure.[33]

Several examples have been found where **1** inserts in a clean reaction into a metal-halogen bond, as described in Scheme 8. Thus, the monosilyl mercury compounds **30** and **31** were obtained from the reaction of **1** with the corresponding mercury dihalides in a 1:1 ratio. A 2:1 ratio led to the formation of the disilylmercury compound **32** in nearly quantitative yield. Similarly, reaction of ferrocenylmercury chloride gave the ferrocenyl(silyl)mercury compound **33**.[34] The molecular structures of **30** and **32** are presented in Fig. 6. Insertion into the Ni–Cl bond under simultaneous elimination of PPh$_3$ was observed in the reaction of **1** with the nickel complex H$_5$C$_5$Ni(PPh$_3$)Cl (see Scheme 8); NMR and X-ray crystal structure data have shown that in the product (Me$_5$C$_5$)$_2$(Cl)SiNiC$_5$H$_5$ (**34**) one of the two pentamethylcyclopentadienyl substituents at silicon is further η$^2$-bonded to the nickel atom, which represents a novel bonding situation in the cyclopentadienyl chemistry of p-block elements.[35] Insertion of **1** into the Au–Cl bond of several (Donor)AuCl complexes led to the formation of the compounds **35–38** of the type (Me$_5$C$_5$)$_2$(Cl)SiAu(Donor) (see Scheme 8).[35] Reaction of **1** with the gold complex (CO)AuCl proceeded under extrusion of CO; the thermally unstable and very air-sensitive reaction product was trapped with the donor molecule PPh$_3$ to give compound **35**.[35]

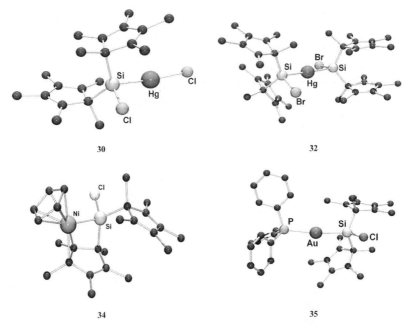

FIG. 6.  Structures of **30**, **32**, **34**, and **35**.

From this result it has been concluded that the reactive intermediate is an insertion product with a structure similar to that of the nickel compound **34** and not a silylene complex as postulated in an earlier publication.[36] The molecular structures of **34** and **35** are presented in Fig. 6.

NMR investigations have shown that **1** reacts in toluene solution with $ZnCl_2$, $ZnBr_2$, and $CdCl_2$ to give the corresponding insertion products of the type $(Me_5C_5)_2(Hal)SiMHal$ (M = Zn, Cd). Concomitant decomposition processes prevent the isolation of the pure compounds.[35]

### 4. Insertion into El–C(alkyl) Bonds

Certain metal alkyl compounds from p- and d-block elements react under very mild conditions with **1** under insertion into the element-carbon bond. Some examples are shown in Scheme 9.

In the low-temperature reaction of **1** with dimethylaluminum chloride insertion into the Al–Cl bond was expected. Instead, insertion into a Al–C bond took place, as concluded from the reaction products. After the insertion process, a reductive elimination step led to the formation of the known aluminum compound **39**[37] and of an undefined polymer of the composition $(Me_5C_5SiMe)_n$.[30] This surprising result has initiated further experiments with other group 13 element alkyls. The low-temperature reaction with trimethylaluminum led to several products, from which only the compound $Me_5C_5AlMe_2$ could be identified unequivocally.[38] In the reaction with trimethylgallium, the highly sensitive Ga–C insertion product $(Me_5C_5)_2MeSiGaMe_2$ was obtained nearly quantitatively.[38] With trimethyl-indium, insertion into two of the three In–C bonds took place to give the air-sensitive disilylindium compound **40**, irrespective of the stoichiometry of the reagents;[39] the molecular structure of **40** is depicted in Fig. 7. A clean insertion was also observed in the reaction of **1** with dimethyl- and diethylcadmium, whereby the air-sensitive silylcadmium compounds **41** and **42** were formed in high yields.[40] Similarly, an insertion took place in the reaction of **1** with dimethyl- and diethylzinc; but in these cases, the mono-insertion products were subject to dyotropic rearrangements involving exchange of an alkyl and of a pentamethylcyclopentadienyl substituent to give the silyl($\eta^5$-pentamethylcyclopentadienyl)zinc complexes **43** and **44**.[40] A dyotropic rearrangement is an uncatalyzed process in which two $\sigma$-bonded substituents migrate simultaneously and intramolecularly. Such processes are rather often observed in the chemistry of cyclopentadienyl compounds of p-block elements.[28]

## 5. Reaction with BrCN, Me₃SiCN, and MeHal

Insertion into a polar carbon-containing bond as present in BrCN, $Me_3SiCN$, and in MeHal (Hal = Br, I) led to the products **45**, **46**,[41] **47**, and **48**[29] (Scheme 10). All transformations described were performed under rather mild conditions. The formation of **45** and **46** most likely proceeds via cycloaddition products as reactive intermediates (see Chapter IV.H). Interestingly, **1** did not react with EtBr, EtI, $i$-PrBr, $i$-PrI, and $t$-BuBr, presumably due to steric reasons. Furthermore, **1** did not insert into

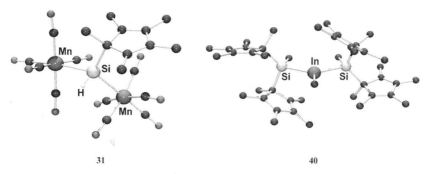

31                                        40

FIG. 7. Structure of **31** and **40**.

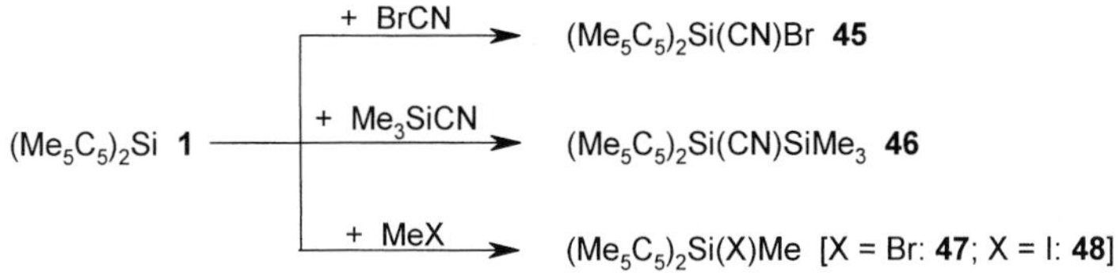

$(Me_5C_5)_2Si$ **1** — 

+ BrCN → $(Me_5C_5)_2Si(CN)Br$ **45**

+ $Me_3SiCN$ → $(Me_5C_5)_2Si(CN)SiMe_3$ **46**

+ MeX → $(Me_5C_5)_2Si(X)Me$  [X = Br: **47**; X = I: **48**]

SCHEME 10.  Reaction of **1** with BrCN, $Me_3SiCN$, and MeHal.

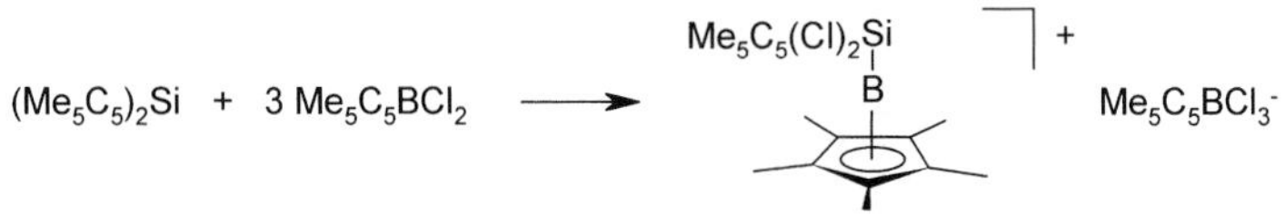

$(Me_5C_5)_2Si$  +  $3\ Me_5C_5BCl_2$  ⟶

$$\left[ \begin{array}{c} Me_5C_5(Cl)_2Si \\ | \\ B \end{array} \right]^+  Me_5C_5BCl_3^-$$

**49**

SCHEME 11.  Reaction of **1** with $Me_5C_5BCl_2$.

(aryl)C–Br and into C–Cl and C–F bonds,[42] in accord with the behavior of many transient silylenes.

The ability to insert in many element–element bonds is an important property of **1**; the $\eta^5$–$\eta^1$ rearrangement of the pentamethylcyclopentadienyl ligands during the reaction is a prerequisite to show a silylene-type reactivity. From a preparative point of view it is worth mentioning that element–silicon bonds which otherwise are difficult to form are easily accessible with the help of **1**. In addition, the leaving group character of the pentamethylcyclopentadienyl substituents allows further chemical transformations (vide infra).

## D. *Reaction with $Me_5C_5BCl_2$, $BHal_3$, $B_2Cl_4$, $AlHal_3$, and $Me_5C_5AlCl_2$*

The pathways in the reaction of **1** with certain halogenoboron and -aluminum compounds are much more complicated than those described in the last chapter. Here, addition and insertion reactions take place in combination with multistep rearrangement and elimination processes. Some reactions are still not fully understood concerning the mechanistic details, but plausible reaction sequences have been suggested.

The reaction of **1** with $Me_5C_5BCl_2$ needed a 1 : 3 stoichiometry to give the boron compounds **49** and $(Me_5C_5)_2BCl$ in high yields (Scheme 11). Species **49** was X-ray structurally characterized and consists of the pentacarba-nido-hexaboronium cation $Me_5C_5(Cl)_2Si–BC_5Me_5^+$ and of the $Me_5C_5BCl_3^-$ anion. The proposed reaction pathway includes several addition, insertion, elimination and rearrangement steps.[30]

$(Me_5C_5)_2Si$ **1** + $\xrightarrow[\substack{CH_2Cl_2 \\ hexane}]{BBr_3}$ $Me_5C_5(X)_2Si\text{-}BC_5Me_5\text{-}Si(X)_2C_5Me_5$ **50**

X = Br

SCHEME 12. Reaction of **1** with $BBr_3$ in $CH_2Cl_2$/hexane.

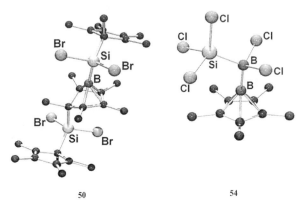

50                              54

FIG. 8. Structure of **50** and **54**.

$(Me_5C_5)_2Si$ **1** + $\xrightarrow[\substack{toluene \\ X = Cl, Br}]{BX_3}$ $[Me_5C_5(X)_2SiB(C_5Me_5)]^+ BX_4^-$ **51**

+ $Me_5C_5(X)_2Si\text{-}BC_5Me_5\text{-}Si(X_2)C_5Me_5)$ **50** + $Me_5C_5BX_2$

SCHEME 13. Reaction of **1** with boron trihalides $BX_3$ in toluene.

The reaction of **1** with the boron trihalides $BCl_3$ and $BBr_3$ turned out to be even more complex. At least three different types of compounds were formed, and the product ratio depended on the polarity of the solvent.[30] In the reaction with $BBr_3$ in dichloromethane/hexane (2:1), the boron compound **50** (X = Br) was isolated as the main product (Scheme 12); X-ray crystal structure analysis revealed the presence of a novel arachno-type cluster possessing a $BC_4$ framework (Fig. 8).

In the reaction with $BCl_3$ or $BBr_3$ in toluene as solvent, ionic compounds of type **51** were isolated as the main products together with the arachno-clusters **50** (X = Cl, Br) and the compounds $Me_5C_5BX_2$ as byproducts (Scheme 13). The nido cluster cation in **51** is identical with that in compound **48**.

Insertion and rearrangement steps were also observed in the reaction of **1** with $B_2Cl_4$[43] (Scheme 14). Compounds **52** and **53** were characterized by NMR spectroscopic or mass-spectrometric data, compound **54** by an X-ray crystal structure analysis. **54** may be regarded as a borane-stabilized boranediyl, consisting of a $Me_5C_5B$: nido-cluster unit (Fig. 8).

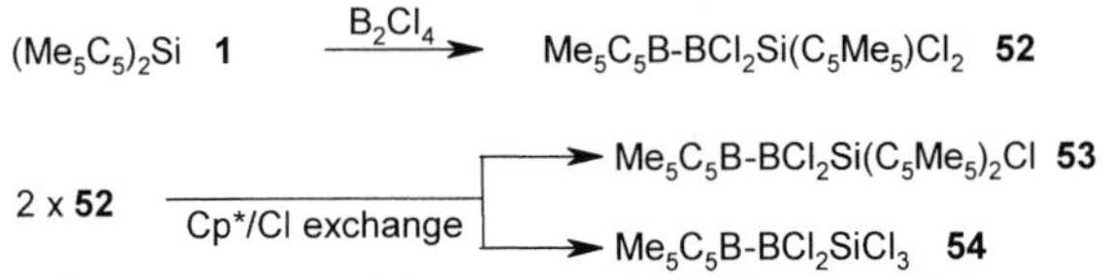

$(Me_5C_5)_2Si$ **1** $\xrightarrow{B_2Cl_4}$ $Me_5C_5B\text{-}BCl_2Si(C_5Me_5)Cl_2$ **52**

2 x **52** $\xrightarrow{\text{Cp*/Cl exchange}}$ 
- $\rightarrow Me_5C_5B\text{-}BCl_2Si(C_5Me_5)_2Cl$ **53**
- $\rightarrow Me_5C_5B\text{-}BCl_2SiCl_3$ **54**

SCHEME 14. Reaction of **1** with $B_2Cl_4$.

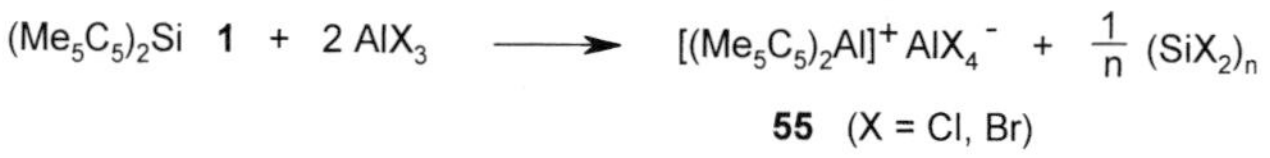

$(Me_5C_5)_2Si$ **1** $+$ $2\,AlX_3$ $\longrightarrow$ $[(Me_5C_5)_2Al]^+ AlX_4^- + \frac{1}{n}(SiX_2)_n$

**55** (X = Cl, Br)

SCHEME 15. Reaction of **1** with aluminum trihalides $AlX_3 (X = Cl, Br)$.

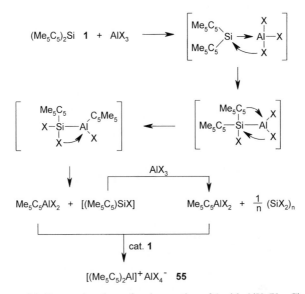

SCHEME 16. Proposed pathway for the reaction of **1** with $AlX_3(X = Cl, Br)$.

Compounds **53** and **54** presumably arise from a $Me_5C_5/Cl$ exchange between two molecules **52** (Scheme 14).

The reaction of **1** with $AlCl_3$ or $AlBr_3$ in a 1:1 stoichiometry led to a mixture of unidentified products. In the reaction with two equivalents of the aluminum trihalides, the known decamethylaluminocenium salts **55**[44] were formed in high yield together with a silicon-containing polymer of unknown structure (Scheme 15).[30] A proposal for the reaction sequence is given in Scheme 16; once more, insertion and elimination reactions as well as rearrangement processes have to be discussed. Interestingly enough, decamethylsilicocene (**1**) acts as a catalyst in the dismutation of $Me_5C_5AlCl_2$. This effect has been proved in an independent experiment (Scheme 17).[30]

$(Me_5C_5AlCl_2)_2 \xrightarrow{\text{cat. 1}} [(Me_5C_5)_2Al]^+AlCl_4^-$  **55** (X = Cl)

SCHEME 17. Catalytic dismutation of $Me_5C_5AlCl_2$.

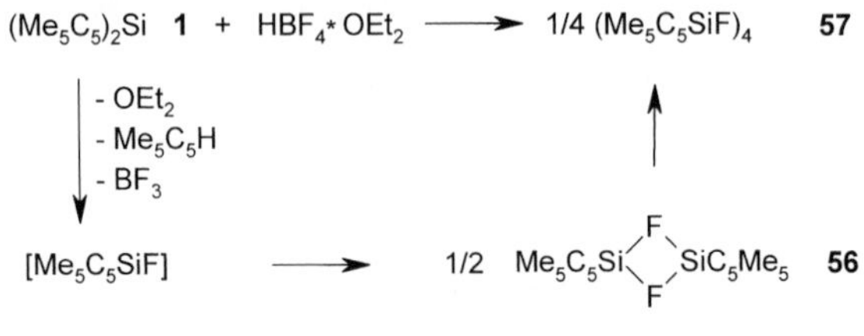

$(Me_5C_5)_2Si$ **1** + $HBF_4 \cdot OEt_2 \longrightarrow 1/4 (Me_5C_5SiF)_4$  **57**

$\downarrow$
- OEt$_2$
- Me$_5$C$_5$H
- BF$_3$

$[Me_5C_5SiF] \longrightarrow 1/2\ Me_5C_5Si\underset{F}{\overset{F}{\diamond}}SiC_5Me_5$  **56**

SCHEME 18. Reaction of **1** with $HBF_4 \cdot OEt_2$

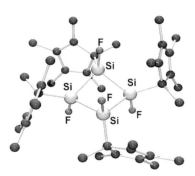

FIG. 9. Structure of **57**.

## E. Reaction with HBF₄ · OEt₂

The decamethylmetallocenes of the heavier congeners germanium, tin, and lead are attacked by electrophiles exclusively at the π-system of one of the pentamethylcyclopentadienyl ligands.[7] Thus, reaction with protic substrates HX leads to stable compounds of the type $Me_5C_5ElX$ or $Me_5C_5El^+\ X^-$ with simultaneous formation of $Me_5C_5H$. In contrast, attack at the lone-pair is the preferred pathway in the chemistry of **1**, whereby oxidative addition products are formed (see previous section). The only one exception so far observed is the reaction of **1** with etheral tetrafluoroboric acid, $HBF_4 \cdot OEt_2$, which leads under elimination of $Me_5C_5H$ to the cyclotetrasilane $(Me_5C_5SiF)_4$ (**57**) as the final product (see Scheme 18).[45] The molecular structure of **57** is presented in Fig. 9. In the reaction sequence, the ionic half-sandwich compound $Me_5C_5Si^+\ BF_4^-$ most likely is a short-lived intermediate, which loses boron trifluoride to give the highly reactive silylene $Me_5C_5SiF$, which dimerizes to the more stable and NMR-spectroscopically characterized fluorine-bridged silylene

$Me_5C_5Si(\mu–F)_2SiC_5Me_5$ (**56**). Calculations for the parent systems have shown that the bridged structure is a minimum on the potential $Si_2H_2F_2$ surface and is less than 10 kcal mol$^{-1}$ higher in energy than the classical disilene. Once formed, the bridged species is kinetically stabilized against interconversion.[46] Kinetic stabilization might be the reason for the observation of **56** at low temperatures. In the last step of the reaction sequence, **56** dimerizes to the stable cyclotetrasilane **57**.

It is interesting to note that another reaction pathway is observed when the pyridine adduct of $HBF_4$ is used as substrate instead of the ether adduct. In this case exclusively oxidative addition of HF takes place to give compound **15**.[32]

The experiments described here and many others performed in our group show clearly, that it will be very difficult to find non-nucleophilic anions which allow the isolation of compounds containing the highly electrophilic half-sandwich cation $Me_5C_5Si^+$. This species so far has been observed only in the gas-phase; stabilization in the condensed phase is still an attractive target.

### F.   *Formation and Characterization of the $(Me_5C_5)_2SiH^+$ Cation*

Simple attack of a proton at the lone-pair in **1** should lead to the bis(pentamethylcyclopentadienyl)silicon cation, $(Me_5C_5)_2SiH^+$, in analogy to the easy protonation of for instance ferrocene. But protic species of different kind find other pathways like oxidative addition reactions or the elimination of $Me_5C_5H$, as described in the previous chapters. Even a reactive intermediate containing the desired cation has never been observed. It was an unexpected reaction sequence which finally led to the formation of the $(Me_5C_5)_2SiH^+$ cation stabilized by an unusual non-nucleophilic anion.[47] As already described in Chapter IV.C and shown in Scheme 6, 4-methyl-catechol reacted with **1** by insertion into one of the two OH bonds to give product **25**. Analogous but less stable compounds were obtained in the reaction of **1** with the parent catechol and with 4-tert.butyl-catechol. Such insertion products decompose in toluene solution already at room-temperature or after gentle warming to give as the main component the extremely air- and moisture-sensitive ionic species of type **58**, as described in Scheme 19. The parent compound **58a** (R = H) was characterized in more detail by several spectroscopic and analytical tools.[47] A structure with fluxional π-bonded pentamethylcyclopentadienyl ligands and with an sp$^2$ hybridized silicon atom was deduced from NMR and IR data. The deuterated cation $(Me_5C_5)_2SiD^+$ was prepared by the reaction of **1** with deuterated catechol $C_6H_4(OD)_2$. How species of type **58** arise from the

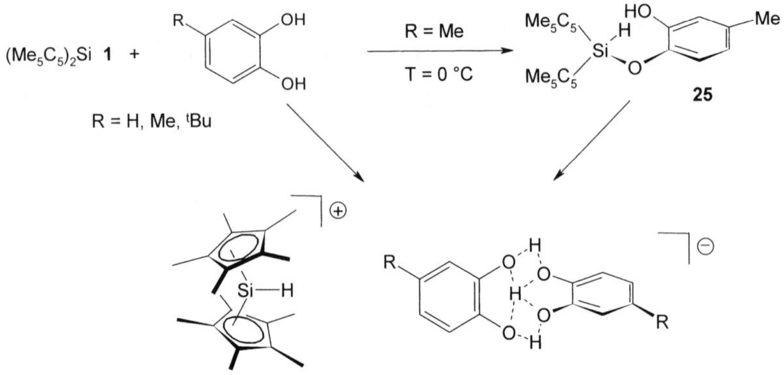

SCHEME 19. Reaction of 1 with catechol derivatives.

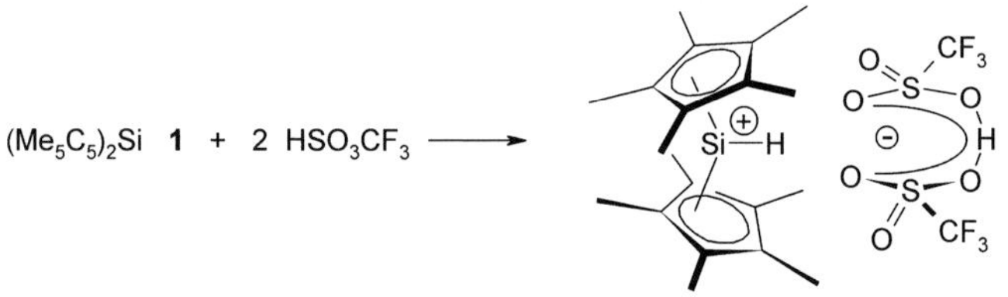

SCHEME 20. Reaction of 1 with trifluoromethane sulfonic acid.

precursors **25** is not yet understood. The cation in **57a** was most likely also generated by the reaction of **1** with two equivalents of trifluormethanesulfonic acid, as described in Scheme 20. The compound [(Me$_5$C$_5$)$_2$SiH] [(F$_3$CSO$_3$)$_2$H] (**59**) was obtained as an extremely air-sensitive oil, which was characterized by NMR and IR data only.[47] It is still an open question to what extent the stability of the cation (Me$_5$C$_5$)$_2$SiH$^+$ depends on cation–anion interactions. Detailed quantum mechanical calculations on the (Me$_5$C$_5$)$_2$SiH$^+$ cation and on the parent species (H$_5$C$_5$)$_2$SiH$^+$ support the proposed π-structure.[48]

## G. *Reactions with Chalcogene-Atom Sources*

Combination of **1** with electron-sextet fragments X might offer an approach to kinetically stabilized pπ–pπ systems of the type (Me$_5$C$_5$)$_2$Si = X, provided that two σ-bonded pentamethylcyclopentadienyl substituents at

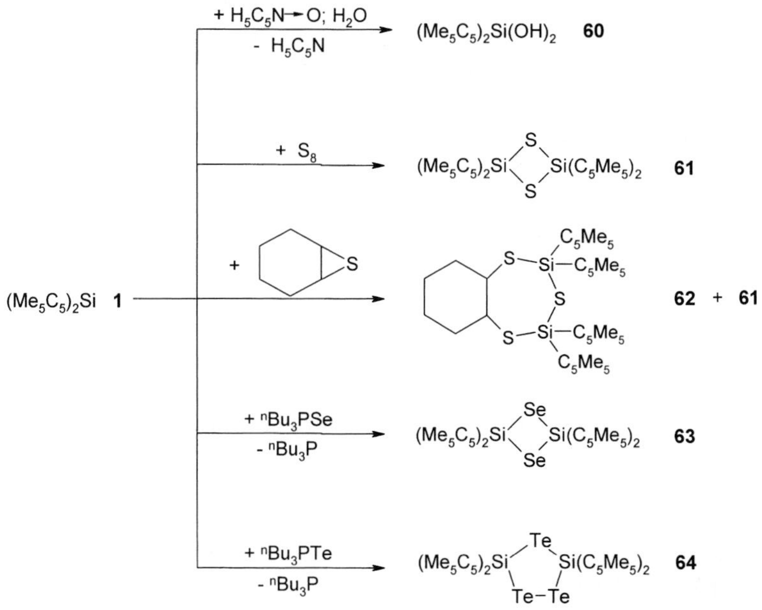

SCHEME 21. Reactions of 1 with chalcogene-atom sources.

silicon are bulky enough to prevent oligomerization or polymerization. The following experiments with oxygen, sulfur, selenium, and tellurium sources will demonstrate that this is not the case (see Scheme 21).

In the reaction of 1 with hydrated pyridine-N-oxide, the silanediol 60 was formed in good yield.[49] The formation of 60 most likely involves a pathway with the silanone $(Me_5C_5)_2Si = O$ as a reactive intermediate, which is transformed to 60 by the addition of a water molecule. The presence of the silanone was proved by an "ene-type" trapping reaction with pinacolone.[39] The dithiadisiletane 61 was formed nearly quantitatively in the reaction of 1 with sulfur; a conceivable silathione as an intermediate in the reaction could not be detected. With cyclohexene sulfide as the sulfur source, higher reaction temperatures were required, and, in addition to 61, the bicyclic compound 62 was formed.[50] Reaction of 1 with tri-n-butylphosphane selenide led nearly quantitatively to the diselenadisiletane 63; an intermediate silaselenone $(Me_5C_5)_2Si = Se$ was trapped with 2,3-dimethylbutadiene by [2 + 4] cycloaddition to give compound 65 (Scheme 22).[50] Finally, the X-ray characterized tritelluradisilole 64 was obtained from the reaction of 1 with tri-n-butylphosphane telluride; reaction intermediates were not detected.[50]

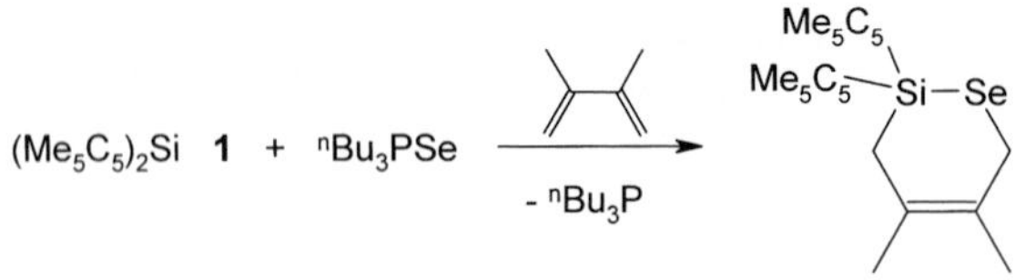

$(Me_5C_5)_2Si$ **1** + $^nBu_3PSe$  $\xrightarrow{\phantom{mm}}$  **65**
$- ^nBu_3P$

SCHEME 22. Trapping of the silaselenone $(Me_5C_5)_2Si = Se$.

### H. Cycloaddition Reactions: Stable Products, Reactive Intermediates and Alternative Pathways

Several compounds containing π-bonds show reactions with **1** which most likely proceed via [2 + 1] or [4 + 1] cycloaddition processes, but no detailed mechanistic studies have been performed so far. Not unexpectedly, the electron-rich species **1** preferentially reacts with electron-poor substrates, and ring-strained or dipolar intermediates rearrange or react further to more stable products in a sometimes rather complicated and surprising fashion. In a few cases even the pentamethylcyclopentadienyl substituents at silicon are involved in the reaction pathways.

#### 1. Reaction with Activated Triple-Bond Systems

In the reaction of **1** with alkynes possessing electron-withdrawing substituents, the corresponding silacyclopropene derivatives **66** and **67** are formed, as described in Scheme 23.[29] An unexpected pathway was observed in the reaction with the electron-poor hexafluorobutyne(2): the X-ray characterized heterocycle **68** was most likely obtained by nucleophilic attack of **1** at the triple bond. A subsequent shift of a fluorine atom from carbon to silicon creates an allene-type molecule which was stabilized by a [2 + 2] cycloaddition process involving a double bond from the pentamethylcyclopentadienyl unit, as described in Scheme 24.[33]

Compound **1** also reacted with several types of activated CN triple-bond systems under cycloaddition. The pathways depend on the further substituents Y at the CN unit. With compounds of the type YCN with $Y = MeS$, $Me_2N$, and $2,4\text{-}Me_2C_6H_3O$, formation of the corresponding diazasilole derivatives **69–71** took place, independent of substrate stoichiometry.[51] Azasilirenes were postulated as intermediates, which were attacked regiospecifically at the silicon–carbon bond by a further substrate molecule to yield the final insertion products in high yields (see Scheme 25); the steric requirements of the pentamethylcyclopentadienyl substituents might prevent other routes of stabilization. The C–C bond formation in these reactions, which proceeds under mild conditions, is of interest for the synthetic chemist. With compounds like BrCN and $Me_3SiCN$ reaction in a

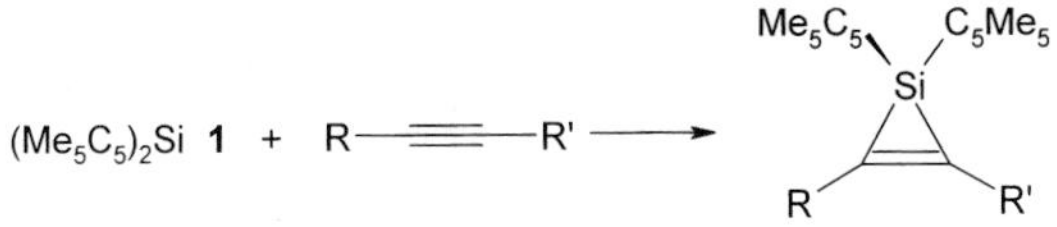

R = R' = CO$_2$Me:    **66**
R = SiMe$_3$, R' = SO$_2$Ph: **67**

SCHEME 23. Cycloaddition of **1** with electron-poor alkynes.

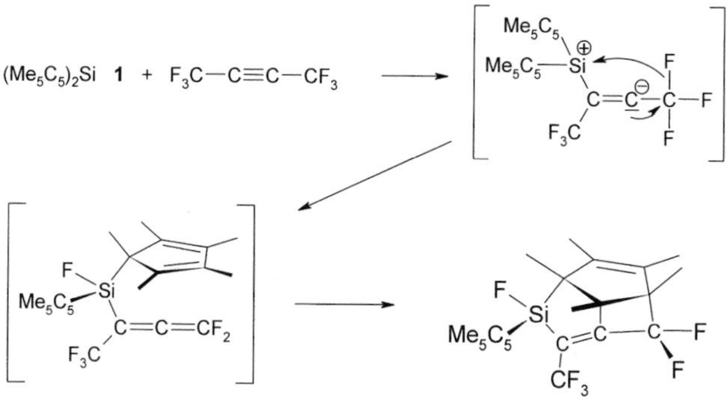

**68**

SCHEME 24. Reaction of **1** with hexafluorobutyne(2).

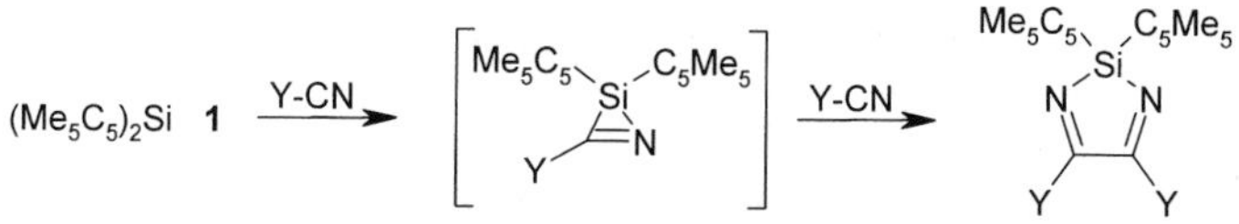

| Y | MeS- | Me$_2$N- | 2,4-Me$_2$C$_6$H$_3$O- |
|---|------|----------|------------------------|
|   | **69** | **70** | **71** |

SCHEME 25. Reaction of **1** with CN triple-bond systems.

1 : 1 stoichiometry was observed and the compounds **13** and **14** were formed (see Scheme 5). Azasilirenes resulting from cycloaddition processes were postulated as reactive intermediates.[41]

## 2. Reaction with the Heterocumulenes CO$_2$, COS, RNCS and CS$_2$

Unprecedented multistep processes were observed in the reaction of **1** with heterocumulenes of the type X = C = Y, which in most cases proceed via several highly reactive intermediates.[52] The products which were

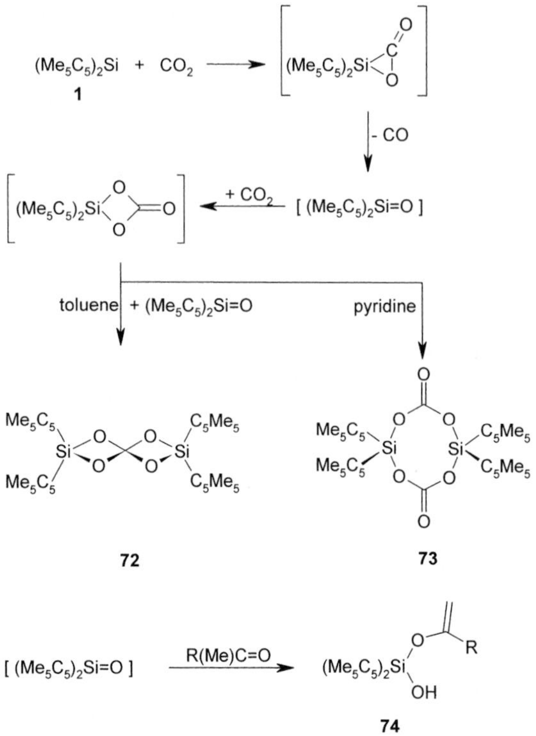

SCHEME 26. Reaction of **1** with $CO_2$ and trapping of the silanone $(Me_5C_5)_2Si=O$.

obtained from the reaction with carbon dioxide depend on the solvent. In toluene the spiro heterocyclic compound **72** was formed, whereas in pyridine the eight-membered ring system **73** was obtained (see Scheme 26). The first intermediate in both cases is a $[2+1]$ cycloaddition product or its ring-opened isomer, which easily loses carbon monoxide to give the silanone $(Me_5C_5)_2Si=O$. This intermediate is transformed by $CO_2$, which is present in the solution, to the $[2+2]$ cycloaddition product $(Me_5C_5)_2SiO_2CO$, which reacts with a second molecule of silanone $(Me_5C_5)_2Si=O$ to give the final product **72**, once more in a $[2+2]$ cycloaddition step. In pyridine as solvent, the silanone $(Me_5C_5)_2Si=O$ is deactivated; as a result, the intermediate $(Me_5C_5)_2SiO_2CO$ does not react with the silanone, but forms the dimerization product **73** after ring opening at one of the Si–O bonds. The intermediate silanone was trapped in ene-type reactions and the compounds **74** were formed.

In the reaction of **1** with carbon oxysulfide, the dithiadisiletane **61** was isolated. Here, the corresponding intermediate loses carbon monoxide to

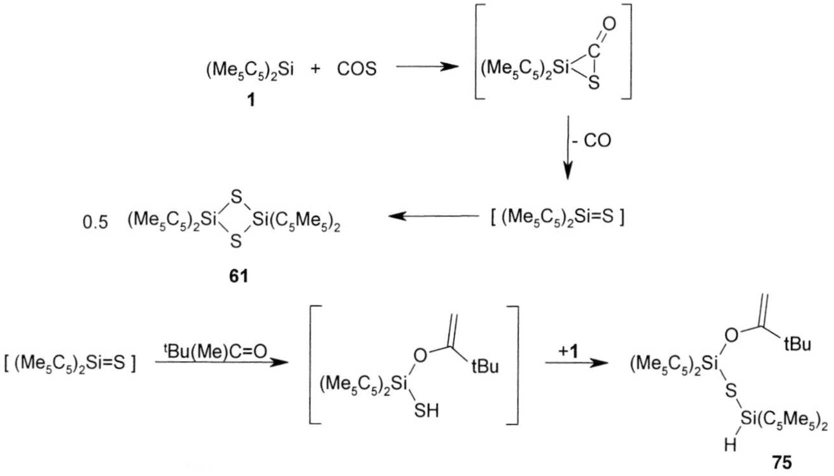

$$(Me_5C_5)_2Si + COS \longrightarrow \left[ (Me_5C_5)_2Si \underset{S}{\overset{O}{<}}C^{\diagdown O} \right]$$
**1**

$\downarrow$ - CO

$$0.5 \quad (Me_5C_5)_2Si \underset{S}{\overset{S}{<}} Si(C_5Me_5)_2 \longleftarrow [ (Me_5C_5)_2Si=S ]$$
**61**

$$[ (Me_5C_5)_2Si=S ] \xrightarrow{{}^tBu(Me)C=O} \left[ (Me_5C_5)_2Si \underset{SH}{\overset{O}{\diagup}} \overset{}{\diagdown} tBu \right] \xrightarrow{+1} (Me_5C_5)_2Si \underset{Si(C_5Me_5)_2}{\overset{O}{\diagup}}tBu$$
H
**75**

SCHEME 27. Reaction of **1** with COS and trapping of the silathione $(Me_5C_5)_2Si=S$.

$$(C_5Me_5)_2Si + RNCS \longrightarrow \left[ (Me_5C_5)_2Si \underset{S}{\overset{NR}{<}}C^{\diagdown} \right]$$
**1**

$\downarrow$ - RNC

$$(C_5Me_5)_2Si \underset{S}{\overset{S}{<}} \rangle=NR \xleftarrow{+ RNCS} [ (Me_5C_5)_2Si=S ]$$

R = Me: **76**
R = Ph: **77**      $\begin{array}{c} + 1 \\ \xrightarrow{\quad\quad} \\ R = Ph \end{array}$      $(Me_5C_5)_2Si \underset{S}{\overset{S}{<}} \underset{NR}{\diagdown} Si(C_5Me_5)_2$

**78**

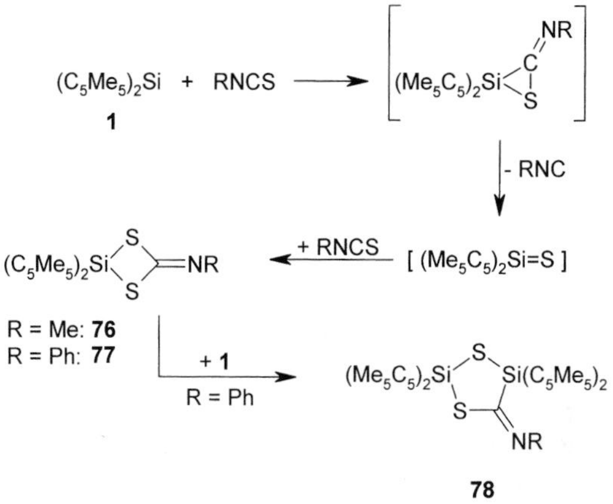

SCHEME 28. Reaction of **1** with isothiocyanates RNCS.

give the silathione $(Me_5C_5)_2Si = S$, which dimerizes to **61**. The intermediate silathione could be trapped and **75** was generated. An ene-type addition product reacted by insertion of **1** into the S–H bond to give the final product **75** (see Scheme 27).

In the reaction of **1** with methyl or phenyl isothiocyanate, the dithiasiletanes **76** or **77** were formed, following a pathway comparable to that observed for the reaction of **1** with carbon dioxide (see Scheme 28). In the case of the reaction with phenyl isothiocyanate, more drastic conditions

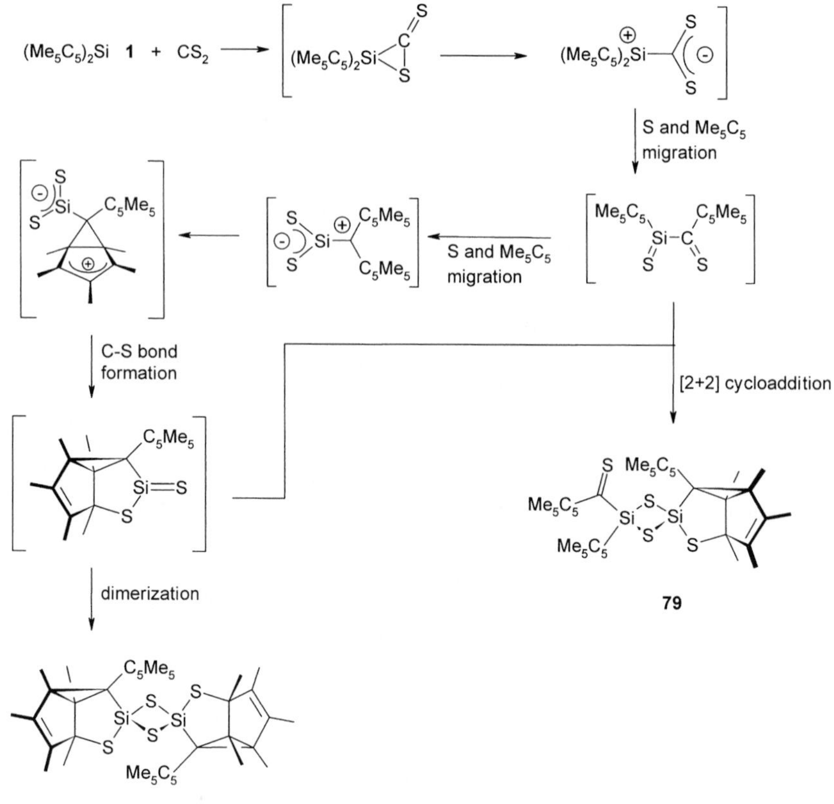

SCHEME 29. Reaction of **1** with CS$_2$.

led to the formation of the heterocyclic compound **78** as the result of the insertion of **1** into the Si–S bond in **77**.

A highly surprising multistep sequence was observed in the reaction of **1** with excess carbon disulfide (see Scheme 29). The symmetrically substituted and X-ray characterized dithiadisiletane **80** was obtained in high yield and the unsymmetrically substituted and also X-ray characterized dithiasiletane **79** was formed as a low-yield by-product. The reaction pathway is tentatively described as shown in Scheme 29. A [2 + 1] cycloaddition product or its ring-opened isomer is the first intermediate. In a subsequent multistep rearrangement, the substituents at carbon and at silicon have to be completely exchanged. Two sulfur atoms migrate from carbon to silicon and two pentamethylcyclopentadienyl groups migrate from silicon to carbon, presumably dictated by the rules valid for dyotropic rearrangement processes. The resulting product once more is highly reactive and gains stabilization by dimerization and by the transformation of a classical carbenium ion into an allyl-type cation. In the last step, C–S bond formation leads to the final product **80**. The formation of **79** as the result of the combination of two postulated intermediates supports the proposed reaction sequence.

## 3. Reaction with Aldehydes, Ketones and Diketones

Extensive studies were performed to investigate the reactions of **1** with organic substrates containing carbonyl groups. Several types of heterocyclic compounds were obtained in the reaction with aldehydes and ketones. Most likely these reactions proceed via $[2+1]$ cycloaddition products of the oxasilirane type as reactive intermediates (see Scheme 30).[53] In the reaction with aldehydes such as benzaldehyde and trans-cinnamaldehyde regio- and stereospecific formation of the respective 1,3,2-dioxasilolane derivatives **81** took place. With acetone as substrate, further reaction under elimination of pentamethylcyclopentadiene finally led to the vinyloxysubstituted dioxasilolane **82**. Rearrangement of the transient oxasiliranes into the 1,2-oxasilapentene(4) derivatives **83** was observed in the reaction with acetophenone or with benzophenone. The 1,3,2-dioxasilole **84** was the product of the reaction of **1** with benzil. In the reaction of **1** with 1,3-diketones such as acetylacetone or hexafluoroacetylacetone which are capable of keto-enol tautomerism, heterocycles **85** of the 2.6-dioxa-1-silacyclohexene(3) type were formed (see Scheme 31).[32] To account for the final products, the formation of intermediate oxidative addition products is assumed, in which migration of a pentamethylcyclopentadienyl substituent from silicon to a carbonyl carbon atom under concomitant Si–O bond formation takes place.

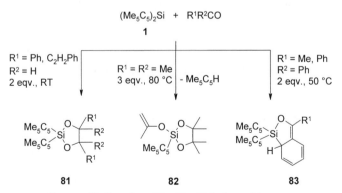

SCHEME 30. Reaction of **1** with aldehydes and ketones.

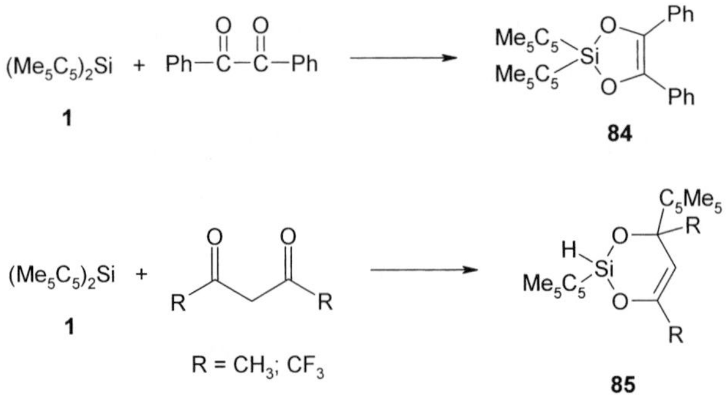

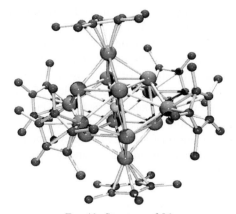

SCHEME 31. Reaction of **1** with diketones.

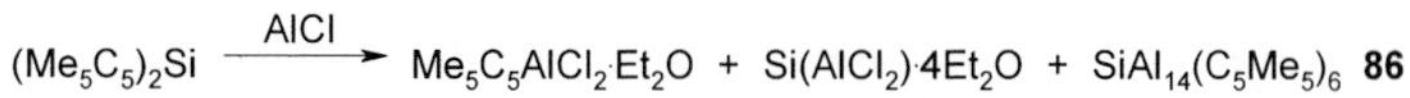

$$(Me_5C_5)_2Si \xrightarrow{\text{AlCl}} Me_5C_5AlCl_2 \cdot Et_2O + Si(AlCl_2) \cdot 4Et_2O + SiAl_{14}(C_5Me_5)_6 \text{ } \mathbf{86}$$

SCHEME 32. Reaction of **1** with AlCl.

FIG. 10. Structure of **86**.

## I. *Formation of the Cluster SiAl$_{14}$(C$_5$Me$_5$)$_6$*

A metastable AlCl solution obtained by co-condensation of the high-temperature molecule AlCl with a mixture of toluene/diethylether reacted with **1** to give compound **86** that bears six $\eta^5$-bonded pentamethylcaclo-pentadienyl substituents that protect a SiAl$_{14}$ cluster fragment from disproportionation and from formation of the bulk material. Furthermore, the compounds Me$_5$C$_5$AlCl$_2$·OEt$_2$ and Si(AlCl$_2$·OEt$_2$)$_4$ were isolated (Scheme 32).[54] The unique molecular structure of **86** is presented in Fig. 10.

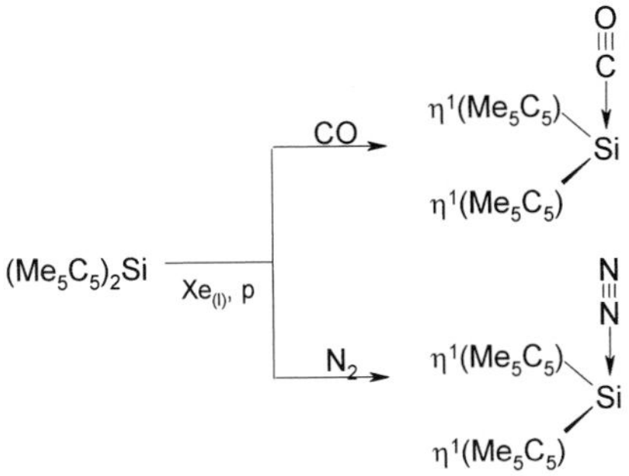

$$
\begin{array}{c}
\overset{\displaystyle O}{\underset{\displaystyle \|}{C}} \\
\eta^1(Me_5C_5)\!\diagdown\!\!\downarrow \\
Si \quad \textbf{87} \\
\eta^1(Me_5C_5)\!\diagup
\end{array}
$$

$(Me_5C_5)_2Si \quad \overline{\phantom{Xe}} Xe_{(l)}, p$

$$
\begin{array}{c}
\overset{\displaystyle N}{\underset{\displaystyle \|}{N}} \\
\eta^1(Me_5C_5)\!\diagdown\!\!\downarrow \\
Si \quad \textbf{88} \\
\eta^1(Me_5C_5)\!\diagup
\end{array}
$$

SCHEME 33. Adduct formation of **1** with CO and $N_2$.

## J. Adduct Formation with CO and $N_2$

In all reactions described in the previous sections, **1** behaved as a nucleophilic reagent. The weak electrophilic character of **1** was demonstrated in the reaction with the donor-molecules CO and $N_2$. Under rather sophisticated reaction conditions, i.e. in liquid xenon or nitrogen as solvent or reactant, the formation of the monocarbonyl (**87**) and of the mono-dinitrogen complex (**88**) was observed; **87** was synthesized by two independent routes (Scheme 33).[55] These reactions were highly incomplete under a few bar of CO or $N_2$ and reversible when the pressure was released. Complex **87** was characterized by three isotopomers, the observed CO and $N_2$ stretching frequencies of $2065\,cm^{-1}$ in **87** and of 2046 [LXe] or $2053\,cm^{-1}$ [$LN_2$] in **88** have been compared with those in the matrix species $Me_2SiCO$ $(1962\,cm^{-1})$,[56] and $SiN_2$ $(1731\,cm^{-1})$.[57] As a result, weak backbonding from the lone-pair at silicon is discussed for **87** and **88**. Calculations[58] support this interpretation and reveal a $\eta^1$-bonding mode of the pentamethylcyclopentadienyl substituents and a bonding situation as portrayed in Scheme 33.

# V

## CONCLUSIONS AND PERSPECTIVES

After a decade of research the basic principles in the chemistry of decamethylsilicocene (**1**) seem to be understood. This compound shows the reactivity of a nucleophilic silylene due to the fact that the $\pi$-bonded pentamethylcyclopentadienyl ligands are easily transferred to $\sigma$-bonded substituents during the reaction. The steric requirements of these substituents permit reactions with bulky substrates. The migratory aptitude and the leaving-group character of the pentamethylcyclopentadienyl groups are the reason for sometimes rather complicated reaction pathways. From there, further surprising results in the chemistry of 1 can be predicted.

Up to date, compound 1 represents the only fully characterized silicocene derivative. Very recent experiments[22] indicate, that it should be possible to prepare also other peralkylated or -arylated derivatives; hydrogen substituents at the cyclopentadienyl rings have to be avoided because they do not withstand the reaction conditions during the reduction of the silicon(IV) precursor. Future work should concentrate on the development of strategies for the synthesis of other silicocene derivatives. The synthesis and characterization of a compound containing the highly electrophilic pentamethylcyclopentadienylsilicon cation, $Me_5C_5Si^+$, is another interesting target; an appropriate counter-anion with very low nucleophilicity has to be found. Experiments to prepare novel silicon(II) compounds by nucleophilic substitution of one or both pentamethylcyclopentadienyl groups in 1 has been unsuccessful so far, and future work should concentrate also on this topic.

ACKNOWLEDGEMENTS

The dedication and expertise of the co-workers cited in the references is gratefully acknowledged. We thank Dr. A. Mix for preparing the drawings and Dr. T. Braun for improving the English. Funding was kindly provided by the Deutsche Forschungsgemeinschaft, by the Fonds der Chemischen Industrie and by the University of Bielefeld.

REFERENCES

(1) Kealy, T. J.; Pauson, P. L. *Nature* **1951**, *168*, 1039; Fischer, E. O.; Pfab, W. *Z. Naturforsch.* **1952**, *7b*, 377; Wilkinson, G.; Rosenblum, M.; Whiting, M. C.; Woodward, R. B. *J. Amer. Chem. Soc.* **1952**, *74*, 215.

(2) Fischer, E. O.; Grubert, H. *Z. Naturforsch.* **1956**, *B11*, 423.

(3) Fischer, E. O.; Grubert, H. *Z. Anorg. Allg. Chem.* **1956**, *286*, 237.

(4) Scibelli, J. V.; Curtis, M. D. *J. Am. Chem. Soc.* **1973**, *95*, 924.

(5) Jutzi, P.; Kohl, F.; Hofmann, P.; Krüger, C.; Tsay, Y. H. *Chem. Ber.* **1980**, *113*, 757.

(6) Atwood, J. L.; Hunter, W. E.; Cowley, A. H.; Jones, R. A.; Stewart, C. A. *J. Chem. Soc., Chem. Commun.* **1984**, 925.

(7) Jutzi, P. *Adv. Organomet. Chem.* **1986**, *26*, 217; Jutzi, P.; Burford, N. *Chem. Rev.* **1999**, *99*, 969.

(8) Jutzi, P.; Hielscher, B. *J. Organomet. Chem.* **1985**, *C25*, 291.

(9) Review on silylenes: P. Gaspar, R. West, in "The Chemistry of Organosilicon Compounds"; Z. Rappoport, Y. Apeloig, Eds.; John Wiley and Sons. Chichester, **1998**; Vol.2, Part 3, pp 2463–2567.

(10) Jutzi, P.; Kanne, D.; Krüger, C. *Angew. Chem. Int. Ed. Engl.* **1986**, *25*, 164; Jutzi, P.; Holtmann, U.; Kanne, D.; Krüger, C.; Blom, R.; Gleiter, R.; Hyla-Kryspin, I. *Chem. Ber.* **1989**, *122*, 1629.

(11) Karsch, H. H.; Keller, U.; Gamper, S.; Müller, G. *Angew. Chem. Int. Ed. Engl.* **1990**, *29*, 295.

(12) Denk, M.; Lennon, R.; Hayashi, R.; West, R.; Belyakov, A. V.; Verne, H. P.; Haaland, A.; Wagner, M.; Metzler, N. *J. Am. Chem. Soc.* **1994**, *116*, 2691.

(13) Reviews on stable silylenes: 1) Haaf, M.; Schmedake, T.A.; West, R. *Acc. Chem. Res.* **2000**, *33*, 704; 2) Gehrhus, B.; Lappert, M.F. *J. Organomet. Chem.* **2001**, *617–618*, 209.

(14) Gehrhus, B.; Lappert, M. F.; Heinecke, J.; Boese, R.; Bläser, D. *J. Chem. Soc. Chem. Commun.* **1995**, 1931.

(15) Heinecke, J.; Oprea, A.; Kindermann, M. K.; Karpati, T.; Nyulaszi, L.; Veszpremi, T. *Chem. Eur. J.* **1998**, *4*, 541.

(16) Denk, M.; Green, J. C.; Metzler, N.; Wagner, M. *J. Chem. Soc. Dalton Trans.* **1994**, 2405.

(17) Kira, M.; Ishida, S.; Iwamoto, T.; Kabuto, C. *J. Am. Chem. Soc.* **1999**, *121*, 9722.

(18) Rees, W. S., Jr.; Schubert, D. M.; Knobler, C. B.; Hawthorne, M. F. *J. Amer. Chem. Soc.* **1986**, *108*, 5369.

(19) Hosmane, N. S.; de Meester, P.; Siriwardane, U.; Islam, M. S.; Chu, S. C. *J. Chem. Soc., Chem. Commun.* **1986**, 1421.

(20) Jutzi, P.; Kanne, D.; Hursthouse, M. B.; Howes, A. *J. Chem. Ber.* **1988**, *121*, 1299.

(21) Evans, W. J.; Ullibarri, T. A.; Jutzi, P. *Inorg. Chim. Acta* **1990**, *168*, 5.

(22) Klipp, A. PhD Thesis, University of Bielefeld, Germany, 1999.

(23) Wrackmeyer, B.; Sebald, A.; Merwin, L. H. *Magn. Reson. Chem.* **1991**, *29*, 260.

(24) Jutzi, P. *Chemie in unserer Zeit* **1999**, *33*, 342.

(25) Jutzi, P.; Burford, N., (Togni, A.; Halterman, R. L. Eds.), *Metallocenes-Synthesis, Reactivity, Applications*, Vol. 1, Wiley-VCH, Weinheim, 1998. p. 3.

(26) Lee, T. J.; Rice, J. E. *J. Am. Chem. Soc.* **1989**, *111*, 2011.

(27) Schoeller, W. W.; Friedrich, O.; Sundermann, A.; Rozhenko, A. *Organometallics* **1999**, *18*, 2099.

(28) Jutzi, P.; Reumann, G. *J. Chem. Soc., Dalton Trans.* **2000**, 2237.

(29) Jutzi, P., (Bassindale, A. R.; Gaspar, P. P. Eds.) Frontiers of Organosilicon Chemistry, The Royal Society of Chemistry, 1991. pp. 307.

(30) Holtmann, U.; Jutzi, P.; Kühler, T.; Neumann, B.; Stammler, H. G. *Organometallics* **1999**, *18*, 5531.

(31) Tacke, M.; Plaggenburg, L.; Schnöckel, H. *Z. Anorg. Allg. Chem.* **1991**, *604*, 35.

(32) Jutzi, P.; Bunte, E. A.; Holtmann, U.; Neumann, B.; Stammler, H. G. *J. Organomet. Chem.* **1993**, *446*, 139.

(33) Jutzi, P.; Eikenberg, D., (Auner, N.; Weis, J. Eds.) Organosilicon Chemistry III, Wiley-VCH, Weinheim, 1998. pp. 76.

(34) Theil, M.; Jutzi, P.; Neumann, B.; Stammler, A.; Stammler, H. G. *Organometallics* **2000**, *19*, 2937.

(35) Theil, M.; Jutzi, P.; Neumann, B.; Stammler, A.; Stammler, H. G. *J. Organomet. Chem.* **2002**, *662*, 34.

(36) Jutzi, P.; Möhrke, A. *Angew. Chem.* **1990**, *102*, 913; *Angew. Chem. Int. Ed. Engl.* **1990**, *29*, 893.

(37) Schonberg, P. R.; Paine, R. T.; Campana, C. F.; Duesler, E. N. *Organometallics* **1982**, *1*, 799.

(38) Kühler, T. Dissertation Univ. Bielefeld **2000**.

(39) Kühler, T.; Jutzi, P.; Stammler, A.; Stammler, H. G. *J. Chem. Soc., Chem. Commun.* **2001**, 539.

(40) Theil, M. PhD Thesis, University of Bielefeld, Germany, **2002**.

(41) Jutzi, P.; Eikenberg, D.; Neumann, B.; Stammler, H. G. *Organometallics* **1996**, *15*, 3659.

(42) Möhrke, A. Dissertation Univ. Bielefeld **1989**.

(43) Greiwe, P.; Bethäuser, A.; Pritzkow, H.; Kühler, T.; Jutzi, P.; Siebert, W. *Eur. J. Inorg. Chem.* **2000**, 1927.
(44) Dohmeier, C.; Schnöckel, H.; Robl, C.; Schneider, U.; Ahlrichs, R. *Angew. Chem.* **1993**, *105*, 1714.
(45) Jutzi, P.; Holtmann, U.; Bögge, H.; Müller, A. *J. Chem. Soc., Chem. Commun.* **1988**, 305.
(46) Maxka, J.; Apeloig, Y. *J. Chem. Soc., Chem. Commun.* **1990**, 737.
(47) Jutzi, P.; Bunte, E.A. *Angew. Chem.* **1992**, *104*, 1636; *Angew. Chem. Int. Ed. Engl.* **1992**, *31*, 1606.
(48) Müller, T.; Jutzi, P.; Kühler, T. *Organometallics* **2001**, *20*, 5619.
(49) Al-Jouaid, S. S.; Eaborn, C.; Hitchcock, P. B.; Lickiss, P.; Möhrke, A.; Jutzi, P. *J. Organomet. Chem.* **1990**, *384*, 33.
(50) Jutzi, P.; Möhrke, A.; Müller, A.; Bögge, H. *Angew. Chem.* **1989**, 101, 1527; *Angew. Chem. Int. Ed. Engl.* **1989**, 28, 1518.
(51) Jutzi, P.; Eikenberg, D.; Neumann, B.; Stammler, H. G. *Organometallics* **1996**, *15*, 3659.
(52) Jutzi, P.; Eikenberg, D.; Möhrke, A.; Neumann, B.; Stammler, H. G. *Organometallics* **1996**, *15*, 753.
(53) Jutzi, P.; Eikenberg, D.; Bunte, E. A.; Möhrke, A.; Neumann, B.; Stammler, H. G. *Organometallics* **1996**, *15*, 1930.
(54) Purath, A.; Dohmeier, C.; Ecker, A.; Köppe, R.; Krautscheid, H.; Schnöckel, H.; Ahlrichs, R.; Stoermer, C.; Friedrich, J.; Jutzi, P. *J. Am. Chem. Soc.* **2000**, *122*, 6955.
(55) Tacke, M.; Klein, Ch.; Stufkens, D. J; Oskam, A.; Jutzi, P.; Bunte, E.A. *Z. Anorg. Allg. Chem.* **1993**, *619*, 865.
(56) Arrington, C. A.; Petty, J. T.; Payne, S.E.; Haskins, W. C. K. *J. Am. Chem. Soc.* **1988**, *110*, 6240; Pearsall, M.; West, R. *J. Am. Chem. Soc.* **1988**, *110*, 7228; SiCO (1899 cm$^{-1}$).
(57) Lemke, R. R.; Ferrante, R. F.; Weltner, W., Jr. *J. Am. Chem. Soc.* **1977**, *99*, 416.
(58) Tacke, M., (Auner, N.; Weis, J. Eds.) Organosilicon Chemistry III, Wiley-VCH, Weinheim, 1998. pp. 70.

# *Organometallic Phosphorous and Arsenic Betaines*

## NIKOLAI N. ZEMLYANSKY and IRINA V. BORISOVA

*A. V. Topchiev Institute of Petrochemical Synthesis,*
*Russian Academy of Sciences (TIPS RAS),*
*29, Leninsky prospect, 119991-GSP-1, Moscow, Russian Federation,*
*E-mail: zemlyan@mail.cnt.ru*

## YURI A. USTYNYUK*

*Department of Chemistry,*
*M. V. Lomonosov Moscow State University,*
*Vorob'evy Gory,*
*119899 Moscow, Russian Federation,*
*E-mail: yust@nmr.chem.msu.su*

*Corresponding author. Fax: 007 (095) 932 8846

35

ADVANCES IN ORGANOMETALLIC CHEMISTRY
VOLUME 49 ISSN 0065-3055/DOI 10.1016/S0065-3055(03)49002-3

©2003 Elsevier Science (USA)
All rights reserved.

# I

## INTRODUCTION

Many zwitterionic organometallic compounds (betaines) are known in which positively and negatively charged centers of different nature are separated by various spacer groups. Substances of this type always exhibit the pronounced specific physical properties and high and unusual reactivity. Several types of organometallic betaines have been reviewed recently.[1-4]

This review is devoted to new betaines containing $^{(+)}E^{15}$–C–$E^{14}$–$X^{(-)}$ (**I**) and $^{(+)}E^{15}$–C–$E^{14(-)}$ (**II**) ($E^{15} = $ P, As; $E^{14} = $ Si, Ge, Sn; X = C, S, O, NR) structural fragments (Table I). The first stable betaines **I** with the phosphonium or arsonium cationic centers separated from the anionic centers localized at the sulfur or carbon atoms by a spacer group containing Group 14 elements (silicon, germanium or tin) have been synthesized and structurally characterized just recently. The first data were also obtained on the formation of betaines **I** with the anionic centers at the nitrogen and oxygen atoms (presented in brackets in Table I) as intermediates in the sulfur substitution reactions of stable betaines **I** with the thiolate group. Only few betaines of the $^{(+)}E^{15}$–C–$E^{14(-)}$ type **II** have been synthesized to date, and only two compounds of this type have been characterized by X-ray data. However, reliable evidence is available for the intermediate formation of these substances in the reactions of phosphorus ylides with silylenes and stannylenes.

### TABLE I

Main Types of Organometallic Betaines **I** and **II** Considered in the Review

| Type $^{(+)}E^{15}$–C–$E^{14}$–$X^{(-)}$ **I** | | Type $^{(+)}E^{15}$–C–$E^{14(-)}$ **II** |
|---|---|---|
| X = C | X = S, O, NR | |
| $^+$P–C–Si–C$^-$ | $^+$P–C–Si–S$^-$ | [$^+$P–C–Si$^-$] |
| $^+$P–C–Ge–C$^-$ | $^+$P–C–Ge–S$^-$ | $^+$P–C–Sn$^-$ |
| | $^+$P–C–Sn–S$^-$ | |
| | $^+$As–C–Si–S$^-$ | |
| | $^+$As–C–Ge–S$^-$ | |
| | [$^+$P–C–Si–O$^-$] | |
| | [$^+$P–C–Si–NR$^-$] | |

Organometallic betaines of type **I** can be considered as the closest structural analogs of carbon betaines of the $^{(+)}P-C-C-X^{(-)}$ type (**IV**), which were regarded for a long time as possible intermediates in classical reactions of carbonyl and thiocarbonyl compounds with phosphorus ylides (Wittig and Corey–Chaykovsky reactions and related processes,[5,6] Scheme 1). Vedejs and coworkers[7,8] proved unambiguously that oxaphosphetanes (**III**) are true intermediates in the reactions of "nonstabilized" phosphorus ylides with carbonyl compounds. The formation of oxabetaines $^{(+)}P-C-C-O^{(-)}$ was detected only in the form of their adducts with lithium salts.[9,10]

The first reliable evidence for the formation of thiabetaines $^{(+)}P-C-C-S^{(-)}$ as true intermediates in thio-Wittig reaction was obtained[11–13] only after the synthesis and complete characterization of the corresponding thiabetaines **I** of the silicon series.[14,15] This evidence was based on the resemblance of the spectral parameters of compounds of both series.[11]

A comparison of the all data accumulated to the present time on the reactivity and structures of organometallic betaines $^{(+)}E^{15}-C-E^{14}-X^{(-)}$ **I** (X = C, S) and corresponding carbon analogs $^{(+)}E^{15}-C-C-X^{(-)}$ **IV** demonstrates a certain resemblance between them. This resemblance opens new challenges for "variations on the Wittig reaction theme" in chemistry of organic derivatives of Group 14 elements. In particular, this suggests the possibility of using betaines **I** as precursors of kinetically unstable compounds with double $E^{14}=X$ bonds, which being generated in solutions can be introduced into further transformations *in situ*.

Evidently Seyferth was the first to recognize the new opportunities of such variations in organosilicon chemistry.[16] He generated dimethylsilylene in the presence of ketones and phosphines. According to authors hypothesis multi-step transformation takes place in the reaction (Scheme 2).

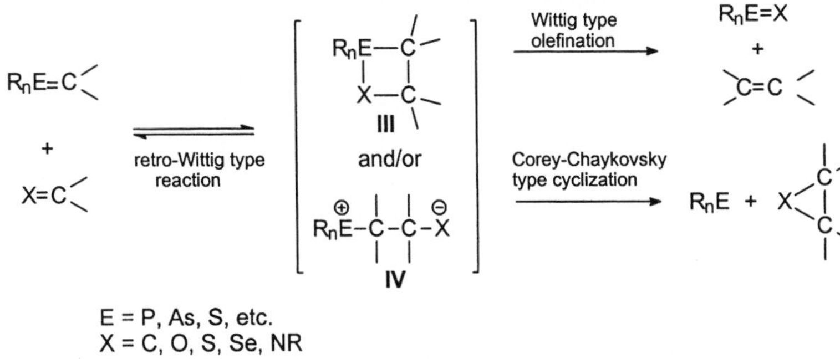

E = P, As, S, etc.
X = C, O, S, Se, NR

SCHEME 1

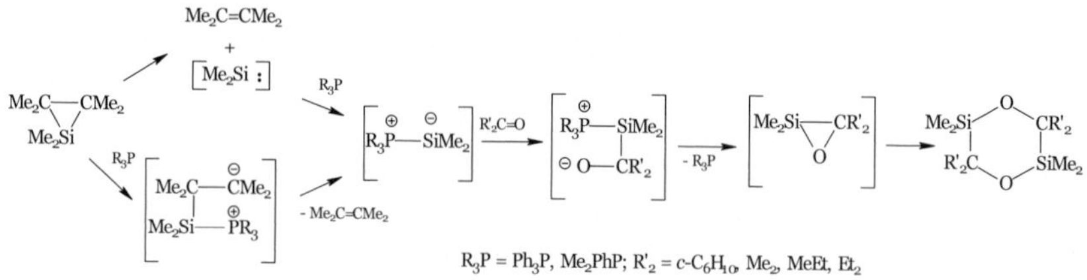

$R_3P = Ph_3P, Me_2PhP; R'_2 = c-C_6H_{10}, Me_2, MeEt, Et_2$

SCHEME 2

Phosphorous silaylide as an initial product reacts first with the ketone generating organophosphorous silabetaine. The latter eliminates $R_3P$ following to Corey–Chaikovsky pathway and resulting silaoxyrane dimerizes.

Unfortunately, this excellent work was not continued and after short communication[16] no full paper with the evidence of reaction scheme was appeared.

Numerous reactions of carbonyl compounds, alcohols, olefins, etc., with compounds bearing $E^{14}=X$ bonds in which the latter act as direct analogs of phosphorus and arsenic ylides have already been accomplished.[17] Recently, an interest in reactivity of compounds with multiple $E^{14}-X$ bonds is increasing due to challenges of important practical applications (see, e.g.,[18]).

This review covers the data on organometallic betaines **I** and **II** to December, 2000. The material concerning the synthesis, structure, and properties of their carbon analogs **IV** is considered only partially.

## II
## SYNTHESIS OF BETAINES $^{(+)}E^{15}-C-E^{14}-X^{(-)}$ (I) AND
## $^{(+)}E^{15}-C-E^{14(-)}$ (II) ($E^{15} = P, As; E^{14} = Si, Ge, Sn; X = C, S$)

Most of the presently known betaines of the both named types were obtained by the reaction of phosphorus or arsenic ylides with stable compounds containing $E=X$ bonds ($E = C$, Si; $X = C$, S), cyclic oligomers of these compounds $(R_2EX)_n$ ($E = Si$, Ge, Sn; $X = S$, $n = 2$, 3), three- and four-membered silacarbocycles, carbenes or their organoelement analogs. The reactions of phosphorus thiabetaines of silicon and germanium with the $(R_3Sn)_2X$ ($X = O$, NMe) compounds ($X = O$, NMe) that occur via intermediate silicon and germanium organophosphorus betaines with oxide or alkylamide anionic centers are considered in Section 5.

## A. *Synthesis of Betaines* I *with Carbanionic Centers*

It is commonly accepted[5,6,19] that unstable betaines **IV** (X = C) are intermediates of the cyclopropanation of olefins with the polar C=C bond by phosphorus ylides. However, only one compound of this type, viz., $Me_3P^{(+)}-CH_2-CMe_2-C_5H_4^{(-)}$ (**1**), synthesized in the reaction of dimethylfulvene with methylenetrimethylphosphorane, was isolated and characterized by multinuclear NMR spectroscopy.[20]

The assumption about an increased (compared to that of standard silenes) kinetic stability of 6,6-dimethyl-6-silafulvene (**2**) (R = Me), which is the silicon analog of dimethylfulvene, has been advanced first on the basis of quantum-chemical calculation by the semiempirical MINDO/2 method.[21] The formation of **2** under gas-phase pyrolysis of allyldimethylsilylcyclopentadiene and (dimethylmetoxysilyl)trimethylsilylcyclopentadiene was experimentally confirmed by several authors[22,23] using the isolation of its dimerization products and reactions with various trapping agents. Chemistry of organoelement analogs of fulvenes, particularly, dibenzosila-,[24,25] -germa-,[26–31] and -stannafulvene is still under intensive study.[32]

Silicon (**5**) and germanium (**6**) betaines with the cyclopentadienyl[20] and/ or fluorenyl[33–35] anionic centers were prepared in 80–90% yields in the reaction of methylenetrimethyl- or ethylidenetriethylphosphorane with cyclopentadienyl- or (fluorenyl)dialkylchlorosilanes and -germanes (**3**). The first stage of this process affords highly reactive 6,6-dialkyl-6-silafulvenes or the corresponding dibenzoelementafulvenes **2**, which further react immediately with an excess of phosphorane as a trapping agent (Scheme 3).

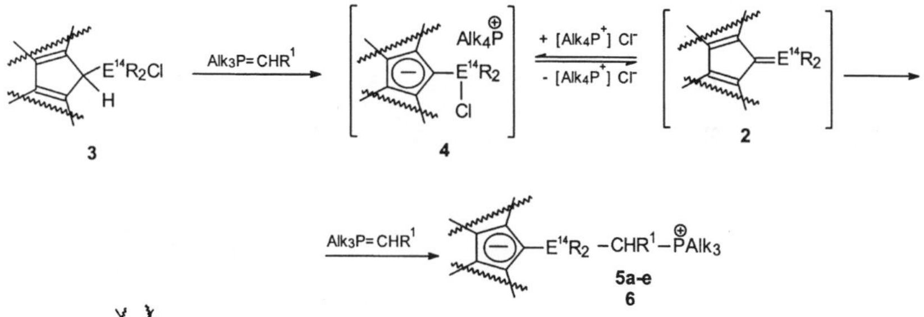

| N |  | $E^{14}$ | $R_2$ | Alk | $R^1$ | yield, % |
|----|-----|-----|----------|-----|-----|--------------|
| 5a | Cp | Si | Me | Me | H | quantitative |
| 5b | Cp | Si | Me | Et | Me | 87 |
| 5c | Cp | Si | $Me_3SiCH_2$ | Me | H | 91 |
| 5d | Fl | Si | $Me_3SiCH_2$ | Me | H | 83 |
| 5e | Fl | Si | Me, *s*-Bu | Me | H | 91 |
| 6 | Fl | Ge | *i*-Bu | Me | H | 87 |

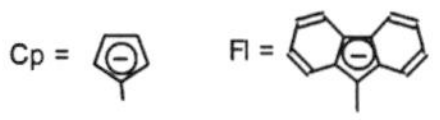

SCHEME 3

Betaines **1**, **5**, and **6** are crystalline white (for the cyclopentadiene series) or yellow (for the fluorene series) substances, which are very sensitive to traces of atmospheric oxygen and moisture. They are poorly soluble in low-polarity solvents and virtually insoluble in nonpolar solvents.

The first kinetically stable dibenzosilafulvene (**7**), whose structure and properties should more correctly be described by the resonance hybrid **7a ⇔ 7b** with a great contribution of the ylide form **7a**, reacts with phosphorus ylide to form betaine (**8**), which is rearranged, under thermodynamically controlled conditions, into the salt (**9**) (Scheme 4).[24,25]

We failed to prepare the stable silicon, germanium, and tin betaines by the reactions of cyclopentadienyl and fluorenylchlorosilanes, germanes, and stannanes with less nucleophilic phosphorous ylides $Ph_3P = CR_2^1$.[36] These reactions lead to elementafulvene cyclooligomers (**10**, **11**)[20,37–43] or to phosphonium salts of dialkylchlorosilyl- or dialkylchlorogermylfluorenes (**3**) regardless of the molar reagent ratio[33,34] (Scheme 5). Based on experiments with chemical traps,[20] we rejected the alternative route of silafulvene dimer formation in solutions by the reaction of phosphorus ylides with chloro (dialkyl)cyclopentadienylsilanes without the intermediate formation of silafulvenes **2**, proposed by Jones *et al.*[44]

The double bond in silenes is strongly polarized. They react with phosphorus ylides, as shown by Brook and MacMillan,[45] like alkenes with the strongly polar C=C bond. Therefore, it is reasonable to suggest that the reaction also occur through the betaine intermediate (**12**) (Scheme 6).

The intermediate formation of betaines with the carbanionic center is also postulated in the reactions of permethylsilirane, sila- and disilacyclobutanes with phosphorus ylides. For data on these betaines isomerized *in situ* to silylated phosphorus ylides, see Section 5.4.

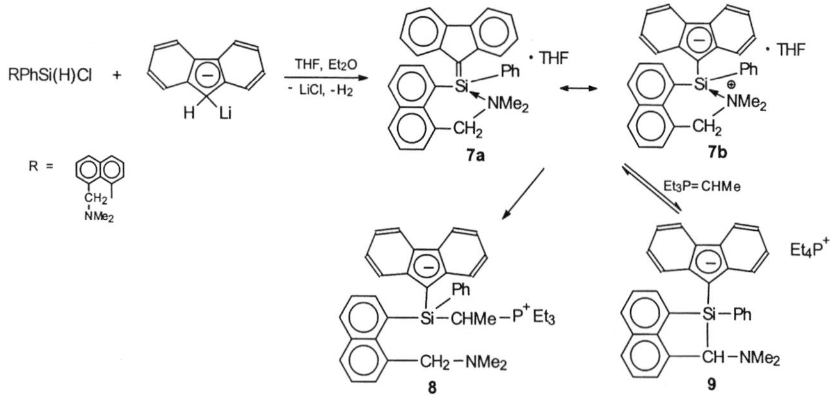

SCHEME 4

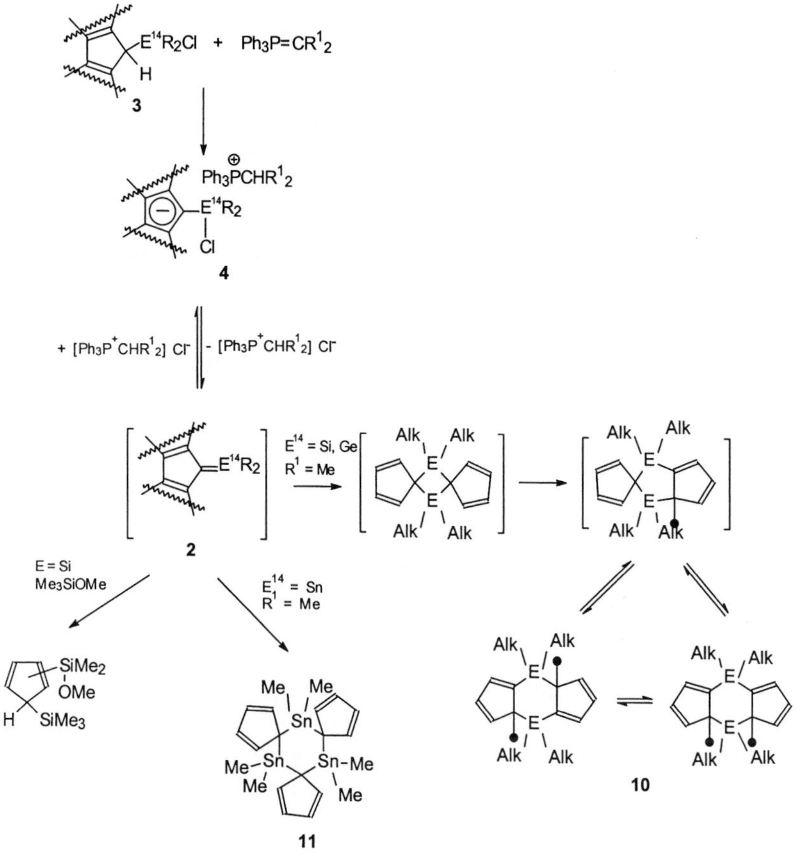

SCHEME 5

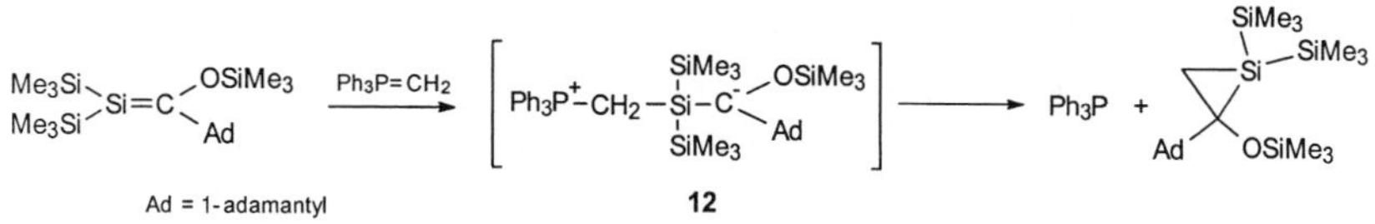

SCHEME 6

## B. Synthesis of Betaines I with Thiolate Centers

### 1. Reactions of Phosphorus Ylides with Carbon Disulfide, Stable Thiocarbonyl Compounds, and Thiosilanones

The first stable organophosphorus betaines with thiolate centers in the carbon series of the type (13) were prepared by the reaction of phosphorus

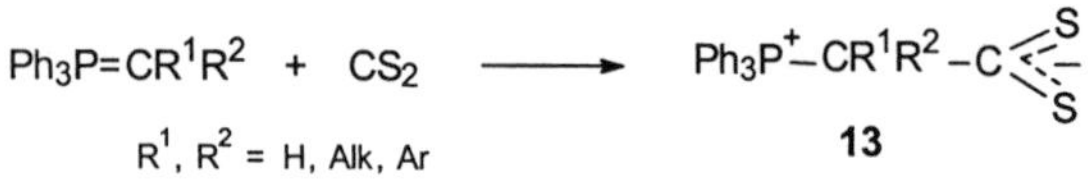
Ph3P=CR1R2 + CS2
R1, R2 = H, Alk, Ar
Ph3P+–CR1R2–C
S
S
13

SCHEME 7

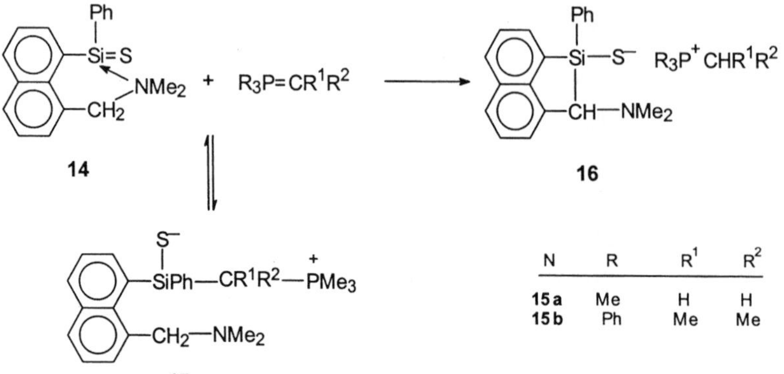
Ph
Si=S
NMe2
CH2
+ R3P=CR1R2
14
Ph
Si—S−
R3P+CHR1R2
CH—NMe2
16
S−
SiPh—CR1R2—PMe3
CH2—NMe2
15a

| N | R | $R^1$ | $R^2$ |
|---|---|---|---|
| 15a | Me | H | H |
| 15b | Ph | Me | Me |

SCHEME 8

ylides with carbon disulfide (Scheme 7). This reaction was studied in detail by independent authors.[46–50] The delocalization of the negative charge over two sulfur atoms of the anionic center increases the resistance of these compounds toward further decomposition. X-ray data were obtained for some of them.[47,50]

The first silicon-organophosphorus betaine with a thiolate center (**15a**) was synthesized by the reaction of stable silanethione (**14**) with trimethylmethylenephosphorane (Scheme 8) and characterized by multinuclear NMR spectroscopy.[14] Compound **15a** is formed under kinetic control and is transformed, under the thermodynamically controlled conditions, into the silaacenaphthene salt (**16**). The processes presented in this scheme reflect the competition of the basicity and nucleophilicity of phosphorus ylides. Betaine **15b** prepared from less nucleophilic and less basic ylide with phenyl substituents at the phosphorus atom is much less resistant toward retrodecomposition compared to the alkyl analog. Its equilibrium concentration does not exceed 6%.

The spectral parameters of **15a** and other silicon-organophosphorus betaines described henceforth in Section 2.2.2 allowed us to show reliably that the reaction of phosphorus ylides with thiocarbonyl compounds, unlike the classical Wittig reaction, occurs through the intermediate formation of betaines (**17**)[11] (Scheme 9). Erker and coworkers performed a more detailed

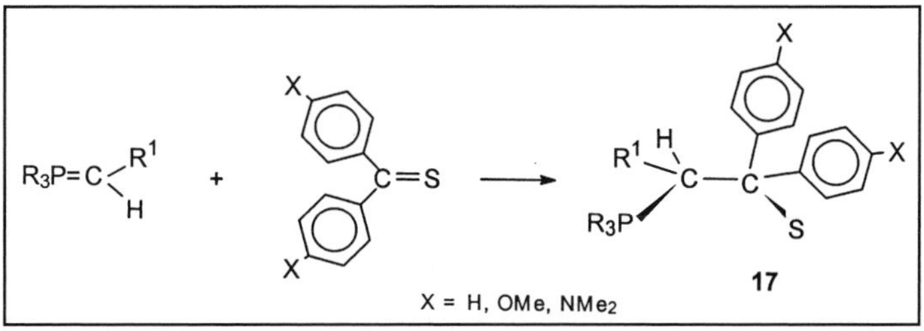

**a:** $Et_3P^+CHMeC(C_6H_4\text{-}p\text{-}NMe_2)_2S^-$

**b:** $Ph_3P^+CHMeC(C_6H_4\text{-}p\text{-}NMe_2)_2S^-$

**c:** $Ph_3P^+CH_2CPh_2S^-$

**d:** $Ph_3P^+CH_2C(C_6H_4\text{-}p\text{-}OMe)_2S^-$

**e:** $MePh_2P^+CH_2C(C_6H_4\text{-}p\text{-}OMe)_2S^-$

**f:** $Me_2PhP^+CH_2C(C_6H_4\text{-}p\text{-}OMe)_2S^-$

**g:** $Me_3P^+CH_2C(C_6H_4\text{-}p\text{-}OMe)_2S^-$

**h:** $c\text{-}PrPh_2P^+CH_2C(C_6H_4\text{-}p\text{-}OMe)_2S^-$

**i:** $c\text{-}Pr_2PhP^+CH_2C(C_6H_4\text{-}p\text{-}OMe)_2S^-$

**j:** $c\text{-}Pr_3P^+CH_2C(C_6H_4\text{-}p\text{-}OMe)_2S^-$

**k:** $c\text{-}Pr_3P^+CH_2C(C_6H_4\text{-}p\text{-}NMe_2)_2S^-$

**l:** $EtPh_2P^+CHMeC(C_6H_4\text{-}p\text{-}OMe)_2S^-$

**m:** $Et_2PhP^+CHMeC(C_6H_4\text{-}p\text{-}OMe)_2S^-$

**n:** $Et_3P^+CHMeC(C_6H_4\text{-}p\text{-}OMe)_2S^-$

SCHEME 9

study of the reactions of thiobenzophenone and its *p*-substituted derivatives with various phosphorus ylides $R_3P=CR'R''$. They obtained a large series of betaines **17** and characterized betaines **17j, n** by X-ray analysis.[12,13,51] The direction of the reaction depends strongly on the polarity of the medium and temperature. For example, at 253 K $Ph_3P$ and thiirane are the main reaction products, whereas olefin and $Ph_3PS$ are formed with the temperature increase. The authors believe that in the Wittig reaction in the series of thiocarbonyl compounds the true structure of the intermediate is determined by the temperature, solvent polarity, and other factors. In their opinion, a continuum of structures intermediate in character exists between thio-betaines and thiaphosphetanes.

The limited range of kinetically stable (under standard conditions) compounds with multiple $E^{14}=X$ bonds and comparatively drastic conditions of generation of the majority of these compounds as intermediates[52–56] forced us to search for detours of betaine **I** preparation, among which the most successful was the reaction of phosphorus and arsenic ylides with organocyclosilthianes and their organogermanium and -tin analogs described in the next section.

## 2. Reactions of Phosphorus and Arsenic Ylides with Organocyclosilthianes and their Organogermanium and -tin Analogs

Organocyclodisilthianes (**18**) and organocyclotrisilthianes (**19**), which from the formal point of view can be considered as dimers and trimers of silanethiones, react readily with "nonstabilized" phosphorus ylides in various solvents at room temperature to form betaines $R_3P^+-CR^1R^2-SiR^3R^4-S^-$ (**20**) in high yields (Scheme 10, Table II). Thioacetone trimer does not react with $Et_3P=CHMe$ under the same conditions.[57,58] According to published data,[59] cyclooligomers of thioaldehydes react with "semistabilized" phosphorus ylides to form olefins, however, experimental data are not presented.

The nature of the substituent at the silicon atom affects substantially the course of this reaction, and the steric effect plays, most likely, the main role. For example, a mixture of oligomers $(Me_2SiS)_n$ ($n = 2, 3$) interacts with $Ph_3P=CMe_2$ for several hours to form **20a** in 90% yield. Oligomers $(Ph_2SiS)_n$ in the same reaction give **20g** in 25% yield. Oligomers $(i\text{-}Pr_2SiS)_n$ do not react with phosphorus ylides under similar conditions.

The nature of ylide is also important in this reaction. The higher the nucleophilicity of the ylide, the faster is the process and the higher is the stability of the betaines formed in solutions. Trialkylalkylidene- and tris(dialkylamino)alkylidenephosphoranes possess the highest reactivity with respect to **18** and **19**. Charge delocalization in the ylide fragment decreases the reactivity. "Semistabilized" phosphorus ylides, e.g., $Ph_3P=CHPh$, do not react with $(Me_2SiS)_n$ under the conditions described above.

All betaines **20** are white or light-yellow crystalline substances, which can be stored for any long time in an inert atmosphere but are very sensitive to atmospheric oxygen and moisture. They are poorly soluble in benzene and ether, moderately soluble in THF and acetonitrile, and are highly soluble in pyridine.

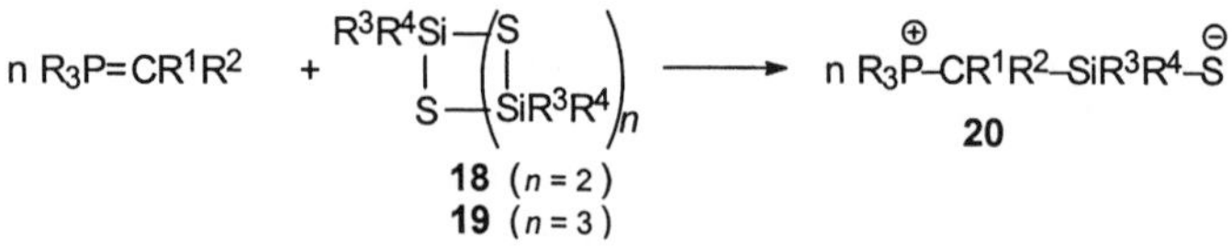

$$n\ R_3P=CR^1R^2 \quad + \quad \begin{array}{c} R^3R^4Si\!-\!S \\ | \qquad | \\ S\!-\!SiR^3R^4 \end{array}\!\Bigg)_n \quad \longrightarrow \quad n\ R_3\overset{\oplus}{P}\!-\!CR^1R^2\!-\!SiR^3R^4\!-\!\overset{\ominus}{S}$$

$$\textbf{20}$$

**18** $(n = 2)$
**19** $(n = 3)$

SCHEME 10

TABLE II
BETAINES I WITH THIOLATE CENTERS

| | Betaine | Solvent | Yield (%) | m.p. (°C) |
|---|---|---|---|---|
| **20a** | $Ph_3P^+CMe_2SiMe_2S^-$ | $Et_2O$ | 91 | |
| | | THF | 77 | 152–153[a] |
| | | $C_6H_6$ | 80 | |
| **20b** | $Ph_3P^+CMe_2Si(CD_3)_2S^-$ | $Et_2O$ | 91 | |
| **20c** | $Ph_3P^+C(CD_3)_2SiMe_2S^-$ | $C_6H_6$ | 78 | |
| **20d** | $Ph_3P^+C(CD_3)_2Me_2Si(CD_3)_2S^-$ | $C_6H_6$ | 86 | |
| **20e** | $Ph_3P^+CMe_2SiMeBzS^-$ | $Et_2O$ | 40 | 80–130[a] |
| **20f** | $Ph_3P^+CMe_2SiMePhS^-$ | $Et_2O$ | 85 | 150–164[a] |
| **20g** | $Ph_3P^+CMe_2SiPh_2S^-$ | $Et_2O$ | 25 | 140–150[a] |
| **20h** | $Ph_3P^+CMe_2SiPh(H)S^-$ | $Et_2O$ | 59 | 145–150[a] |
| **20i** | $Ph_3P^+CMe_2SiEt(H)S^-$ | $Et_2O$ | 95 | |
| **20j** | $Ph_3P^+CMe_2SiMe(OEt)S^-$ | $Et_2O$ | 57 | 60–80[a] |
| **20k** | $Ph_3P^+CHMeSiMe_2S^-$ | $Et_2O$ | 72 | 80–100[a] |
| | | $C_6H_6$ | 38 | |
| **20l** | $Ph_3P^+CH_2SiMe_2S^-$ | $Et_2O$ | 74 | 94–115[a] |
| **20m** | $Ph_3P^+CH_2SiMePhS^-$ | THF-$d_8$ | 59 | |
| | | $C_5D_5N$ | ~100 | |
| **20n** | $Et_3P^+CHMeSiMe_2S^-$ | $Et_2O$ | 92 | 135–136 |
| | | THF | 69 | |
| **20o** | $Et_3P^+CHMeSiPh_2S^-$ | $Et_2O$ | 81 | 139–140 |
| **20p** | $Me_3P^+CH_2SiPh_2S^-$ | THF-$d_8$ | 70 | |
| **20q** | $(Me_2N)_3P^+CMe_2SiMe_2S^-$ | $Et_2O$ | 74 | 130–131 |

[a]With decomposition.

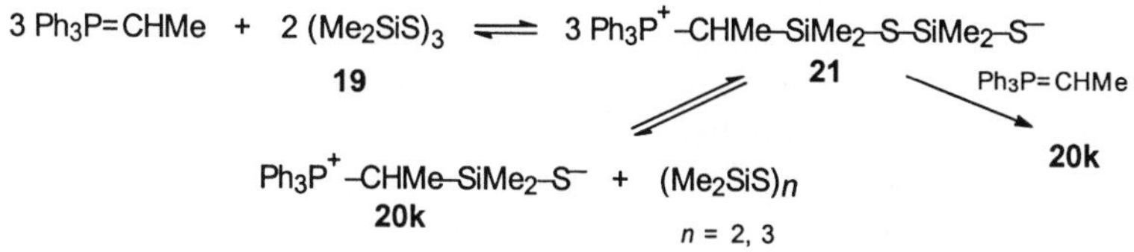

SCHEME 11

The reaction of **18** and **19** with phosphorus ylides occurs as a stepwise process. Betaine (**21**) can be isolated when $(Me_2SiS)_3$ reacts with $Ph_3P=CHMe$ in a 3:2 ratio of the reactants (Scheme 11). This substance is quite stable in the solid state but on dissolving in pyridine it is reversibly transformed into a mixture of **20k** and $(Me_2SiS)_3$. The equilibrium concentration of **21** in a solution at room temperature is at most 28% according to the NMR data, and the addition of one more equivalent of $Ph_3P=CHMe$ to the solution results in the quantitative transformation of **21** into **20k**.

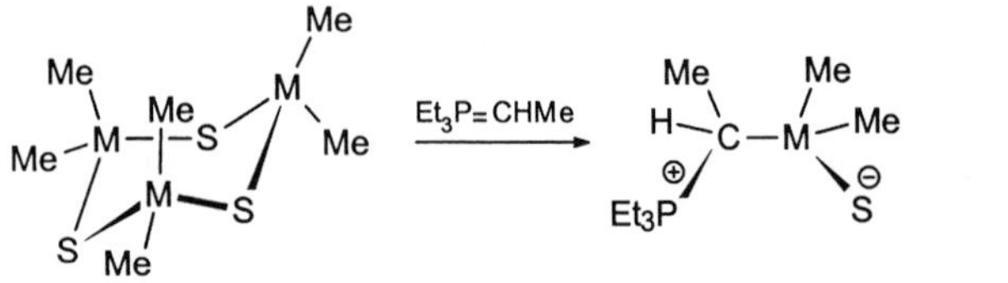

**22** (M = Ge)
**23** (M = Sn)

**24** (M = Ge); yield 84 %
**25** (M = Sn); yield 50-95 %

SCHEME 12

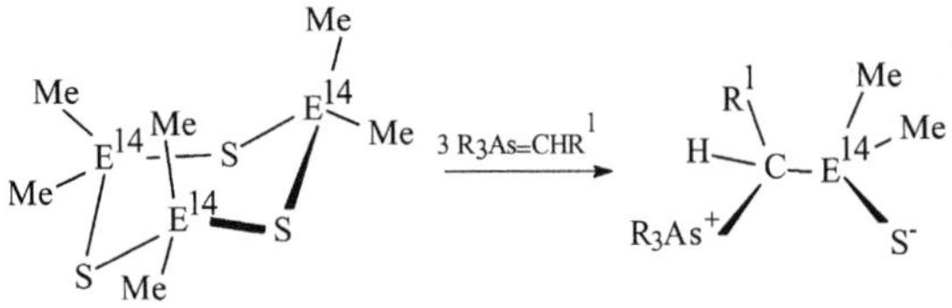

**19** ($E^{14}$ = Si)
**22** ($E^{14}$ = Ge)

**26** ($E^{14}$ = Si)
**27** ($E^{14}$ = Ge)

| N | $E^{14}$ | R | $R^1$ | yield |
|------|------|------|------|------|
| 26a: | Si | Et; | Ph | 45 % |
| 26b: | Si | Et | Ph | ~95 % |
| 26ñ: | Si | Et | Me₃Si | 84 % |
| 27: | Ge | Et | Me₃Si | 76 % |

SCHEME 13

Hexamethylcyclotrigermathiane (**22**) and hexamethylcyclotristannathiane (**23**) also react easily with $Et_3P=CHMe$, which enabled us to obtain the first betaines with the thiolate center in the germanium (**24**) and tin (**25**) series (Scheme 12).[60,61] Both betaines are solid finely crystalline white substances, whose solubility and stability in the solid state and in solutions are similar to those of the silicon analog **20n**.

The resistance of the $E^{14}$–S bond in cyclotrimetallathianes toward nucleophilic reagents, for example, water and alcohol,[62] increases on going from the silicon compounds to the corresponding germanium and tin derivatives. This is due, most likely, to the fact that the reaction of less nucleophilic ylides with phenyl groups at the phosphorus atom with trithianes $(R_2MS)_3$ (M = Ge, Sn) occurs slowly and is impeded by several side processes.

Nucleophilicity of arsenic ylides $R_3As=CR^1R^2$ is much higher than that of phosphorus ylides.[63,64] Therefore, "nonstabilized" arsenic ylides react with cyclothianes $(R_2MS)_n$ (M = Si, Ge) much more vigorously. "Semistabilized" arsenic ylides $R_3As=CHPh$ (R = Et, Ph) and $Et_3As=CHSiMe_3$ also react readily with $(Me_2SiS)_n$ and $(Me_2GeS)_n$ to form betaines (**26**) and (**27**) in a high yield[65] (Scheme 13). Betaines **26a**, **26c**, and **27** are stable crystalline white compounds resembling in properties and solubility

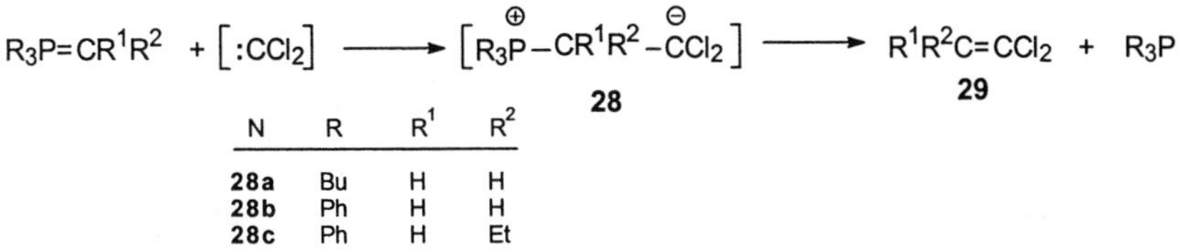

$$R_3P=CR^1R^2 + [:CCl_2] \longrightarrow \left[ \overset{\oplus}{R_3P}-CR^1R^2-\overset{\ominus}{CCl_2} \right] \longrightarrow R^1R^2C=CCl_2 + R_3P$$

28      29

| N | R | $R^1$ | $R^2$ |
|---|---|---|---|
| 28a | Bu | H | H |
| 28b | Ph | H | H |
| 28c | Ph | H | Et |

SCHEME 14

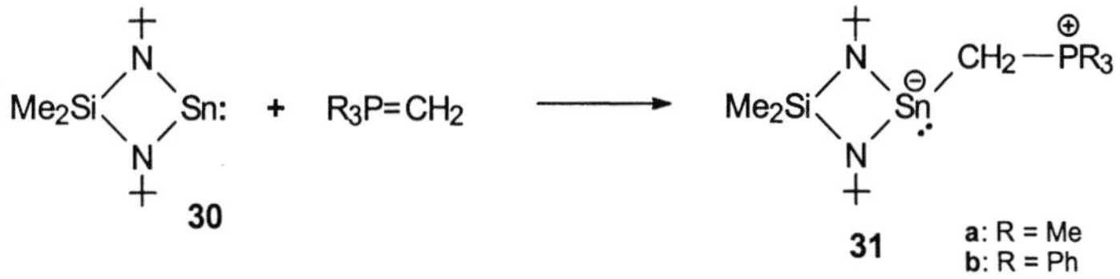

SCHEME 15

organophosphorus betaines **20**. Unlike them, **26b** being stored in a solution for several hours is transformed into disilolane (see Section 5.1).

## C. Synthesis of Betaines II

The reactions of dichlorocarbene with phosphorus ylides result in the corresponding olefins and phosphines.[66–68] In the reaction of dichlorocarbene generated in situ with tributyl- and triphenylmethylenephosphoranes or triphenylethylidenephosphorane, the olefin yield increases as the nucleophilicity of phosphorus ylide increases. According to,[67] the reaction starts from the electrophilic attack of carbene at the $\alpha$-C atom of phosphorus ylide. Then the intermediately formed betaine (**28**) (Scheme 14) decomposes to eliminate the phosphine molecule and form dichloroolefin (**29**).

The first organophosphorus betaines $^{(+)}E^{15}$–$C$–$E^{14(-)}$ (**31**) with the negative charge on the atom of the Group 14 element were prepared by Veith and Huch[69] in the reaction of cyclic stannylene (**30**) with phosphorus ylides (Scheme 15).

Many of the presently known stable silylenes, germylenes, and stannylenes, among which are cyclic diaminosilylene (**32**),[70] dialkoxygermylene $Ge(OCH_2CH_2NMe_2)_2$ (**33**),[71,72] and diphenoxystannylene $Sn(OC_6H_2CH_2NMe_2-2,4,6)$ (**34**),[73] contain electron-donating substituents at the element atom, and hence can be classified as nucleophilic analogs of carbenes. Due to this, their reactivity in the reaction with phosphorus ylides at the element atom is decreased. We found that the reaction of **32** with trimethylmethylenephosphorane occurs with the formation of silylated ylide (**36**) as the

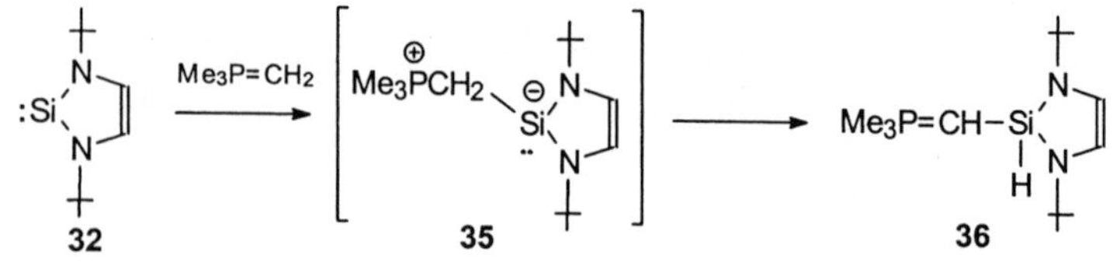

SCHEME 16

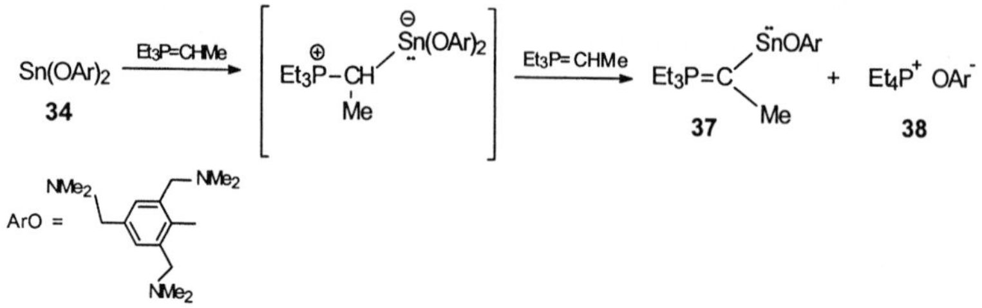

SCHEME 17

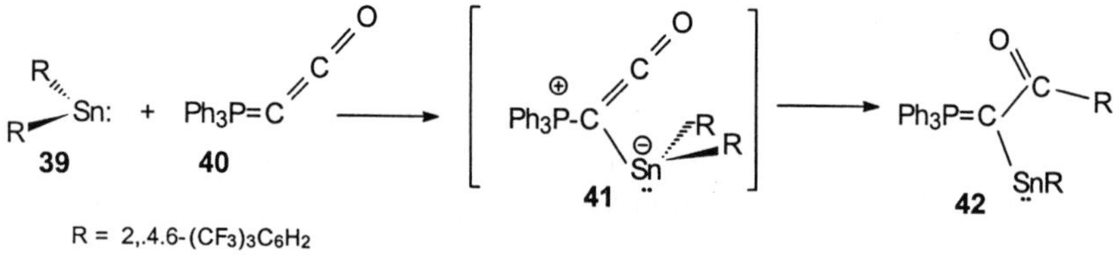

SCHEME 18

final product.[74] Evidently, the first step of the reaction gives betaine (35), which is further isomerized to 36 (Scheme 16). Similar isomerization processes are considered in more detail in Section 5.4.

The reaction of 34 with triethylethylidenephosphorane is more complex. According to the multinuclear NMR data, the reaction occurs at the 1 : 2 ratio of the reactants. The Sn–O bond is cleaved to give phosphonium phenoxide (38) and stannylene (37) in which the tin atom is also bound to the ylide carbon atom of phosphorane (Scheme 17).[61] Metallation reactions of this type are well known.[61,75]

According to the data of Grützmacher *et al.*, the isomerization of the proposed intermediate betaine 41, which is formed in the reaction of bis[2,4,6-tris(trifluoromethyl)phenyl]stannylene 39 with phosphorus ylide 40, affords stannylene 41 (Scheme 18).[76]

## III

## STRUCTURE OF BETAINES $^{(+)}E^{15}$–C–$E^{14}$–$X^{(-)}$ I AND $^{(+)}E^{15}$–C–$E^{14(-)}$ II ACCORDING TO X-RAY DATA

The most important geometric parameters obtained by the X-ray diffraction study of betaines **I** and **II** are presented in Table III. The structure of several compounds is shown in Figs. 1 and 2.

The strong intramolecular Coulomb interaction between the differently charged centers $^{(+)}E^{15}$ and $S^{(-)}$ in molecules of betaines **I** (see Section 6) results in the situation when all of them have a *gauche*-conformation of the main chain. In the organophosphorus betaines the dihedral $E^{15}$–C–$E^{14}$–S angles range from 38 to 56°, and the nonvalent $P \cdots S$ distances are longer than the sum of the covalent radii of phosphorus and sulfur (2.3 Å), but for betaines of the carbon series they are much shorter than the sum of their van der Waals radii (3.9 Å). This results in considerable steric strains, which appear as the elongation of all bonds of the main $E^{15}$–C–$E^{14}$–S chain, especially of the central C–$E^{14}$ bonds. For example, in thiabetaines of the carbon series **17** studied by Erker and coworkers[12,13,51] and betaines of the silicon series **20** studied by us,[15,77] the C–C and C–Si bond lengths exceed by more than 0.1 Å the statistical mean values. The C–Si–$S^-$ and $^+$P–C–Si bond angles are increased compared to the ideal tetrahedral angle of 109.5°, and the phosphorus and arsenic atoms gain a distorted tetrahedral configuration. The Si–$S^{(-)}$ bond lengths in **20** are shorter than the typical ordinary bonds (2.145 Å[78]) but longer than the corresponding double Si=S bonds (1.948(4) Å[79]). By contrast, in carbon analogs **17** the C–$S^{(-)}$ bond length coincides or even is a little longer than the statistical mean value of the C–S ordinary bond (1.820 Å).[78] These specific features of the betaine structure reflect their reactivity (see Section 5).

The ethylation of betaine **20a** at the sulfur atom results in the disappearance of the anionic center, which eliminates the strong Coulomb interaction, and a molecule of salt **43** takes the most favorable *trans*-conformation.

The introduction of more bulky groups to the silicon atom and, by contrast, a decrease in steric strains at the carbon atom in betaines of the silicon series decreases the dihedral P–C–Si–S angle and shortens the $^+P \cdots S^-$ nonvalent contact and S–Si bond length. A comparison of the geometric parameters for betaines **20o** and **20a, q** indicates explicitly a tendency for four-membered cycle closure when such changes are introduced.

The geometric structure of organogermanium betaine **24** is similar, as a whole, to that of the corresponding silicon analog **20o**. The main distinctions are an additional shortening of the P–C bond lengths and a further

## TABLE III

Some X-ray Parameters for Betaines of the Type $^{(+)}E^{15}{-}C{-}E^{14}{-}S^{(-)}$ **I**, $^{(+)}E^{15}{-}C{-}E^{14(-)}$ **II** and the Salt $^{(+)}Ph_3P{-}CMe_2{-}SiMe_2{-}SEt\ Br^{(-)}$ (**43**)

| Compound | $E^{15}{-}C$ | $C{-}E^{14}$ | $E^{14}{-}S$ | $E^{15}{\cdots}S$ | $E^{15}{-}C{-}E^{14}$ | $C{-}E^{14}{-}S$ | $E^{15}{-}C{-}E^{14}{-}S$ | Ref. |
|---|---|---|---|---|---|---|---|---|
| *P–C–C–S betaines* | | | | | | | | |
| $Ph_3P^+{-}CMe_2{-}CS_2^-$ **13** | 1.880(3) | 1.526(4) | 1.682(3) 1.676(3) | 3.247 | 111.7(2) | 118.3(2) | 40.9 | 50 |
| $Et_3P^+{-}CHMe{-}C(C_6H_4OMe\text{-}p)_2{-}S^-$ **17n** | 1.845(7) | 1.554(8) | 1.833(6) | 3.109(5) | 113.0(4) | 105.9(4) | 47.7(5) | 13 |
| $c\text{-}Pr_3P^+{-}CH_2{-}C(C_6H_4NMe_2p)_2{-}S^-$ **17j** | 1.806(3) | 1.557(4) | 1.841(3) | 3.312(2) | 117.6(2) | 107.6(2) | 52.9(3) | 14 |
| *P–C–Si–S betaines* | | | | | | | | |
| $Ph_3P^+{-}CMe_2{-}SiMe_2{-}S^-$ **20a** | 1.825(4) | 1.986(4) | 2.048(2) | 3.988(4) | 115.6(2) | 114.6(1) | 56.1(2) | 16, 77 |
| $Et_3P^+{-}CHMe{-}SiPh_2{-}S^-$ **20o** | 1.811(3) | 1.934(4) | 2.044(2) | 3.681(4) | 112.8(1) | 114.7(2) | 38.2(2) | 77 |
| $(Me_2N)_3P^+{-}CMe_2{-}SiMe_2{-}S^-$ **20q** | 1.830(4) | 1.979(3) | 2.037(2) | 3.980(4) | 115.7(2) | 116.9(1) | 50.4(2) | 16, 77 |
| *P–C–Ge–S betaine* | | | | | | | | |
| $Et_3P^+{-}CHMe{-}GeMe_2{-}S^{-a}$ **24** | 1.783(3) 1.785(2) | 2.032(3) 2.041(3) | 2.140(7) 2.140(7) | 3.774(2) 3.810(2) | 114.35(13) 114.88(12) | 114.86(7) 115.47(7) | 25.9(2) 27.3(2) | 60 |
| *As–C–Si–S betaine* | | | | | | | | |
| $Et_3As^+{-}CHPh{-}SiMe_2{-}S^-$ **26a** | 1.936(2) | 1.946(4) | 2.0455(11) | 3.518(1) | 111.44(17) | 109.97(8) | 33.13(19) | 65 |
| *P–C–M⁻ betaine* | | | | | | | | |
| $Ph_3P^+\ CH_2Sn^-{:}(NBu\text{-}t)_2SiMe_2$ **31** | 1.753(8) | 2.442(6) | – | – | 123.3(3) | – | – | 81 |
| $Ph_3P^+\ CH_2Sn^-{:}(NBu\text{-}t)_2SiMe_2\cdot C_6H_6$ **31 · C₆H₆** | 1.74(1) | 2.40(1) | – | – | 121.9(4) | – | – | 69, 81 |
| **44** | 1.710(5) | 2.278(5) 2.293(5) | – | – | 105.1(2) 119.2(2) | – | – | 81 |
| *P,Si,S-salt* | | | | | | | | |
| $Ph_3P^+{-}CMe_2{-}SiMe_2{-}S{-}Et]Br^-$ **43** | 1.839(5) | 1.933(6) | 2.140(3) | 5.010(5) | 116.6(3) | 107.3(2) | 177.0(2) | 16, 77 |

Distances (Å) and angles (°).
aFor two independent molecules.

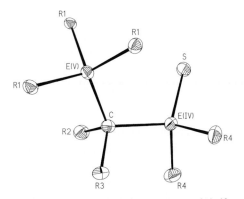

FIG. 1. Geometry of main structure fragment of thiabetaines $^{(+)}E^{15}$–C–$E^{14}$–$S^{(-)}$ **17j, n, 20a, o, q, 24,** and **26a.**

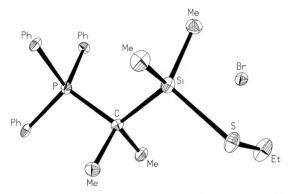

FIG. 2. Geometry of main fragment of salt $Ph_3P^+$–$CMe_2$–$SiMe_2$–SEt $Br^-$ **43.**

decrease in the torsion angle of the main chain with a simultaneous increase in the $^{(+)}P \cdots S^{(-)}$ distance. Thus, steric strains in a molecule of **20** are substantially weakened, which is reflected as an increase in the thermal stability of the compound (see Section 5).

In silicon-arsonium betaine **26a** the dihedral angle of the main chain increases compared to the above considered betaine of the germanium series but a very short nonvalent As$\cdots$S contact (3.518 Å) is retained in the molecule, which is much shorter than the sum of van der Waals radii of the corresponding atoms ($\sim 4.5$ Å for S and As), although it much exceeds the As–S ordinary bond lengths (2.20–2.50 Å[80]). The nonvalent contact of the hydrogen atom in the *o*-position of the phenyl ring with the As atom (As(1)$\cdots$H(9) 3.23 Å) indicates considerable steric strains in the molecule. The coordination of the As atom in it can be considered as either distorted tetrahedral or distorted trigonal-bipyramidal with a significant shift of the arsenic atom (0.655 Å) from the equatorial plane of the C(1)–C(2)–C(6) trigonal bipyramid. These specific features of the structure of arsonium betaine **26a** are manifested in the easiness of its selective decomposition according to the Corey–Chaykovsky type and reflected in its NMR spectral parameters.

Betaines of type **II** are presented by only three structures **31**, **31** · $C_6D_6$, and **44** and, therefore, no conclusion about the regularities in their structure can be drawn. Note that in **31** the P–C distances are shortened compared to those in betaines **I** and the standard P–C bond length (1.800 Å).[78] Crystal solvate **31** · $C_6D_6$ is a rotamer of betaine **31**. Their main structural distinction is the existence of a short nonvalent contact between the hydrogen atom in the *o*-position of one of the Ph ring at the phosphorus atom and the nitrogen atom in the four-membered heterocycle $Me_2Si(NBu$-$t)_2Sn$, which is considered by the authors of [69] as a bridging hydrogen bond C–H · · · N.

Polycyclic compound **44**, which is formed from **31**, can also be considered as a dibetaine with an unusual structure. Its two five-membered rings are arranged perpendicularly to the central four-membered cycle due to the orthogonal arrangement of ligands typical of tin atoms with the trigonal-pyramidal configuration.[81] This betaine can be considered by the type of tin bonds as a triorganostannate. Note again the unusually short P–C bond in this compound, which is by 0.04 Å shorter than that in adduct **31** and by only 0.05 Å longer than that in $Ph_3P=CH_2$.

It follows from the above X-ray data that betaines **I** and **II** have some structural peculiarities. Two main peculiarities are especially pronounced for betaines **I**. These compounds have the sterically strained *gauche*-conformation of the main chain due to the intramolecular Coulomb interaction of the cationic and anionic centers and noticeable distortions of the bond lengths in it. In Section 5 we discuss how these peculiarities of the betaine structure reflect their reactivity.

# IV

## NMR PARAMETERS OF BETAINES $^{(+)}E^{15}$–C–$E^{14}$–$X^{(-)}$ I AND

## $^{(+)}P$–C–$E^{14(-)}$ II

Multinuclear NMR spectroscopy is a very informative and reliable method for the identification and study of betaines of types **I** and **II** in solutions. The main NMR parameters of these compounds are presented in Table IV. The data for some of their carbon analogs **17** are presented for comparison.

## TABLE IV

SOME NMR PARAMETERS FOR BETAINES $^{(+)}E^{15}$—C—$E^{14}$—$S^{(-)}$ **I**, $^{(+)}E^{15}$—C—$E^{14(-)}$ **II**, AND SALT $^{(+)}Ph_3P$—$CMe_2$—$SiMe_2$—$SEt$ $Br^{(-)}$ (**43**)

| Betaine | $\delta_H$ (P$^+$—CH$_n$), ppm ($^2J_{PH}$; $^3J_{HH}$, Hz) | $\delta_C$ (E$^{15+}$CE$^{14}$), ppm ($^1J_{CP}$, Hz) | $\delta_P$, ppm | $\delta_{Si}$, ppm ($^2J_{SiP}$, Hz)/or $\delta_{PCC(Ar)}$ ($^2J_{CP}$, Hz) | Ref. |
|---|---|---|---|---|---|
| *P—C—C* betaine | | | | | |
| **1** Me$_3$P$^+$CH$_2$CMe$_2$C$_5$H$_4^-$ | 2.16 d (12.5) | — | — | — | 20 |
| *P—C—E$^{14}$—C$^-$ betaines* | | | | | |
| **5a** Me$_3$P$^+$CH$_2$SiMe$_2$C$_5$H$_4^-$ | 1.21 d (17.8) | — | — | — | 20 |
| **5c** Me$_3$P$^+$CH$_2$Si(CH$_2$SiMe$_3$)$_2$C$_5$H$_4^-$ | 1.16 d (17.8) | — | — | — | 20 |
| **5d** Me$_3$P$^+$CH$_2$Si(CH$_2$SiMe$_3$)$_2$C$_{13}$H$_9^-$ | — | 13.35 d (45.1) | 22.1 | −13.11 d (5.0) 0.03 d (1.1) | 33, 34 |
| **5e** Me$_3$P$^+$CH$_2$SiMe(Bu − *s*)C$_{13}$H$_9^-$ | — | 9.68 d (46.4) | 21.8 | −10.6 d (5.3) | 33, 34 |
| **6** Me$_3$P$^+$CH$_2$Ge(Bu − *i*)$_2$C$_{13}$H$_9^-$ | — | 8.26 d (44.4) | — | — | 35 |
| *P—C—C—S$^-$ betaines* | | | | | |
| **17a** Et$_3$P$^+$CHMeC(C$_6$H$_4$-*p*-NMe$_2$)$_2$S$^-$ | 2.20 dq (7.7; 12.3) | 49.0 d (67.6) | | | 12, 58 |
| **17b** Ph$_3$P$^+$CHMeC(C$_6$H$_4$-*p*-NMe$_2$)$_2$S$^-$ | 2.26 m, br (obscured) | | 27.33 | | 12, 58 |
| **17c** Ph$_3$P$^+$CH$_2$CPh$_2$S$^-$ | — | 52 (82) | 5 | | 13, 14 |
| **17d** Ph$_3$P$^+$CH$_2$C(C$_6$H$_4$-*p*-OMe)$_2$S$^-$ | 5.18 d (9.8) | 51.8 d (82.1) | 0.4 | | 14 |
| **17e** MePh$_2$P$^+$CH$_2$C(C$_6$H$_4$-*p*-OMe)$_2$S$^-$ | 5.01 d (10.0) | 51.6 d (78.8) | 1.9 | 51.9 d (2.5) | 14 |
| **17f** Me$_2$PhP$^+$CH$_2$C(C$_6$H$_4$-*p*-OMe)$_2$S$^-$ | 4.37 d (10.6) | 50.9 d (73.9) | 9.2 | 52.2 d (2.7) | 14 |
| **17g** Me$_3$P$^+$CH$_2$C(C$_6$H$_4$-*p*-OMe)$_2$S$^-$ | 4.05 d (10.5) | 49.1 d (75.1) | 11.5 | 51.9 d (3.3) | 14 |
| **17h** c-PrPh$_2$P$^+$CH$_2$C(C$_6$H$_4$-*p*-OMe)$_2$S$^-$ | 5.16 d (10.0) | 52.1 d (77.9) | 13.9 | 51.9 d (2.1) | 14 |
| **17i** c-Pr$_2$PhP$^+$CH$_2$C(C$_6$H$_4$-*p*-OMe)$_2$S$^-$ | 4.59 d (9.9) | 48.8 d (81.2) | 16.8 | 51.9 d (2.2) | 14 |
| **17g** c-Pr$_3$P$^+$CH$_2$C(C$_6$H$_4$-*p*-OMe)$_2$S$^-$ | 3.89 d (10.6) | 46.0 d (71.7) | 29.9 | 51.1 d (1.9) | 14 |
| **17k** c-Pr$_3$P$^+$CH$_2$C(C$_6$H$_4$-*p*-NMe$_2$)$_2$S$^-$ | 3.89 d (10.2) | 46.4 d (72.2) | 27.7 | 50.8 d (1.8) | 14 |
| **17l** EtPh$_2$P$^+$CHMeC(C$_6$H$_4$-*p*-OMe)$_2$S$^-$ | 5.50 q (0; 6.6) | 57.4 d (70.1) | 14.4 | 58.3 d (0) | 14 |
| **17m** Et$_2$PhP$^+$CHMeC(C$_6$H$_4$-*p*-OMe)$_2$S$^-$ | 4.83 q (0; 6.7) | 53.8 d (67.4) | 25.3 | 58.2 d (0) | 14 |

*(Continued)*

## TABLE IV

Some NMR Parameters for Betaines $^{(+)}E^{15}$–C–$E^{14}$–S$^{(-)}$ **I**, $^{(+)}E^{15}$–C–$E^{14}$–S$^{(-)}$ **II**, and Salt $^{(+)}$Ph$_3$P–CMe$_2$–SiMe$_2$–SEt Br$^{(-)}$ (**43**)

| | Betaine | $\delta_H$ (P$^+$–CH$_n$), ppm ($^2J_{PH}$; $^3J_{HH}$, Hz) | $\delta_C$ (E$^{15}$+CE$^{14}$), ppm ($^1J_{CP}$, Hz) | $\delta_P$, ppm | $\delta_{Si}$, ppm ($^2J_{SiP}$, Hz)/or $\delta_{PCC(Ar)}$ ($^2J_{CP}$, Hz) | Ref. |
|---|---|---|---|---|---|---|
| **17n** | Et$_3$P$^+$CHMeC(C$_6$H$_4$-$p$-OMe)$_2$S$^-$ | 4.65 dq (2.4; 6.8) | 48.5 d (67.7) | 27.5 | 57.6 (0) | 14 |
| | *P–C–Si–S$^-$ betaines* | | | | | |
| **15a** | Me$_3$P$^+$CH$_2$SiPhNft–S$^-$ | — | 18.1 (45.5) | 26.5 | −7.1 (2.0) | 15 |
| **20a** | Ph$_3$P$^+$CMe$_2$SiMe$_2$S$^-$ or (Ph$_3$P$^+$CMe$_2$SiMe$_2$S$^-$)$_2$·LiBr | — | 27.7 d (22.9) | 39.7 | 13.8 d (2.2) | 58 |
| **20e** | Ph$_3$P$^+$CMe$_2$SiMeBzS$^-$ | | 28.7 d (23.2) | 38.8 | 13.6 d (1.5) | 58 |
| **20f** | Ph$_3$P$^+$CMe$_2$SiMePhS$^-$ | | 28.9 d (21.7) | 40.4 | 8.0 d (2.9) | 58 |
| **20g** | Ph$_3$P$^+$CMe$_2$SiPh$_2$S$^-$ | | 30.9 d (23.3) | 40.4 | 1.3 d (3.6) | 58 |
| **20h** | Ph$_3$P$^+$CMe$_2$SiPh(H)S$^-$ | — | 25.2 d (22.9) | 39.8 | — | 58 |
| **20i** | Ph$_3$P$^+$CMe$_2$SiEt(H)S$^-$ | — | 24.7 d (23.8) | 39.2 | 9.5 d (1.3) | 58 |
| **20j** | Ph$_3$P$^+$CMe$_2$SiMe(OEt)S$^-$ | — | 26.7 d (20.7) | 40.1 | — | 58 |
| **20k** | Ph$_3$P$^+$CHMeSiMe$_2$S$^-$ | 3.17 dq (18.3; 7.2) | 17.8 d (32.1) | 32.9 | 6.7 d (2.4) | 58 |
| **20l** | Ph$_3$P$^+$CH$_2$SiMe$_2$S$^-$ | 2.90 d (17.9) | 15.2 d (38.1) | 26.7 | 0.2 d (6.3) | 58 |
| **20m** | Ph$_3$P$^+$CH$_2$SiMePhS$^-$ | $\delta_A$ 2.85, $\delta_M$ 2.72 ($J_{AX}=J_{MX}=17.5$) ($J_{AM}=14.5$) | 15.1 d (39.5) | 26.5 | −4.3 d (5.6) | 58 |
| **20n** | Et$_3$P$^+$CHMeSiMe$_2$S$^-$ | 1.54 dq (18.1; 7.5) | 14.9 d (35.9) | 43.7 | 4.4 d (2.2) | 58 |
| **20o** | Et$_3$P$^+$CHMeSiPh$_2$S$^-$ | 2.32–2.51 m$^b$ | 13.4 d (35.1) | 44.7 | −3.2 d (1.0) | 58 |
| **20p** | Me$_3$P$^+$CH$_2$SiPh$_2$S$^-$ | 2.09 d (17.6) | 14.2 d (45.5) | 26.1 | −9.6 d (4.4) | 58 |
| **20q** | (Me$_2$N)$_3$P$^+$CMe$_2$SiMe$_2$S$^-$ | — | 30.2 d (83.9) | 74.1 | 14.1 d (1.8) | 58 |
| **21** | Ph$_3$P$^+$CHMeSiMe$_2$SSiMe$_2$S$^-$ | ~3.10 (obscured) | 15.9 d (23.4) | 31.7 | 16.1 (SSiS) 17.2 br (PCSi) ($\nu_{1/2}=5.35$ Hz) | 58 |
| | *P–C–Ge–S$^-$ betaine* | | | | | |
| **24** | Et$_3$P$^+$CHMeGeMe$_2$S$^-$ | 1.56 dq (15.6; 7.6) | 14.1 d (38.2) | 43.5 | — | 60 |

| | | | | | |
|---|---|---|---|---|---|
| **25** | *P–C–Sn–S⁻ betaine* Et$_3$P$^+$CHMeSnMe$_2$S$^-$ | br | 7.89 (37.0) | 43.6 | – | 61 |
| **26a** | *As–C–Si–S⁻ betaines* Et$_3$As$^+$CH(Ph)SiMe$_2$S$^-$ | 3.69 s | 39.52 | – | – | 65 |
| **26c** | Et$_3$As$^+$CH(SiMe$_3$)SiMe$_2$S$^-$ | 1.41 s | 15.93 | – | 0.24 (SiMe$_2$); 1.88 (SiMe$_3$) | 65 |
| **27** | *As–C–Ge–S⁻ betaine* Et$_3$As$^+$CH(SiMe$_3$)GeMe$_2$S$^-$ | 1.37 s | 14.75 | – | – | 65 |
| **31** | *P–C–Sn betaines* | 0.3 d (12) | – | 18 | – | 69 |
| **31 · C₆H₆** | | 1.62 d (10) | – | 24 | – | 69 |

The parameters of organophosphorus betaines **5, 6, 15, 17, 20, 21, 24, 25,** and **31** and the corresponding phosphonium cations $R_3P^+-CHR^1R^2$, as well as cations of silylated phosphonium salts $R_3P^+-CHR^1-SiR_3^2$,[82–84] are quite similar. The $^{31}P$ NMR signals of the betaines lie in the region typical of tetracoordinated phosphorus.[85,86]

The downfield shift of the $^{13}C$ signal of the main $P-C-E^{14}-S$ chain relative to those of the corresponding phosphonium cation and parent phosphorus ylide is a specific feature of all phosphorus betaines. It is especially significant for thiabetaines **17** of the carbon series (by 37–43 ppm relative to those of phosphonium cations and by 56–72 ppm relative to those of ylides) and is much lower than those of organometallic betaines **15, 20, 21, 24,** and **25.** All direct spin–spin coupling constants $^1J_{PC}$ for betaines **17** are much higher (by 11–25 Hz) than those in the corresponding phosphonium cations (49–61 Hz)[13] but somewhat lower than those in the preceding phosphorus ylides (96–117 Hz).[85,86] By contrast, in Si-, Ge-, and Sn-betaines of the thiolate and carbanionic series $^1J_{PC}$ decrease by 10–26 Hz[58] compared to their values in cations of phosphonium salts $R_3P^+-CHR^1R^2$.

The geminal coupling constants $^2J_{PH}$ for betaines of the silicon series, silylated phosphonium cations,[87] and silicon-organophosphorus betaines **5** and **6** with the cyclopentadieneylide or fluoreneylide anionic centers lie in the region of 17.5–18.3 Hz, i.e., they are by $\sim$4–6 Hz higher than those in the corresponding unsubstituted phosphonium cations. By contrast, for betaines of the carbon series **17** they are somewhat lower (0–12.3 Hz). The solvent polarity and the presence of lithium salts have no substantial effect on the spectral parameters of thiabetaines of the silicon series in solutions.[58] According to the $^{13}C$ and $^{31}P$ CP MAS NMR spectra, the structures of these betaines in solution and crystal are the same.[58] The phosphorus thiabetaine of the germanium **24** and tin **25** series resembles in spectral parameters silicon analogs **20**.[60,61] Erker and coworkers[13,51] studied in detail the NMR spectra of a large set of thiabetaines of the carbon series **17** and established that their spectral parameters change significantly in different solvents. According to the opinion advanced in,[13] the structures of betaines **17** in low-polarity solvents and in the crystalline state are also identical.

The spectra of arsonium thiabetaine **26a** at room temperature exhibits a noticeable broadening of the resonance $C_o$ and $C_m$ signals in the $^{13}C$ NMR spectrum and of the $H_o$ signal in the $^1H$ NMR spectrum, which indicates, most likely, restricted rotation around the C–Ph bond.[65] The presence of the chiral carbon atom in a molecule of thiabetaines results in the diastereotopic doubling of signals of the organic groups at the silicon, germanium or tin atom. The doubling of signals of the substituents at the $\alpha$-C atom of $E^{15}-C-E^{14}$ is observed when the substituents at the $E^{14}$ atom are different in nature.

## V

## REACTIVITY OF BETAINES $^{(+)}E^{15}$–C–$E^{14}$–$X^{(-)}$ I AND

## $^{(+)}E^{15}$–C–$E^{14(-)}$ II ($E^{15}$ = P, As; $E^{14}$ = Si, Ge, Sn; X = S)

Betaines **I** and **II** contain several reaction centers, which predetermines their potentially rich and diverse reactivity. The reactivity of silicon organophosphorus betaines **I** bearing the thiolate center was studied in most detail.

### A. *Photo- and Thermodecomposition of Thiobetaines $^{(+)}E^{15}$–C–Si–$S^{(-)}$ I ($E^{15}$ = P, As)*

As mentioned above (see Scheme 1), three main directions of the decomposition of intermediates that formed are possible when phosphorus and arsenic ylides react with compounds bearing C=X bonds:[5,6,19,63,64,88] (i) elimination of $R_3E^{15}$=X to form olefins (Wittig type reaction); (ii) retro-Wittig type decomposition; and (iii) elimination of $R_3E^{15}$ and formation of three-membered cycles (Corey–Chaykovsky type reaction). According to the data of Erker and coworkers,[12,13,51] under kinetic control, the reaction of phosphorus ylides with thiocarbonyl compounds also affords phosphines and thiiranes, whose further transformations lead to olefins and $R_3PS$ under thermodynamic control.

Available experimental data suggest that the decomposition of betaines **I** occurs via direction (iii) with $R_3E^{15}$ elimination giving three-membered heterocycles or via retro-Wittig type (ii) to eliminate $R_3E^{15}$=$CR^1R^2$ leading to the compounds with an $E^{14}$=X bond (Scheme 19).

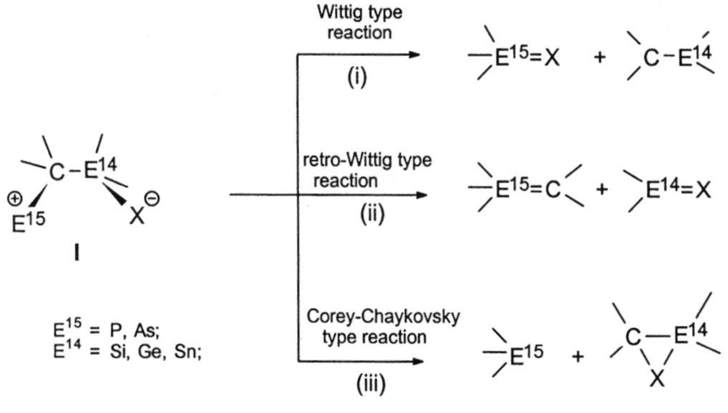

SCHEME 19

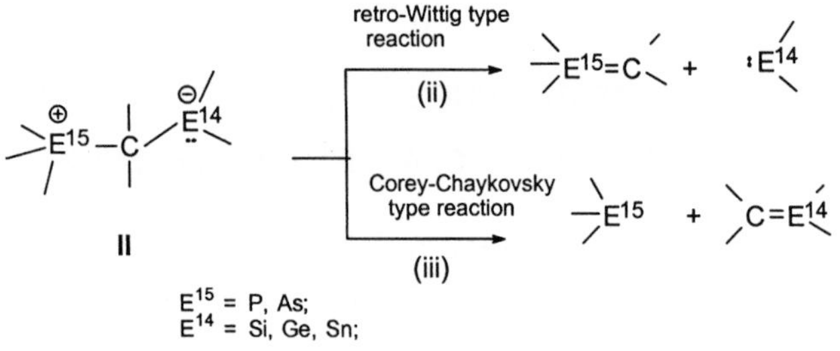

$E^{15}$ = P, As;
$E^{14}$ = Si, Ge, Sn;

SCHEME 20

According to quantum-chemical calculations, decompositions of two last types are possible for betaines of type **II** (see Section 6). Retro-Wittig decomposition (ii) is the process inverse to their formation. The direction (iii) resulting in the formation of elementaolefins is much more interesting (Scheme 20).

### 1. Photodecomposition of Thiobetaines I

UV irradiation stimulates the intramolecular charge transfer $S^- \rightarrow {}^{(+)}E^{15}$, which results in the ${}^{(+)}E^{15}$–C bond cleavage and decomposition of betaines ${}^{(+)}E^{15}$–C–Si–$S^{(-)}$ ($E^{15}$ = P, As) by the Corey–Chaykovsky type reaction. When a suspension of betaines **20a** in benzene is irradiated by a medium-pressure mercury lamp at 20 °C, they decompose to form triphenylphosphine and silathiiranes (**45a**)[89] (Scheme 21).

Unlike carbon analogs, silathiirane **45a** is kinetically unstable. Similarly, to other sterically nonhindered elementathiiranes of Group 14,[90–93] it forms cyclodimer (**46a**) in a high yield or reacts with "chemical traps" if the latter are present in the reaction medium. GC/MS analysis of the photolysis products of a suspension of **20a** in benzene with an addition of acetone shows (**47a**), the product of the reaction of silathiirane **45a** with acetone, along with dimer **46a**, which is the main reaction product. The starting betaine **20a** can also act as a trapping agent with respect to silathiiranes **45a** to form the 1 : 1 adduct (**48a**). Further elimination of $Ph_3P=CMe_2$ from **48a** results in dithiadisilolane (**49a**).[89] Under homogeneous conditions, the concentration of **20a** is rather high, and dithiadisilolane **49a** becomes the main product of **20a** photolysis in pyridine; the yield of **49a** increases to 57% and that of **46a** decreases to 24%.

Photodecomposition under homogeneous conditions of betaine $Et_3As^+$–CHPh–$SiMe_2S^-$ **26a** occurs similarly and gives selectively $Et_3As$ and corresponding dithiadisilolane **49b** (Scheme 22).[65]

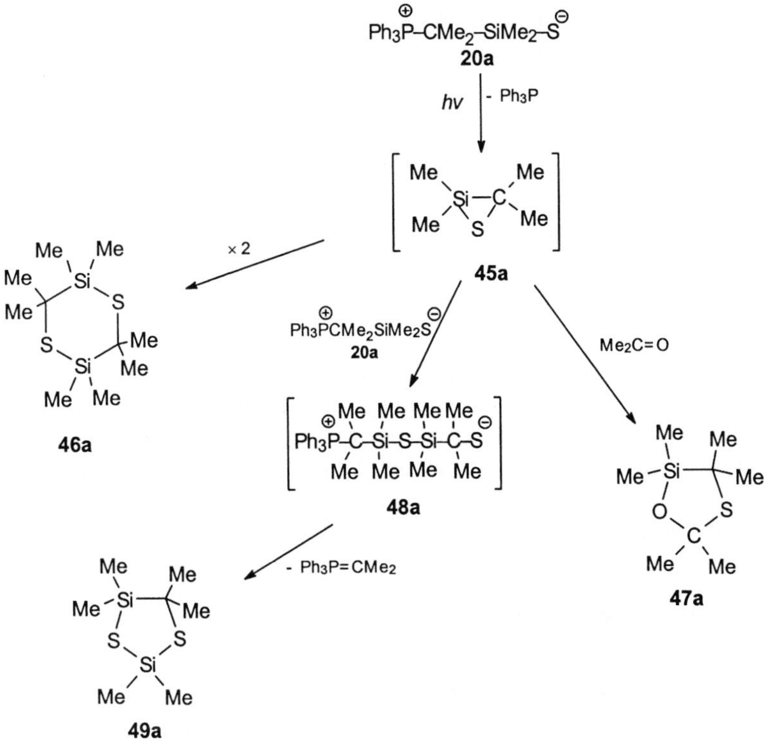

SCHEME 21

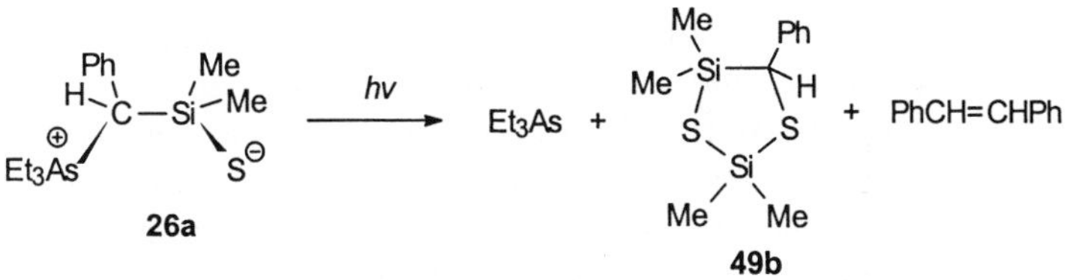

SCHEME 22

## 2. Thermodecomposition of Thiobetaines I

Organosilicon betaines with the thiolate center containing alkyl or dialkylamino groups at the phosphorus atom are rather resistant toward thermolysis. For example, betaines $Alk_3P^+-CR^1R^2-SiR_2^3-S^-$ **20n, o, p, q** in pyridine solutions remain stable on heating to $150\,^\circ$C under anaerobic conditions for several hours.[89] Betaines with phenyl substituents at the phosphorus atoms are less thermoresistant. On heating a solution of **20a** in

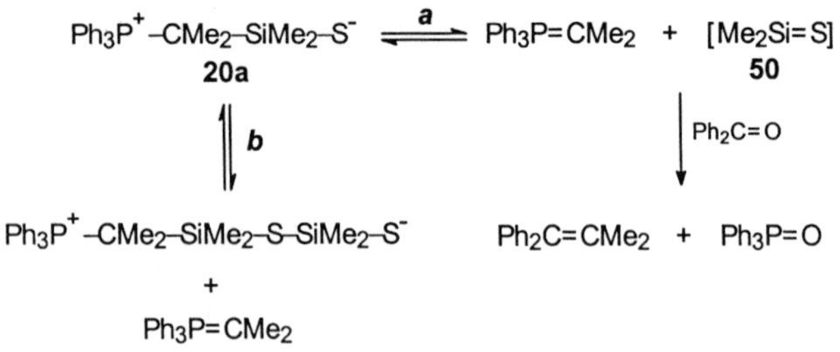

$$Ph_3P^+-CMe_2-SiMe_2-S^- \xrightleftharpoons{\textbf{a}} Ph_3P=CMe_2 \;+\; [Me_2Si=S]$$

$$\textbf{20a} \hspace{6cm} \textbf{50}$$

$$\Bigg\Updownarrow \textbf{b} \hspace{6cm} \Bigg\downarrow Ph_2C=O$$

$$Ph_3P^+-CMe_2-SiMe_2-S-SiMe_2-S^- \hspace{1cm} Ph_2C=CMe_2 \;+\; Ph_3P=O$$

$$+$$

$$Ph_3P=CMe_2$$

SCHEME 23

$C_5D_5N$ above 80 °C, it gains the characteristic color of phosphorus ylide, and signals of phosphorus ylides and cyclosilathianes appear in the $^1H$, $^{13}C$, and $^{31}P$ NMR spectra. In the presence of an equivalent amount of benzophenone in this solution, 1,1-dimethyl-2,2-diphenylethylene and $Ph_3PO$ are formed in 53% yield at 100 °C for 10 min. This indicates that the retro-Wittig decomposition of **20a** occurs in the solution (Scheme 23, equilibrium **a**). Probably, phosphorus ylide is also formed in the equilibrium bimolecular reaction between two betaine molecules (Scheme 23 equilibrium **b**). The ratio of the contributions of these two reactions is strongly determined by the solvent and temperature.

An increase in the temperature of the solution shifts both equilibria to the right. When betaine **20a** is dissolved in $C_5D_5N$ at room temperature, only signals of the starting betaine are observed in the $^1H$, $^{13}C$, $^{29}Si$, and $^{31}P$ NMR spectra. On heating (90 °C, $\sim 30$ min) of this solution, doublets of methyl groups (characteristic of $Ph_3P=CMe_2$) at 2.00 ppm ($^3J_{PH} = 16.5\,Hz$) in the $^1H$ NMR spectrum, and 20.9 ppm ($^2J_{CH} = 13.6\,Hz$) in the $^{13}C$ NMR spectrum, respectively, and a singlet at $\sim +10$ ppm in the $^{31}P$ NMR spectrum are detected. All signals of phosphorus ylides are broadened. The ratio of concentrations **20a**:$Ph_3P=CMe_2$ is 0.8:1. When the temperature decreases to 50 °C, the ratio increases to 2.9:1, and subsequent repeated heating to 75 °C results in its decrease to 1.46:1. The $^1H$ and $^{29}Si$ NMR spectra of heated solutions of **20a** in $C_5D_5N$ also exhibit somewhat broadened signals of cyclooligomers $(Me_2SiS)_n$ in the region of 0.7–1.0 ppm ($^1H$) and at $\sim 21.0$ ($n = 3$) and $\sim 16$ ppm ($n = 2$) ($^{29}Si$), respectively. Several irreversible processes, which will be considered below, occur in solutions in parallel with the retro-Wittig decomposition. Therefore, the ratio presented above of concentrations of betaine and phosphorane cannot be considered equilibrium.

Thermolysis of betaines **20** is not so selective as their photolysis.[89] Decomposition of the Corey–Chaykovsky mode, which is predominant under irradiation, occurs in parallel with retro-Wittig type fragmentation.

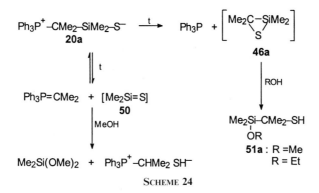

SCHEME 24

The composition of thermolysis products in ethanol, which acted as a chemical trap, allows the estimation of the contributions of these processes under different conditions (Scheme 24). On heating of an alcohol solution of **20a** in a sealed tube, the short-lived silanethione $Me_2Si=S$ (**50**) and silathiirane **46a** are trapped immediately by ethanol. The first compound forms $Me_2Si(OEt)_2$, and the second one gives silylated mercaptan $Me_2Si(OEt)$ $CMe_2SH$ (**51**). At 150 °C these products are formed in a ratio of 1:1. The Corey–Chaykovsky type reaction becomes predominant at 245 °C. The $Me_2Si(OEt)_2$:**51** ratio at this temperature is already 1:3.

During pyrolysis (150 °C) of betaines $Ph_3P^+$–$CR^1R^2$–$SiR^3R^4$–$S^-$ **20a, f, k** in $C_6H_6$, $C_5D_5N$ or melt, the corresponding silathiiranes **46** formed in the Corey–Chaykovsky type reaction are transformed into the expected dithia-disilolanes **49** as in photolysis under homogeneous conditions (Scheme 21). Another main product of betaines **20** pyrolysis are salts $[Ph_3P^+CHR_2]_2$ $[(R^3R^4SiS^-)_2S]^{2-}$ **52**.

We succeeded to isolate and characterize by NMR and X-ray data the $[Et_4P^+]_2[S–SiMe_2–S–SiMe_2–S]^{2-}$ salt **53**, which was prepared by the treatment of the salt $[Ph_3P^+CHMe_2]_2[(Me_2SiS^-)_2S]^{2-}$ **52a** with $Et_3P=CHMe$. This ylide possesses a higher basicity than $Ph_3P=CMe_2$ and, hence, $[Ph_3P^+CHMe_2]$ is deprotonated and the cation is replaced in **52a**. In crystal the planar $[S–SiMe_2–S–SiMe_2–S]^{2-}$ anion has the W-configuration (Fig. 3). The terminal Si–S$^-$ bonds are somewhat longer than the bond in the initial thiabetaines, and the Si–S distances agree well with the statistical mean value.[78]

The mechanism of salts **52** formation is yet unclear. Based on available data, we can assume that precursors of $[(R^3R^4SiS^-)_2S]^{2-}$ anions are betaines with the $^+P–C–(Si–S–)_xSi–S^-$ skeleton formed due to the insertion of short-lived silathiones $[R^3R_4Si=S]$ into the initial betaines **20** or to the bimolecular reaction via direction **b** (Scheme 23). This is indirectly indicated by the fact

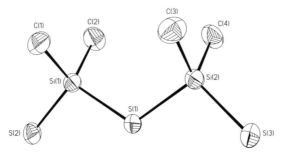

FIG. 3. Geometry of anion part of salt **53**.

that the pyrolysis of betaine $Ph_3P^+$–CHMe–SiMe$_2$–S–SiMe$_2$–S$^-$ **21**, which already contains this structural fragment, gives the salt [$Ph_3P^+$ CH$_2$Me]$_2$[(Me$_2$SiS$^-$)$_2$S]$^{2-}$ **52k** in 70% yield, and the yield of dithiadisilolane Me$_2$Si–CHMe–S–Me$_2$Si–S **49k** is only 5.5%, whereas, under the same conditions, pyrolysis of betaine $Ph_3P^+$–CHMe–SiMe$_2$–S$^-$ **20k** gives this salt in a yield of only 49% and the yield of **49k** increases to 23%.

An arresting fact is the participation of all hydrogen atoms of **20a** in the formation of cations [$Ph_3P^+$CHMe$_2$]. The values of the cation ratio [$Ph_3P^+$CHMe$_2$]/[$Ph_3P^+$CDMe$_2$] of the salts formed upon thermolysis of **20a** and its selectively deuterated analogs are listed in Table V. Clearly, the hydrogen atoms of Me$_2$C groups, Me$_2$Si groups, and pyridine are the sources of protons for [$Ph_3P^+$CHMe$_2$] formation. [$Ph_3P^+$CHMe$_2$] is also formed upon **20d** thermolysis in pyridine-$d_5$, which unambiguously indicates that hydrogen atoms of the phenyl groups at the phosphorus atom also participate in the formation of **52**.

Methyl groups at Si are engaged in proton transfer most actively. The enhanced acidity of the MeSi groups was demonstrated in reactions of various methylsilanes with RLi.[94,95]

An increase in the CH-acidity of the substituents at the Si atom results in an enhanced yield of **52**. Thus, even in the course of betaine $Ph_3P^+$CMe$_2$ SiMeBnS$^-$ **20e** synthesis at 20 °C up to 50% of salt **52e** is formed. Betaine $Ph_3P^+$CMe$_2$SiMePhS$^-$ **20f** under the thermolysis conditions gives salt **52f** in 4% yield (yield of 2,4,5,5-tetramethyl-2,4-diphenyl-1,3-dithia-2,4-dis-ilolane (**49f**) about 80%). Hence, the yield of dithiadisilolanes **49** formed upon thermodecomposition of **20** (Corey–Chaykovsky type reaction) decreases, whereas the yield of **52** increases in the order **20f**, **20a**, **20k**, and **20e**. It follows from these results that the thermodecomposition of betaines **20** in solutions under drastic conditions results, most likely, in a complex system of equilibrium and nonequilibrium processes involving phosphorus ylides, silanethiones **50**, their cyclodimers **18**, **19**, and betaines of the type $^+$P–C–(Si–S–)$_x$Si–S$^-$.

TABLE V

RATIO OF $[Ph_3PCHMe_2]^+/[Ph_3PCDMe_2]^+$ CATIONS OBTAINED UPON THERMOLYSIS OF **20a–d** (100°C, 5 h, APPROXIMATELY 80% CONVERSION) FOR PYRIDINE-$d_5$ AND PYRIDINE-$H_5$ (IN PARENTHESIS)*

| | | Relative yield | |
|---|---|---|---|
| Betaine | | $Ph_3P^+CHMe_2$ | $Ph_3P^+CDMe_2$ |
| **20a** | $Ph_3P^+CMe_2SiMe_2S^-$ | 64 (100) | 36 (0) |
| **20b** | $Ph_3P^+CMe_2Si(CD_3)_2S^-$ | 35 (57) | 65 (43) |
| **20c** | $Ph_3P^+C(CD_3)_2SiMe_2S^-$ | 46 (58) | 54 (42) |
| **20d** | $Ph_3P^+C(CD_3)_2Si(CD_3)_2S^-$ | 36 | 64 |

*The ratio of $Ph_3P$ : **49a**:**52a** is 1:0.5:0.5. The temperature increase to 150°C promotes the conversion up to 100% but does not change the product ratio in the reaction mixture.

## B. Alkylation and Acetylation of Thiobetaines $^{(+)}P–C–Si–S^{(-)}$ I

A suspension of betaine **20a** in THF was heated with ethyl bromide to give the expected salt **43** in 70% yield[57,84,96] (Scheme 25).

The X-ray diffraction data for this compound are presented in Section 3. Betaines containing a hydrogen atom in the $\alpha$-position to the phosphonium center and capable of reversible isomerization to silylated ylides are alkylated by ethyl bromide in a different ways. This reaction resulting in a complex mixture of products is considered below in Section 5.4.

The reactions of betaines **20a, n** with acetyl chloride are unusual.[84,97] The structure and composition of the products formed depend on the molar ratio of reactants. At equimolar amounts of the reactants in THF, the cyclic compound **54** is formed in 96–98% yield (Scheme 26).

Probably, the primarily formed S-acetyl derivative **54** undergoes rearrangement with the migration of the organosilicon fragment to the oxygen atom. This process is thermodynamically favorable. The second thiabetaine molecule, acting as a base, deprotonates thioacetate **55**. This type of deprotonation of thiocarbonyl compounds is known.[98] The acetylation of the thioenolate ion **56** leads to the $\beta$-thiodicarbonyl compound **57**. The subsequent enolization and formation of the very strong Si–O bond complete the cycle closure and formation of **59**. Phosphonium salt **58**, as should be expected,[99] decomposes to cleave the Si–C bond.

According to X-ray data (Fig. 4), a molecule of 2,2,6-trimethyl-1,3-dioxa-2-silacyclohex-5-ene-4-thione has a very strained planar cycle in which the tetrahedral geometry of the silicon atom is strongly distorted. The O–Si–O (103.7(1)°) and C–Si–C (115.1(1)°) bond angles differ considerably from the ideal tetrahedral angle (109.5°), and the Si–O bonds (1.668(2) and 1.675(2) Å) are much longer than the average value of 1.645 Å.[78]

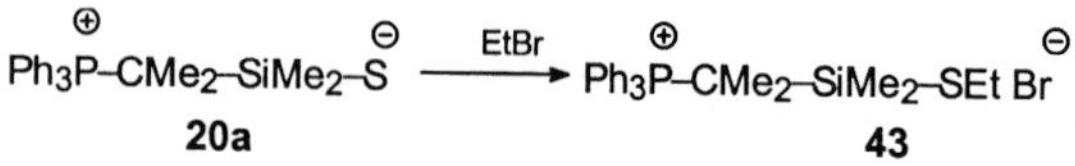

$$\overset{\oplus}{Ph_3P}-CMe_2-SiMe_2-\overset{\ominus}{S} \xrightarrow{EtBr} \overset{\oplus}{Ph_3P}-CMe_2-SiMe_2-SEt \ \overset{\ominus}{Br}$$

**20a**　　　　　　　　　　　　　　　　　**43**

SCHEME 25

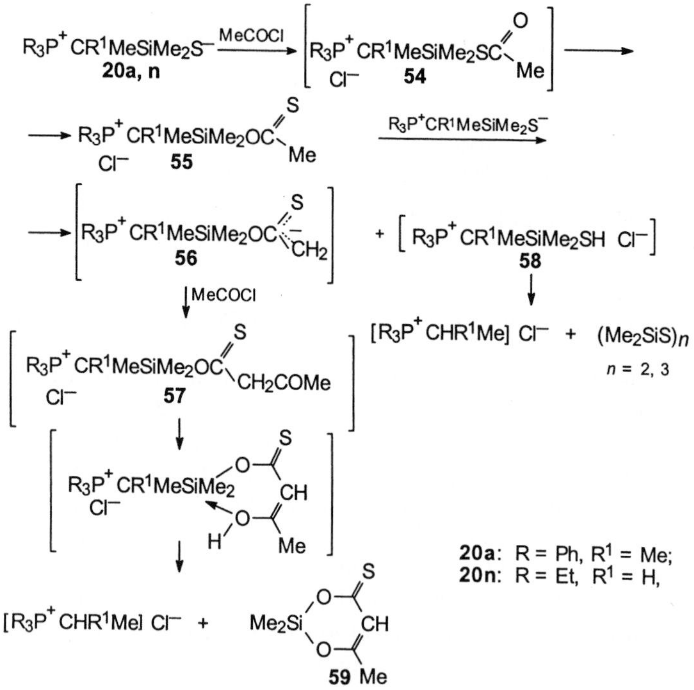

$$R_3\overset{+}{P}CR^1MeSiMe_2S^- \xrightarrow{MeCOCl} \left[ R_3\overset{+}{P}CR^1MeSiMe_2SC\overset{O}{\underset{Me}{\diagdown}} \right] \longrightarrow$$

**20a, n**　　　　　　　　　　　　Cl⁻　**54**

$$\longrightarrow R_3\overset{+}{P}CR^1MeSiMe_2OC\overset{S}{\underset{Me}{\diagdown}} \xrightarrow{R_3\overset{+}{P}CR^1MeSiMe_2S^-}$$

Cl⁻　**55**

$$\longrightarrow \left[ R_3\overset{+}{P}CR^1MeSiMe_2OC\overset{S}{\underset{CH_2}{\diagdown\diagup}} \right] + \left[ R_3\overset{+}{P}CR^1MeSiMe_2SH \ Cl^- \right]$$

**56**　　　　　　　　　　　　　　　**58**

↓ MeCOCl

$$\left[ R_3\overset{+}{P}CR^1MeSiMe_2OC\overset{S}{\underset{CH_2COMe}{\diagdown}} \right]$$

Cl⁻　**57**

$[R_3\overset{+}{P}CHR^1Me] \ Cl^- \ + \ (Me_2SiS)_n$

$n = 2, 3$

$$\left[ R_3\overset{+}{P}CR^1MeSiMe_2 \underset{Cl^-}{\phantom{x}} \begin{array}{c} O-C\overset{S}{\diagdown} \\ \diagdown CH \\ O-C \end{array} \underset{Me}{H} \right]$$

**20a:** R = Ph, R¹ = Me;
**20n:** R = Et, R¹ = H,

↓

$[R_3\overset{+}{P}CHR^1Me] \ Cl^- \ + \ Me_2Si \begin{array}{c} O-C\overset{S}{\diagdown} \\ \diagdown CH \\ O-C \end{array} Me$

**59** Me

SCHEME 26

Probably, these are precisely the distortions that are responsible for the very high reactivity of **59**, in particular, with respect to oxygen and moisture.

Heating of betaine **20n** with a high excess of acetyl chloride in ether gives salt **60** in an almost 100% yield (Scheme 27).[84,97]

## C. *Reactions of Thiobetaines* $^{(+)}P–C–Si–S^{(-)}$ I *with Compounds* $(R_3Sn)_2X$ (X= O, NMe)

As known, compounds of the $(R_3Sn)_2X$ type enter readily into various exchange reactions.[62] No intermediate silicon organophosphorus betaine **61** with the oxide anionic center was detected in the reactions of thiobetaines **20a,** with $(R_3Sn)_2O$ in ether at room temperature.[84,96,100] In solutions they

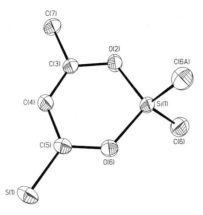

Fig. 4. Geometry of cyclic skeleton of compound **59**.

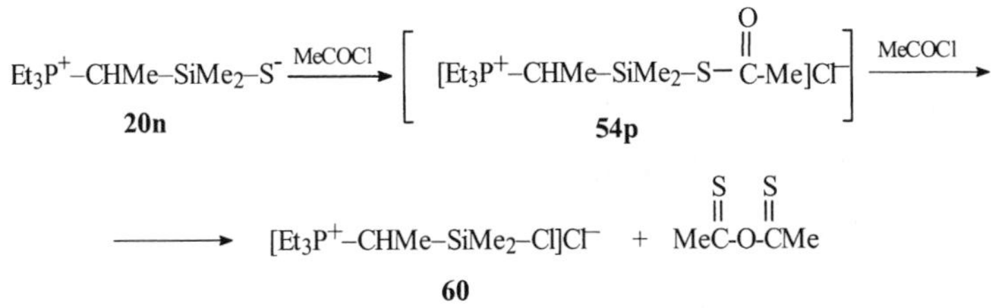

$$Et_3P^+\text{--}CHMe\text{--}SiMe_2\text{--}S^- \xrightarrow{MeCOCl} \left[ [Et_3P^+\text{--}CHMe\text{--}SiMe_2\text{--}S\text{--}\overset{\overset{O}{\parallel}}{C}\text{-Me}]Cl^- \right] \xrightarrow{MeCOCl}$$

**20n**                                      **54p**

$$\longrightarrow [Et_3P^+\text{--}CHMe\text{--}SiMe_2\text{--}Cl]Cl^- + Me\overset{\overset{S}{\parallel}}{C}\text{-O-}\overset{\overset{S}{\parallel}}{C}Me$$

**60**

SCHEME 27

$$Ph_3\overset{+}{P}\text{--}CMe_2\text{--}SiMeR^1\text{--}S^- \xrightarrow[- (Et_3Sn)_2S]{(Et_3Sn)_2O} \left[ Ph_3\overset{+}{P}\text{--}CMe_2\text{--}SiMeR^1\text{--}O^- \right] \longrightarrow$$

**20a, f**                                            **61**

$$\xrightarrow{\phantom{xxx}} Ph_3P\text{=}CMe_2 + [R^1MeSi\text{=}O]$$

**62a, f**

Me₃SiOMe → Me₃SiOSiMeR¹(OMe)   **64a, f**

→ (R¹MeSiO)ₙ   **63a, f**

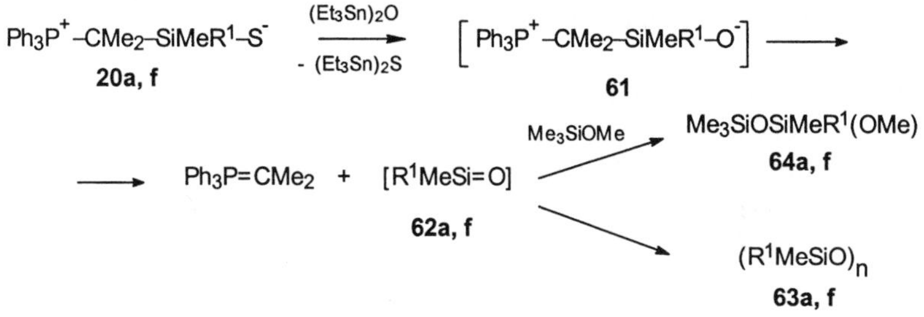

SCHEME 28

undergo the retro-Wittig *in situ* decomposition to form highly reactive silanones **62**, whose chemistry has been intensely developed in recent years.[53,101,102] The existence of **62** was detected by the formation of cyclooligomers **63** and methoxydisiloxanes **64** in the reaction with trimethyl-methoxysilane, which is an efficient chemical trap for silanones (Scheme 28). Thus, oxabetaines **I** with the oxide anionic center possess a much lower thermal stability than their thiaanalogs, whose decomposition under similar

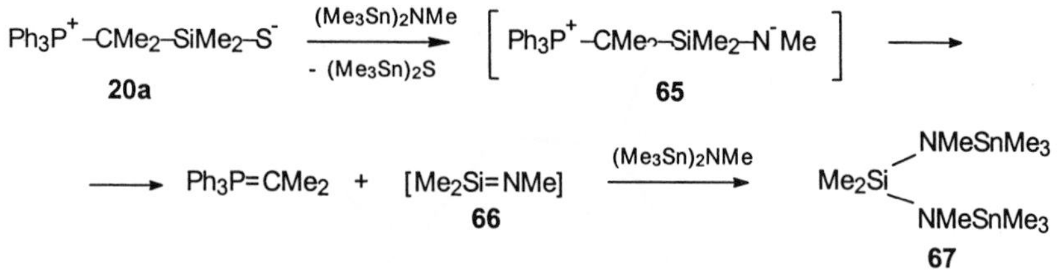

Ph$_3$P$^+$–CMe$_2$–SiMe$_2$–S$^-$  $\xrightarrow[\text{– (Me}_3\text{Sn)}_2\text{S}]{\text{(Me}_3\text{Sn)}_2\text{NMe}}$  $\left[ \text{Ph}_3\text{P}^+\text{–CMe}_2\text{–SiMe}_2\text{–N}^-\text{Me} \right]$  $\longrightarrow$

**20a**                                                                        **65**

$\longrightarrow$  Ph$_3$P=CMe$_2$  +  [Me$_2$Si=NMe]  $\xrightarrow{\text{(Me}_3\text{Sn)}_2\text{NMe}}$  Me$_2$Si$\begin{array}{c}\diagup\text{NMeSnMe}_3\\\diagdown\text{NMeSnMe}_3\end{array}$

**66**                                                              **67**

SCHEME 29

Ph$_3$P=CHR  +  CS$_2$  $\longrightarrow$  Ph$_3$P$^+$–CHR–C$\begin{array}{c}\diagup\text{S}\\\diagdown\text{S}^-\end{array}$  $\rightleftharpoons$  Ph$_3$P=CR–C–SH  $\overset{\text{Ph}_3\text{P=CHR}}{\rightleftharpoons}$

                                       **13**                            **68**  ‖S

$\rightleftharpoons$  Ph$_3$P$^+$CH$_2$R $\left[ \text{Ph}_3\text{P=CR–C}\begin{array}{c}\diagup\text{S}\\\diagdown\text{S}^-\end{array} \right]$

R = Alk, Ar, CN                    **69**

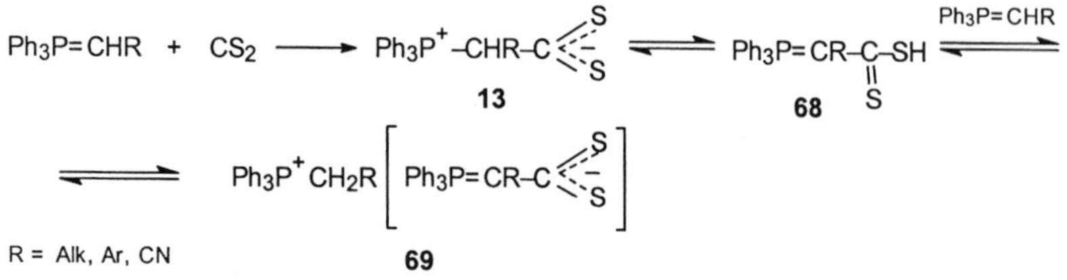

SCHEME 30

conditions occurs with a noticeable rate only at temperatures higher than 50 °C.

The reaction of silicon organophosphorus betaines with bis(trimethylstannyl)methylamine occurs similarly.[84,96] Amidobetaine **65**, which is most likely formed as an intermediate, decomposes to form silaneimine **66**, which is trapped immediately by bis(trimethylstannyl)methylamine. When a twofold excess of organostannane was used, the yield of compound **67** was 90% (Scheme 29).

Despite the intermediate formation of betaines with oxide and amide anionic centers in the above reactions is highly probable, more unambiguous additional proofs are needed for a final conclusion.

### D. *Isomerization of Organophosphorus Betaines* I *Containing a Hydrogen Atom in the α-Position to the Phosphonium Center*

Betaines **13** containing hydrogen atoms in the α-position to the triphenylphosphonium center are reversibly isomerized in a solution to form phosphoranylidenealkane dithiocarboxylic acids **68** or their salts **69** (Scheme 30).[48,49]

Similar isomerization of silicon organophosphorus betaines **20k–m** to silylated phosphorus ylides **70** (Scheme 31)[84,96] is distinctly detected by $^1$H, $^{13}$C, and $^{31}$P NMR from the appearance of signals characteristic of

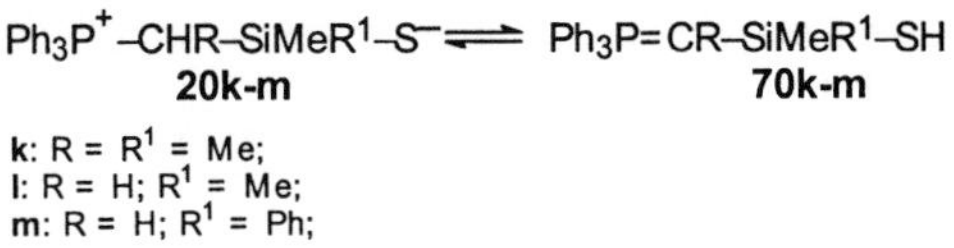

$$Ph_3P^+-CHR-SiMeR^1-S^- \rightleftharpoons Ph_3P=CR-SiMeR^1-SH$$

$$\textbf{20k-m} \qquad\qquad\qquad \textbf{70k-m}$$

**k**: R = R$^1$ = Me;
**l**: R = H; R$^1$ = Me;
**m**: R = H; R$^1$ = Ph;

SCHEME 31

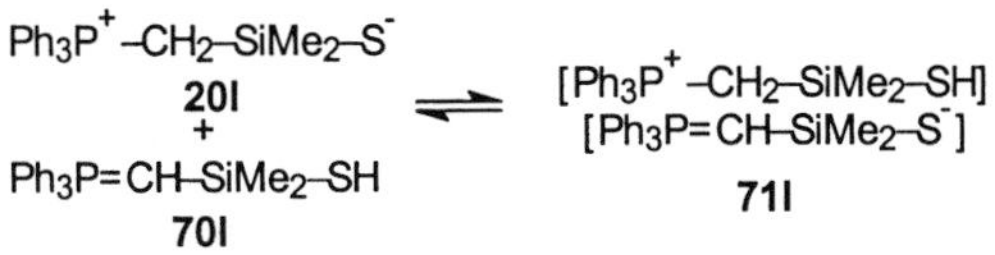

$$Ph_3P^+-CH_2-SiMe_2-S^-$$
$$\textbf{20l}$$
$$+ \qquad\qquad \rightleftharpoons \qquad [Ph_3P^+-CH_2-SiMe_2-SH]$$
$$\qquad\qquad\qquad\qquad\qquad [Ph_3P=CH-SiMe_2-S^-]$$
$$Ph_3P=CH-SiMe_2-SH \qquad\qquad\qquad \textbf{71l}$$
$$\textbf{70l}$$

SCHEME 32

phosphorus ylide,[83,103] including that silylated at the ylide carbon atom.[83,104]

An increase in the temperature or a decrease in the solvent polarity shifts the equilibrium toward ylides **70**.[84] For example, in the synthesis of **20k** in benzene, up to 62% **70k** in a mixture with betaine **20k** were detected in the solution, whereas in pyridine the content of **20k** did not exceed 1.4% at $\sim 20\,^{\circ}C$ and 3.6% at $90\,^{\circ}C$.

Equilibrium processes of intermolecular proton transfer between the thiolate center and SH group are observed in a solution along with the isomerization of betaine **20l** to **70l**. These processes result in salts **71** (Scheme 32) similar to salts **69** in the carbon series, which was proved by $^{29}Si$ NMR spectroscopy.

Isomerization of thiabetaines **20** to silylated ylides was chemically confirmed by the study of the reactions of **20** with ethyl bromide and alcohols.[84] Unlike betaine **20a** based on phosphorus ylide with the quaternary ylide carbon (Scheme 25), the reaction of ethyl bromide with $Ph_3P^+CHMeSiMe_2S^-$ **20k** affords a complex mixture of products. In this case, after the alkylation of the thiolate center, the halide ion as a nucleophile cleaves the C–Si bond in ylide **70k** to form phosphorus ylide $Ph_3P=CHMe$. The latter enters into further transformations, which are described in detail for similar systems in.[99] For betaine **20k** the final products of these reactions are $Ph_3P=C(Me)SiMe_2X$, $Ph_3P^+C(H)MeSiMe_2X\ Y^-$, $Me_2SiXY$ (X, Y = Br, SEt), $Ph_3P^+Et\ Br^-$, and $Ph_3P=CHMe$.[84]

Isomerization of betaines **20k, l** to silylated phosphorus ylides was also confirmed by their reaction with methanol-$d_1$, which occurs rather rapidly at room temperature to give $Me_2Si(OMe)_2$ and mixtures of α-deuterated phosphonium salts **72k, l** (Scheme 33). It is most probable that, in this case, the reaction occurs mainly through the intermediate formation of α-silylated

$$Ph_3P^+-CHR-SiMe_2-S^- \; \rightleftharpoons \; Ph_3P=CR-SiMe_2-SH \xrightarrow{\text{MeOD}}$$

$$\text{20k, l}$$

$$\longrightarrow \; Ph_3P=CDR \; + \; Me_2Si(OMe)(SH) \xrightarrow{\text{MeOD}}$$

$$\longrightarrow \; Me_2Si(OMe)_2 \; + \; [Ph_3P^+CD_2R] \, SH^- \; (\text{or} \; [Ph_3P^+CHDR] \, SD^-)$$

$$\text{72k, l}$$

k: R = Me;
l: R = H

SCHEME 33

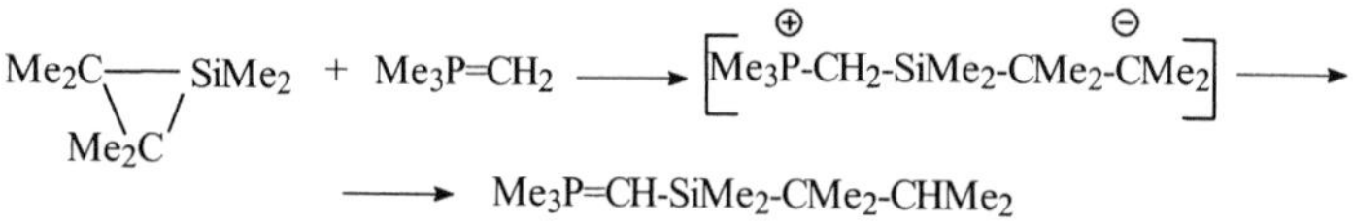

$$Me_2C \!\!\!\!\begin{array}{c} \\ \diagdown \diagup \\ Me_2C \end{array}\!\!\!\! SiMe_2 \; + \; Me_3P=CH_2 \; \longrightarrow \; \left[ \overset{\oplus}{Me_3P}\text{-}CH_2\text{-}SiMe_2\text{-}CMe_2\text{-}\overset{\ominus}{CMe_2} \right] \longrightarrow$$

$$\longrightarrow \; Me_3P=CH\text{-}SiMe_2\text{-}CMe_2\text{-}CHMe_2$$

SCHEME 34

phosphorus ylides **70**, whose further transformations in methanol-$d_1$ occur according to general schemes established for the reactions of alcohols with $R_3P=CHR^1$ ($R^1 = H$, Alk or $Me_3Si$).[105,106]

We failed to detect the formation of noticeable amounts of doubly deuterated salt $[Et_3P^+CD_2Me]SH^-$ during alcoholysis of betaine $Et_3P^+CHMeSiMe_2S^-$ **20n**. This indicates that betaines with the trialkylphosphonium center virtually do not undergo isomerization to silylated ylides, which is due to a lower CH acidity of phosphonium salts of the alkyl series compared to those of the aryl series, and correspondingly, to a higher basicity (nucleophilicity) of the conjugated bases, phosphorus ylides.[106]

The intermediate formation of betaines and their subsequent irreversible isomerization to silylated phosphorus ylides have previously been postulated for the reactions of phosphorus ylides with hexamethylsilirane,[107] sila-[108] and disilacyclobutanes[109] with different substituents at the silicon atom (Schemes 34 and 35).

Gilman and Tomasi proposed the following scheme of the reaction of octaphenylcyclotetrasilane with phosphorus ylide[110] (Scheme 36).

Octaphenylcyclotetrasilane reacts similarly with $Et_3P=CHMe$. At a ratio of reactants of 1 : 4, the composition of the final product remains unchanged. Under the same conditions, $Ph_3P=CMe_2$ does not react with octaphenylcyclotetrasilane.[36]

### E. *Chemical Transformations of Betaines* $^{(+)}P\text{-}C\text{-}Sn^{(-)}$ II

Veith and Huch[81] obtained interesting results by studying betaine $Ph_3P^+-CH_2-Sn(NC_4H_9\text{-}t)_2SiMe_2$ **31** decomposition under heating at $105\,^\circ C$ (Scheme 37).

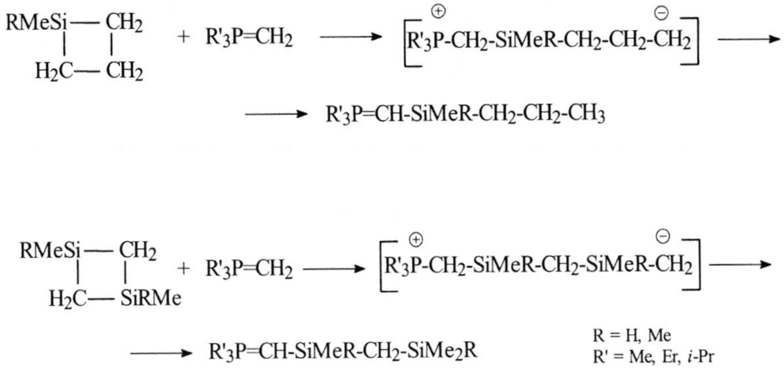

$$\begin{array}{c} RMeSi\!-\!CH_2 \\ |\qquad\quad| \\ H_2C\!-\!CH_2 \end{array} \;+\; R'_3P{=}CH_2 \;\longrightarrow\; \left[\overset{\oplus}{R'_3P}\text{-}CH_2\text{-}SiMeR\text{-}CH_2\text{-}CH_2\text{-}\overset{\ominus}{CH_2}\right] \;\longrightarrow$$

$$\longrightarrow\; R'_3P{=}CH\text{-}SiMeR\text{-}CH_2\text{-}CH_2\text{-}CH_3$$

$$\begin{array}{c} RMeSi\!-\!CH_2 \\ |\qquad\quad| \\ H_2C\!-\!SiRMe \end{array} \;+\; R'_3P{=}CH_2 \;\longrightarrow\; \left[\overset{\oplus}{R'_3P}\text{-}CH_2\text{-}SiMeR\text{-}CH_2\text{-}SiMeR\text{-}\overset{\ominus}{CH_2}\right] \;\longrightarrow$$

$$\longrightarrow\; R'_3P{=}CH\text{-}SiMeR\text{-}CH_2\text{-}SiMe_2R$$

R = H, Me
R' = Me, Er, *i*-Pr

**SCHEME 35**

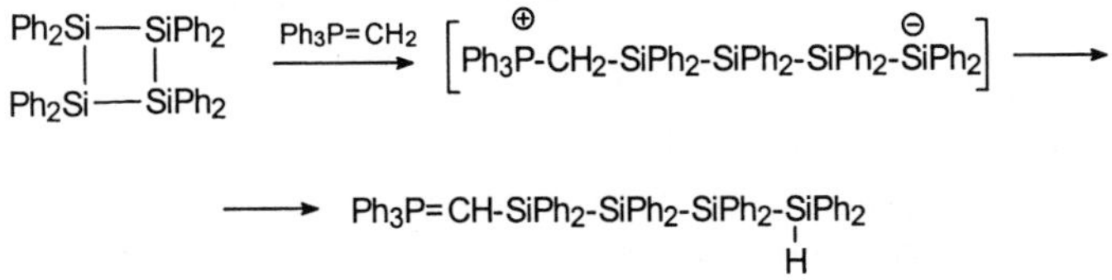

$$\begin{array}{c} Ph_2Si\!-\!SiPh_2 \\ |\qquad\quad| \\ Ph_2Si\!-\!SiPh_2 \end{array} \;\xrightarrow{Ph_3P{=}CH_2}\; \left[\overset{\oplus}{Ph_3P}\text{-}CH_2\text{-}SiPh_2\text{-}SiPh_2\text{-}SiPh_2\text{-}\overset{\ominus}{SiPh_2}\right] \;\longrightarrow$$

$$\longrightarrow\; Ph_3P{=}CH\text{-}SiPh_2\text{-}SiPh_2\text{-}SiPh_2\text{-}\underset{\underset{H}{|}}{SiPh_2}$$

**SCHEME 36**

**31**

**44**

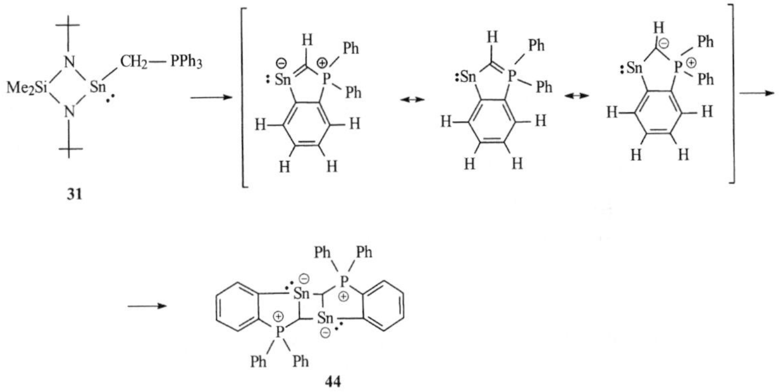

**SCHEME 37**

The scheme proposed by the authors[81] includes the formation of a hydrogen bond between the proton at the *ortho*-carbon atom of the phenyl ring and one of the nitrogen atoms of cyclostannylene followed by the Sn–N bond cleavage and C–Sn bond formation. The second Sn–N bond is cleaved by one of the protons of the $CH_2$ group bound to the phosphorus atom. The scheme of reactions was confirmed by the formation of $Me_2Si(NDBu-t)$ (NHBu-$t$) during pyrolysis of deuterated betaine $Ph_3P^+-CD_2-Sn[N(Bu-t)]_2$ $SiMe_2$ and by kinetic data.

The study of the reactivity of betaines **I** and **II** is yet far from completion. The data presented show that the scope of their reactions is wide and diverse. Therefore, we can expect that compounds of this type will find various synthetic applications, including those for the preparation and *in situ* generation of compounds with multiple $E^{14}=X$ bonds.

# VI

## QUANTUM-CHEMICAL INVESTIGATION OF POTENTIAL ENERGY SURFACES OF BETAINES $^{(+)}E^{15}-C-E^{14}-X^{(-)}$ I AND $^{(+)}E^{15}-C-E^{14(-)}$ II

In the last decades, quantum-chemical studies have become an integral part of modern chemical research due to the appearance of modern quantum-chemical methods and computer programs providing reliable quantitative information on the structure and reactivity of medium-sized organoelement compounds at rather moderate calculation expenses.[111] We studied in detail the potential energy surfaces (PES) of betaines **I** and **II**[60,61,65,112–114] by DFT calculations using the generalized gradient approximation and PBE functional[115] and TZ2p quality basis sets. For the expansion of one-electron wave functions, we used the extended atomic basis sets of grouped functions of the Gaussian type in the form {311/1} for the H atom, {611111/411/11} for C, N, and O atoms, {6111111111/5111111/11} for Si, P, and S atoms, {7111111111111/5111111111/51111} for Ge, As, and Se atoms, and {7111111111111111/5111111111111/61111111} for the Sn atom (figures in braces show the sample of the group of the Gaussian functions for AO with *s*-, *p*-, and *d*-symmetry). According to the algorithm proposed in,[116] to calculate the matrix elements of the Coulomb and exchange–correlation potentials, we used the expansion of the electron density in the auxiliary basis set consisting of sets of centered on atoms nongrouped Gaussian functions of the type (5s1p) for H, (10s3p3d1f) for C, N, and O, (14s7p7d1f1g) for Si, P, and S, (18s5p5d3f3g) for Ge, As, and Se, and (21s6p6d4f1g) for Sn. Stationary points on the PES of the studied systems were identified by analysis of the Hessians. The second derivatives of energy with respect to coordinates were calculated analytically. The correction to energies of zero vibrations was calculated in the harmonic approximation. Calculations were performed by the program PRIRODA written by D.N. Laikov.[117]

### A. Geometry of Betaine Molecules $^{(+)}E^{15}$–$C$–$E^{14}$–$S^{(-)}$. Comparison of Calculated and X-ray Data

Table VI contains the geometric parameters of the main structural $^{(+)}E^{15}$–$C$–$E^{14}$–$S^{(-)}$ fragment found by the calculation for several betaines **I**, which were studied by us using X-ray analysis (see Section 3). The X-ray data are presented in parentheses.

Comparison shows a satisfactory qualitative agreement of the calculated and experimental values. The calculation reproduces well the main structural feature of all betaines: their sterically strained *gauche*-conformation due to a strong Coulomb interaction between the cationic and anionic centers. The calculated covalent bond lengths of the $^{(+)}E^{15}$–$C$–$E^{14}$–$S^{(-)}$ fragment are somewhat overestimated (by less than 0.06 Å) compared to the experimental values. These deviations can be related to effects of crystalline packing, which are usually significant for polar bonds. The C–C, $E^{15}$–C, and C–$E^{14}$ bond lengths at the periphery of the molecules are reproduced with a higher accuracy (deviations not greater than 0.04 Å). The values of the S–$E^{14}$–C and $E^{14}$–C–$E^{15}$ bond angles and dihedral $E^{15}$–C–$E^{14}$–S angles are somewhat underestimated compared to the experimental values. Greater deviations are observed for the dihedral angles. They are related to the competition of the electrostatic intermolecular interactions in crystal with the intramolecular Coulomb interactions of the thiolate anionic and phosphonium (arsonium) cationic centers. The calculation reproduces well the specific feature mentioned in Section 3 of the structure of arsonium betaine **26a**. Due to the short As–S contact, the coordination of the arsenic atom can be considered as intermediate between a distorted tetrahedron and a trigonal bipyramid.

### B. Potential Energy Surfaces of Model Betaines $^{(+)}Me_3E^{15}$–$CH_2$–$E^{14}Me_2$–$X^{(-)}$ ($E^{15} = P$, As; $X = S$, $C_5H_4$, O, NMe, Se)

We studied the PES for betaines **I** using the model compounds $^{(+)}Me_3E^{15}$–$CH_2$–$E^{14}Me_2$–$X^{(-)}$ ($E^{15} = P$, As; $X = S$, $C_5H_4$, O, NMe, Se).[112–114,116,117] The replacement of radicals at atoms of elements by methyl groups in the molecules allows a substantial decrease in the expenses of calculation time retaining an appropriate closeness of the systems to real molecules for which experimental data were obtained.

As mentioned in Sections 1 and 3, one of the most important problems in the study of the Wittig reaction mechanism is the determination of the relative stability of betaines with the open structure and cyclic oxapho-sphetanes as intermediates. The solution to a similar problem in chemistry

## TABLE VI

GEOMETRIC PARAMETERS OF THE $^{(+)}E^{15}$–C–$E^{14}$–$S^{(-)}$ FRAGMENT FOR SEVERAL BETAINES **I** WITH THE THIOLATE CENTER ACCORDING TO DFT CALCULATIONS AND X-RAY DATA GIVEN IN PARENTHESIS)

| Betaines | $r$ (Å) | | | | $\varphi$ (°) | | $\theta$ (°) |
| --- | --- | --- | --- | --- | --- | --- | --- |
| | $E^{15}$–C | C–$E^{14}$ | $E^{14}$–S | $E^{15}$...S | $E^{15}$–C–$E^{14}$ | C–$E^{14}$–S | $E^{15}$–C–$E^{14}$–S |
| $Ph_3P^+$–$CMe_2$–$SiMe_2$–$S^-$ | 1.865 (1.825) | 2.037 (1.986) | 2.064 (2.048) | 3.687 (3.988) | 111.7 (115.60) | 111.9 (114.6) | 36.7 (56.1) |
| $Ph_3P^+$–CHMe–$SiPh_2$–$S^-$ | 1.827 (1.811) | 1.993 (1.934) | 2.063 (2.044) | 3.527 (3.681) | 110.1 (112.8) | 110.9 (114.7) | 32.9(38.2) |
| $Et_3As^+$–CHPh–$SiMe_2$–$S^-$ | 1.986 (1.936) | 1.969 (1.946) | 2.083 (2.0455) | 3.040 (3.518) | 105.1 (111.47) | 102.0 (109.97) | 24.4 (33.13) |
| $Et_3P^+$–CHMe–$GeMe_2$–$S^-$ | 1.815 (1.783) | 2.108 (2.041) | 2.157 (2.140) | 3.642 (3.774) | 110.8 (114.3) | 111.2 (114.9) | 26.0 (25.9) |

72

of organoelement betaines **I** is doubtless of interest. Therefore, we studied in detail the region of PES of model betaines corresponding to the rotation about the central $CH_2$–$E^{14}$ bond and searched for minima corresponding to the isomers with a cyclic structure.

All studied model compounds can distinctly be divided into three groups (Table VII). The first group is composed of substances in which the sulfur, selenium or cyclopentadienyl anion acts as an anionic center. They exist only in open betaine forms, and their PES do not contain local minima corresponding to cyclic isomers. The second group contains compounds with arsonium cationic and oxide anionic centers and silicon and germanium betaines with arsonium and amide centers. They exist as cyclic isomers and their PES have no local minima corresponding to the open forms. Finally, the third group consists of six studied compounds with phosphonium cationic and oxide or amide anionic centers and arsonium-imide betaine. Their PES have minima for both cyclic and open forms separated by low barriers.

The *gauche*-conformation is the most stable for all betaines of the first group and open forms of compounds from the third group. Table VIII contains the most important geometric parameters of these betaines.

The dipole moments of betaines of the first and third groups are presented in Table IX.

The local minima corresponding to the *trans*-conformers are observed on PES for betaines $^{(+)}Me_3E^{15}$–$CH_2$–$E^{14}Me_2$–$X^{(-)}$ ($E^{15} = P$, As; $E^{14} = Si$, Ge; $X = S$, Se, $C_5H_4$). The main geometric parameters of the *trans*-conformations of these betaines and their relative energies and dipole moments are presented in Table X. The following regularities are distinctly seen from the

TABLE VII

BETAINE **A** and CYCLIC **B** FORMS OF MODEL COMPOUNDS $X$–$E^{14}Me_2$–$CH_2$–$E^{15}Me_3$ ($X = S$, Se, $C_5H_4$, O, NMe; $E^{14} = Si$, Ge, Sn; $E^{15} = P$, As)

$$^{-}X\text{-}E^{14}Me_2\text{-}CH_2\text{-}^{+}E^{15}Me_3 \qquad \overline{^{-}X\text{-}E^{14}Me_2\text{-}CH_2\text{-}E^{15}Me_3}$$
$$\textbf{A} \qquad\qquad\qquad \textbf{B}$$

| $E^{14}$ | S–P | S–As | Se–P | Se–As | O–P | O–As | N–P | N–As |
|---|---|---|---|---|---|---|---|---|
| | | | | $X$–$E^{15}$ | | | | |
| Si | b | b | b | b | b + c | c | b + c | c |
| Ge | b | b | b | b | b + c | c | b + c | c |
| Sn | b | b | b | b | b + c | c | b + c | b + c |

*Note*: b, only the betaine form is present on PES; c, only the cyclic form is present on PES; b + c, both betaine and cyclic forms are possible.

TABLE VIII

MOST IMPORTANT GEOMETRIC PARAMETERS OF BETAINES $^{(-)}$X–E$^{14}$Me$_2$–CH$_2$$^{(+)}$E$^{15}$Me$_3$

(X = S, Se, C$_5$H$_4$, O, NMe; E$^{14}$ = Si, Ge, Sn; E$^{15}$ = P, As)

| | r (Å) | | | | | | φ (°) | | θ (°) |
|---|---|---|---|---|---|---|---|---|---|
| | X–E$^{14}$ | E$^{14}$–C | E$^{14}$–C$_{Me}$ | C–E$^{15}$ | X···E$^{15}$ | E$^{14}$···E$^{15}$ | X–E$^{14}$–C | E$^{14}$–C–E$^{15}$ | X–E$^{14}$–C–E$^{15}$ |
| P–C–Si–S | 2.065 | 2.003 | 1.904 | 1.794 | 3.584 | 3.196 | 111.8 | 114.6 | 20.8 |
| P–C–Ge–S | 2.154 | 2.106 | 1.988 | 1.788 | 3.658 | 3.277 | 111.7 | 114.3 | 0.8 |
| P–C–Sn–S | 2.338 | 2.324 | 2.196 | 1.782 | 3.815 | 3.452 | 108.1 | 113.8 | 0.0 |
| As–C–Si–S | 2.071 | 1.974 | 1.905 | 1.928 | 3.403 | 3.213 | 108.4 | 110.9 | 23.2 |
| As–C–Ge–S | 2.159 | 2.078 | 1.987 | 1.920 | 3.530 | 3.312 | 109.3 | 111.8 | 9.7 |
| As–C–Sn–S | 2.341 | 2.304 | 2.196 | 1.912 | 3.740 | 3.504 | 107.0 | 112.1 | 0.2 |
| P–C–Si–Se | 2.211 | 1.997 | 1.907 | 1.796 | 3.733 | 3.214 | 112.3 | 115.7 | 26.0 |
| P–C–Ge–Se | 2.294 | 2.102 | 1.988 | 1.789 | 3.807 | 3.292 | 112.7 | 115.4 | 12.1 |
| P–C–Sn–Se | 2.467 | 2.324 | 2.196 | 1.782 | 3.968 | 3.467 | 109.5 | 115.1 | 0.1 |
| As–C–Si–Se | 2.217 | 1.971 | 1.906 | 1.929 | 3.563 | 3.239 | 109.1 | 112.4 | 27.5 |
| As–C–Ge–Se | 2.299 | 2.074 | 1.987 | 1.920 | 3.689 | 3.328 | 110.1 | 112.8 | 19.2 |
| As–C–Sn–Se | 2.471 | 2.305 | 2.196 | 1.911 | 3.896 | 3.530 | 108.6 | 113.4 | 0.1 |
| P–C–Si–O | 1.599 | 2.012 | 1.907 | 1.794 | 2.994 | 3.070 | 106.0 | 107.4 | 0.2 |
| P–C–Ge–O | 1.721 | 2.114 | 1.989 | 1.789 | 3.097 | 3.152 | 105.1 | 107.4 | 0.6 |
| P–C–Sn–O | 1.931 | 2.325 | 2.198 | 1.787 | 3.218 | 3.332 | 100.4 | 107.6 | 0.1 |
| P–C–Si–N | 1.680 | 1.981 | 1.918 | 1.797 | 3.042 | 3.141 | 102.8 | 112.4 | 6.4 |
| P–C–Ge–N | 1.801 | 2.070 | 2.004 | 1.794 | 3.120 | 3.212 | 101.8 | 112.3 | 0.8 |
| P–C–Sn–N | 2.012 | 2.293 | 2.214 | 1.788 | 3.245 | 3.402 | 97.3 | 112.4 | 1.9 |
| As–C–Sn–N | 2.011 | 2.270 | 2.215 | 1.919 | 3.100 | 3.446 | 94.4 | 110.4 | 0.2 |
| P–C–Si–C$_5$H$_4$ | 1.822 | 1.988 | 1.899 | 1.802 | 3.489 | 3.208 | 109.1 | 115.6 | 32.3 |

TABLE IX

DIPOLE MOMENTS OF THE *CIS*-CONFORMERS OF BETAINES OF THE FIRST AND THE THIRD GROUP

| Betaine | μ (D) | Betaine | μ (D) |
|---|---|---|---|
| P–C–Si–S | 9.7 | As–C–Si–Se | 9.6 |
| P–C–Ge–S | 9.6 | As–C–Ge–Se | 9.8 |
| P–C–Sn–S | 9.6 | As–C–Sn–Se | 9.9 |
| As–C–Si–S | 9.2 | P–C–Si–O | 7.8 |
| As–C–Ge–S | 9.4 | P–C–Ge–O | 7.6 |
| As–C–Sn–S | 9.6 | P–C–Sn–O | 7.5 |
| P–C–Si–Se | 10.1 | P–C–Si–N | 7.5 |
| P–C–Ge–Se | 9.9 | P–C–Ge–N | 7.4 |
| P–C–Sn–Se | 10.0 | P–C–Sn–N | 7.2 |
| P–C–Si–C$_5$H$_4$ | 10.5 | As–C–Sn–N | 7.0 |

### TABLE X

GEOMETRIC PARAMETERS, DIPOLE MOMENTS, AND RELATIVE ENERGIES OF THE *TRANS*-CONFORMERS OF BETAINES OF THE FIRST GROUP

| Betaine | $r$ (Å) | | | | $\alpha$ (°) | | $\theta$ (°) | $\Delta E°$ (kcal/ mol) | $\mu$ (D) |
|---|---|---|---|---|---|---|---|---|---|
| | $X-E^{14}$ | $E^{14}-C$ | $C-E^{15}$ | $X \cdots E^{15}$ | $X-E^{14}-C$ | $E^{14}-C-E^{15}$ | $X-E^{14} -C-E^{15}$ | | |
| P–C–Si–S | 2.034 | 2.015 | 1.783 | 4.891 | 104.8 | 123.4 | 159.2 | 11.8 | 14.1 |
| P–C–Ge–S | 2.119 | 2.127 | 1.772 | 4.967 | 103.8 | 122.4 | 150.4 | 12.3 | 14.0 |
| As–C–Si–S | 2.037 | 1.992 | 1.927 | 4.997 | 103.6 | 122.1 | 163.0 | 12.1 | 14.1 |
| As–C–Ge–S | 2.121 | 2.099 | 1.914 | 5.096 | 102.6 | 121.8 | 159.0 | 12.7 | 14.2 |
| P–C–Si–Se | 2.179 | 2.011 | 1.786 | 5.009 | 104.8 | 123.2 | 158.4 | 11.3 | 14.3 |
| P–C–Ge–Se | 2.259 | 2.121 | 1.775 | 5.089 | 103.9 | 122.3 | 151.7 | 11.7 | 14.3 |
| As–C–Si–Se | 2.182 | 1.986 | 1.934 | 5.118 | 103.3 | 122.0 | 163.4 | 11.6 | 14.4 |
| As–C–Ge–Se | 2.261 | 2.093 | 1.919 | 5.213 | 102.4 | 121.8 | 159.7 | 12.0 | 14.4 |

analysis of the data in Tables VI, VII, VIII, IX and X for betaines of the first group:

- Torsion angles $E^{15}–C–E^{14}–X$ for phosphonium and arsonium betaines decrease regularly in the series $E^{14}$=Si–Ge–Sn. This is unambiguously related to an increase in the length of the central $E^{14}–C$ bond along this series, which results in a decrease in steric strains stipulated by the repulsive interactions H–Me and $X–E^{15}Me_3$ in the eclipsed conformation.
- Central $E^{14}–C$ bonds are longer (about 0.1 Å) than the $E^{14}–C_{Me}$ bonds, and the $E^{14}–C–E^{15}$ angles are greater than the ideal tetrahedral angle of 109.4°, which indicates considerable steric strain due to short nonvalent contacts $X \cdots E^{15}$. For betaines of the tin series, the $E^{14} \cdots E^{15}$ distance approaches the sum of van der Waals radii of atoms (Table XI). This regularity is distinctly seen in the experimental X-ray data (see Section 3).
- It is characteristic that the interatomic $E^{14} \cdots E^{15}$ distances in model betaines is by 0.6–0.8 Å shorter than the sums of the corresponding van der Waals radii (Table XI). This also increases the steric strains in the *gauche*-conformers.
- The values of dipole moments of betaines (Table IX) range from 9.2 to 10.1 D.
- The values of rotation barriers about the central $E^{14}–C$ bond are 11.6–13.0 kcal/mol.
- The *trans*-conformers of betaines $^{(+)}Me_3E^{15}–CH_2–E^{14}Me_2–X^{(-)}$ ($E^{15}$ = P, As; $E^{14}$ = Si, Ge; X = S, Se, $C_5H_4$) lie by 11.3–12.7 kcal/mol higher than the corresponding *cis* forms. The $X–E^{14}–C$ bond

TABLE XI

Sums of van der Waals and Covalent Radii for Pairs
of Atoms $E^{15}$, $E^{14}$, and X

| Pairs of atoms | $\sum R$ coval. | $\sum R$ vdw |
|---|---|---|
| S–P | 2.14 | 3.75 |
| S–As | 2.25 | 3.85 |
| Se–P | 2.27 | 3.90 |
| Se–As | 2.38 | 4.00 |
| O–P | 1.76 | 3.30 |
| O–As | 1.87 | 3.40 |
| N–P | 1.80 | 3.44 |
| N–As | 1.91 | 3.54 |
| Si–P | 2.27 | 3.90 |
| Si–As | 2.38 | 4.00 |
| Ge–P | 2.32 | 3.90 |
| Ge–As | 2.43 | 4.00 |
| Sn–P | 2.50 | 4.10 |
| Sn–As | 2.61 | 4.20 |

angle is decreased to 102.4–104.1°, and the $E^{14}$–C–$E^{15}$ angle is increased to 121.8–123.4°. The dipole moments increase to 14.0–14.4 D, which is related to an increase in the distance between the charged $^{(-)}X \cdots E^{15(+)}$ centers to 4.891–5.213 Å.

The energies of the covalent O–As and N–As bonds are higher than those of the S–As and Se–As bonds by about 25 kcal/mol. This favors the closure of the strained four-membered cycles in arsonium betaines at X = O, NMe because the energy gain due to the formation of the O–As and N–As bonds overcomes the destabilization effect of formation of small cycles. The structural parameters of cycles for the compounds of the second group and cyclic isomers of the third group are presented in Table XII.

In compounds of the second group and cyclic isomers of the compounds of the third group, the configuration at the $E^{15}$ atoms changes compared to their betaine analogs. Their phosphorus and arsenic atoms are in the trigonal-bipyramidal configuration. The $E^{15}$–C bond with the carbon atom of the methyl group in the axial position is by 0.05–0.08 Å longer than the corresponding bond with the carbon atom in the equatorial position. The $E^{15}$ atom is shifted relatively to the equatorial plane toward the axial methyl group by 0.03–0.06 Å. A decrease in the X–$E^{14}$–C bond angle in the Si–Ge–Sn series from 85.1–93.5° for silicon and to 74.6–81.2° for tin is another typical feature of cyclic isomers of betaines.

TABLE XII

GEOMETRIC PARAMETERS OF THE CYCLIC COMPOUNDS OF SECOND AND THIRD GROUPS

| | $r$ (Å) | | | | | | $\varphi$ (°) | | $\theta$ (°) |
|---|---|---|---|---|---|---|---|---|---|
| | $X\text{–}E^{14}$ | $E^{14}\text{–}C$ | $E^{14}\text{–}C_{Me}$ | $C\text{–}E^{15}$ | $X\text{–}E^{15}$ | $E^{14}\cdots E^{15}$ | $X\text{–}E^{14}\text{–}C$ | $E^{14}\text{–}C\text{–}E^{15}$ | $X\text{–}E^{14}\text{–}C\text{–}E^{15}$ |
| P–C–Si–O | 1.654 | 1.910 | 1.894 | 1.867 | 2.006 | 2.739 | 89.6 | 92.9 | 0.0 |
| P–C–Ge–O | 1.798 | 1.976 | 1.972 | 1.885 | 1.901 | 2.808 | 84.1 | 93.3 | 0.0 |
| P–C–Sn–O | 2.008 | 2.176 | 2.181 | 1.894 | 1.860 | 2.994 | 76.6 | 94.4 | 0.1 |
| As–C–Si–O | 1.646 | 1.924 | 1.897 | 1.979 | 2.164 | 2.838 | 93.5 | 93.3 | 0.1 |
| As–C–Ge–O | 1.787 | 1.989 | 1.975 | 1.997 | 2.068 | 2.900 | 88.4 | 93.4 | 0.0 |
| As–C–Sn–O | 1.996 | 2.188 | 2.184 | 2.006 | 2.037 | 3.079 | 81.2 | 94.4 | 0.0 |
| P–C–Si–N | 1.731 | 1.894 | 1.899 | 1.889 | 1.926 | 2.795 | 85.1 | 95.2 | 0.8 |
| P–C–Ge–N | 1.844 | 1.969 | 1.978 | 1.896 | 1.898 | 2.880 | 81.1 | 96.3 | 3.7 |
| P–C–Sn–N | 2.049 | 2.172 | 2.191 | 1.897 | 1.892 | 3.074 | 74.6 | 97.9 | 2.4 |
| As–C–Si–N | 1.727 | 1.901 | 1.902 | 2.006 | 2.075 | 2.890 | 89.0 | 95.4 | 0.2 |
| As–C–Ge–N | 1.843 | 1.973 | 1.979 | 2.016 | 2.048 | 2.967 | 85.1 | 96.1 | 6.7 |
| As–C–Sn–N | 2.047 | 2.176 | 2.193 | 2.017 | 2.043 | 3.156 | 78.6 | 97.6 | 4.7 |

The O–P and N–P bond energies are by approximately 15 kcal/mol lower than those of the O–As and N–As bonds. Therefore, PES of compounds of the third group ($E^{15} = P$, $X = O$, NMe) contain minima corresponding to both possible forms, open betaine and cyclic. This group also includes the compound of the arsonium series with the amide anionic center ($E^{14} = Sn$, $E^{15} = As$, $X = NMe$). Its cyclic isomer is destabilized to a greater extent than the cycles with germanium and silicon due to the too short nonbonding contact of the As and Sn atoms, the distance between which is by 0.75 Å shorter than the sum of their van der Waals radii (see Table XI).

The rotation barriers about the central $E^{14}\text{–}C$ bond in betaines of the third group are 15.3–18.2 kcal/mol. The barrier value increases in the series Si–Ge–Sn and on going from $X = O$ to NMe.

In the compounds of the third group, cyclic isomers lie lower than open betaine forms in all cases. The relative energies and activation barriers are presented in Table XIII. It is noteworthy that the cyclic isomers of betaines of the second and third groups are much less polar than the corresponding betaines. The values of dipole moments range from 4.5 to 5.6 D. The energy gain for the formation of the cyclic form for betaines of this group lies in the range from 0.5 to 10 kcal/mol. Therefore, we can conclude that a dynamic equilibrium between cyclic and open forms of betaines can exist in a solution. The polarity of the medium has to affect strongly its position. An increase in the medium polarity shifts it toward more polar open forms.

TABLE XIII

THE RELATIVE ENERGIES (kcal/mol) OF THE CYCLIC
ISOMERS $\Delta E^\circ$ AND ACTIVATION BARRIERS $E_a^\circ$ FOR
INTERCONVERSION OF OPEN AND CYCLIC ISOMERS OF THE
THIRD GROUP

| Betaine | $E_a^\circ$ | $\Delta E^\circ$ |
|---------|-------------|------------------|
| P–C–Si–O | 0.9 | −0.5 |
| P–C–Ge–O | 0.9 | −4.3 |
| P–C–Sn–O | 2.6 | −3.5 |
| P–C–Si–N | 0.1 | −8.8 |
| P–C–Ge–N | 1.3 | −9.4 |
| P–C–Sn–N | 2.9 | −6.8 |
| As–C–Sn–N | 0.1 | −10.9 |

## C. Main Directions of $^{(+)}Me_3E^{15}–CH_2–E^{14}Me_2–X^{(-)}$ ($E^{15} = P$, As; X = S, $C_5H_4$, O, NMe, Se) Betaine Decomposition and Isomerization

Among four main directions of the chemical transformations of betaines of the first group, which are presented in Scheme 38, isomerization to metallated ylides (direction **A**), retro-Wittig type decomposition (direction **B**), and decomposition according to the Corey–Chaykovsky type (direction **C**) resulting in compounds with the three-membered cycle (see Section 5) were experimentally observed. The Wittig-type decomposition (direction **D**), which has not yet been found under experimental conditions, is of special interest. In studying these processes by experimental and theoretical methods, one should keep in mind that the possibility of their occurrence is determined, to a substantial extent, by subsequent reactions possible for the primarily formed decomposition products. Among such reactions, strongly exothermic dimerization (oligomerization) processes of compounds with $X=E^{14}$ bonds and three-membered heterocycles bearing $E^{14}$ atoms are most significant. The high polarity and affinity for the formation of complexes of the starting betaines and compounds with the $X=E^{14}$ bonds require taking into account solvation effects.

### 1. Isomerization of Betaines to Metallated Phosphorus and Arsenic Ylides (Direction **A**)

In all cases, ylides $HX–E^{14}Me_2–CH=E^{15}Me_3$ (X = S, Se; $E^{14}$ = Si, Ge; $E^{15}$ = P, As) isomeric to betaines lie in energy higher than the corresponding betaines. The differences in energies are almost independent of $E^{14}$ and determined by the type of ylide only. They are 12–16 kcal/mol for

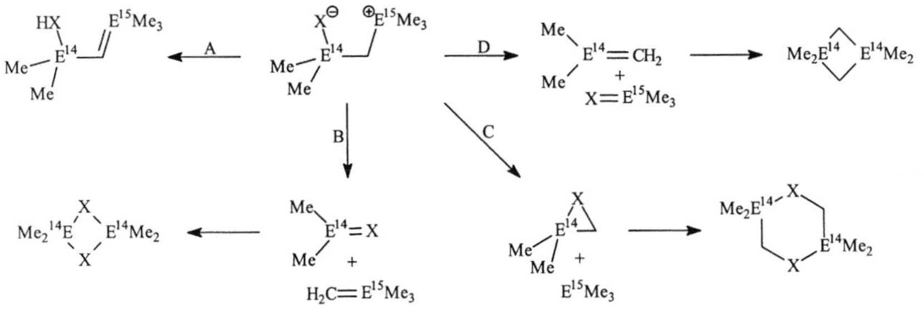

HX    $E^{15}Me_3$                    $X^{\ominus}$  $^{\oplus}E^{15}Me_3$              Me                                    Me_2E^{14}    E^{14}Me_2
Me—$E^{14}$  ←——— A ———  Me—$E^{14}$  ——— D ———  $E^{14}$=CH_2  ———→
Me                                    Me                        Me                    Me
Me                                                              +
                                                                X=$E^{15}Me_3$

B ↓                          C ↘

X                            Me                          X
Me_2$^{14}$E    E^{14}Me_2  ←——  $E^{14}$=X              Me    $E^{14}$           Me_2E^{14}    X
X                            Me                          Me         ———→              E^{14}Me_2
                             +                           Me                          X
                          $H_2C=E^{15}Me_3$              +
                                                        $E^{15}Me_3$

X = S, Se, O, NMe; $E^{14}$ = Si, Ge, Sn; $E^{15}$ = P, As.

**SCHEME 38**

phosphorus ylides and 15–21 kcal/mol for ylides of the arsenic series. In the case of betaines with $X = O$, $E^{15} = P$, ylides are comparable in stability with betaines or even more stable. In the case of $X = O$, $E^{15} = As$, ylides lie in energy by 2–9 kcal/mol higher than the cyclic forms. The ylide isomers of imide betaines $X = NMe$, in all cases, lie lower in energy by 2–20 kcal/mol than the corresponding open or cyclic betaine forms.

In metallated ylides $HX–E^{14}Me_2–CH=E^{15}Me_3$, the central fragment retains the *gauche*-conformation inherent in betaines. The electrostatic attraction between the X atoms bearing partial negative charges and positively charged arsenic and phosphorus atoms is retained, although it is noticeably weakened compared to that in the corresponding betaines. The dipole moments of ylides $HX–E^{14}Me_2–CH=E^{15}Me_3$ ($\mu \sim 3–5\,D$) is much lower than those of the corresponding betaines ($\mu \sim 7–10\,D$). Due to this, the equilibrium between the betaine and ylide forms in low-polar media should be shifted toward ylide, which we observed experimentally for phosphorus–silicon thiabetaines (see Section 5.4).

Transition states for betaine isomerization to ylides via the intramolecular mechanism were not localized. We believe that these processes are intermolecular and involve donor solvent molecules or the second betaine molecule as a proton carrier.

## 2. Retro-Wittig Decomposition of Betaines (Direction **B**)

The monomolecular retro-Wittig type decomposition of betaines $^{(-)}X–E^{14}Me_2–CH_2–E^{15}Me_3^{(+)}$, under gas-phase conditions, occurs as a simple cleavage of the central $E^{14}$–C bond and, hence, is a strongly endothermic process (Table XIV). The high thermal resistance of betaines **20b** and **20p** with alkyl groups at the phosphorus atom (see Section 5) agrees with these data.

However, the $Me_2E^{14}=X$ ($E^{14} = Si$, Ge, Sn; $X = O$, S, NMe, $C_5H_4$) compounds formed during the retro-Wittig type decomposition are

TABLE XIV

Activation Barriers ($E_a$) and Reaction Heats (kcal/mol) for Main Directions of Betaines I Decomposition[a]

| Betaine | Reaction A ylide | Reaction B | | | Reaction C | | | Reaction D | | |
| --- | --- | --- | --- | --- | --- | --- | --- | --- | --- | --- |
| | | $E_a^{\circ}$ | $Me_2E^{14}{=}X$ | Dimer | $E_a^{\circ}$ | $Me_2E^{14}(X)CH_2$ | Dimer | $E_a^{\circ}$ | $Me_2E^{14}{=}CH_2$ | Dimer |
| P–C–Si–S | 12.3 | – (17.8) | 43.0 (21.7) | 9.9 | 32.6 | 19.8 | 0.3 | 37.4 | 34.0 (27.2) | –5.1 |
| P–C–Ge–S | 12.0 | – (–) | 38.2 (22.1) | 8.1 | 35.5 | 20.0 | –1.6 | 32.2 | 28.9 (25.0) | –5.7 |
| P–C–Sn–S | 13.1 | – (–) | 40.2 (22.0) | 5.1 | 41.6 | 24.7 | 1.4 | 35.2 | 38.4 (30.9) | –0.5 |
| As–C–Si–S | 16.4 | – (21.1) | 47.2 (25.9) | 14.0 | 20.7 | 5.9 | –16.6 | 29.7 | 35.3 (28.4) | –3.8 |
| As–C–Ge–S | 15.7 | – (–) | 41.7 (25.6) | 11.6 | 22.9 | 5.3 | –16.2 | 27.1 | 29.6 (25.7) | –5.1 |
| As–C–Sn–S | 16.3 | – (–) | 43.2 (25.0) | 8.1 | 27.9 | 9.6 | –13.6 | 29.0 | 38.5 (31.0) | –0.4 |
| P–C–Si–Se | 14.6 | – (18.5) | 42.9 (22.0) | 12.3 | 30.1 | 19.4 | 1.4 | 35.4 | 35.1 (28.2) | –4.0 |
| P–C–Ge–Se | 15.6 | – (–) | 37.6 (21.9) | 9.2 | 33.1 | 20.5 | 0.3 | 35.6 | 32.5 (28.6) | –2.2 |
| P–C–Sn–Se | 18.0 | – (–) | 39.2 (21.4) | 6.3 | 39.8 | 26.5 | 4.3 | 40.1 | 44.1 (36.6) | 5.2 |
| As–C–Si–Se | 18.9 | – (21.9) | 47.2 (26.3) | 16.6 | 18.2 | 5.5 | –12.5 | 28.4 | 32.9 (26.0) | –6.2 |
| As–C–Ge–Se | 19.3 | – (–) | 41.3 (25.6) | 12.9 | 20.7 | 6.0 | –14.2 | 27.8 | 29.7 (25.9) | –4.9 |
| As–C–Sn–Se | 21.2 | – (–) | 42.3 (24.5) | 9.4 | 26.1 | 11.6 | –10.7 | 31.6 | 40.7 (33.3) | 1.9 |

| | | | | | | | | | |
|---|---|---|---|---|---|---|---|---|---|
| P–C–Si–O | 0.8 | —(–) | 44.6 (21.8) | −2.9 | 42.3 | 29.3 | −1.0 | — | 28.2 (21.4) | −10.9 |
| P–C–Ge–O | −4.6 | —(–) | 39.9 (22.5) | −0.7 | 42.3 | 21.2 | −8.7 | 5.5 | 8.5 (4.6) | −26.2 |
| P–C–Sn–O | −8.7 | —(–) | 43.5 (23.9) | −3.1 | 46.3 | 16.0 | −10.7 | 1.2 | 8.5 (1.0) | −30.4 |
| As–C–Si–O* | 8.8 | —(–) | 52.4 (29.6) | 4.9 | 33.5 | 19.1 | −11.3 | — | 50.2 (43.4) | 11.1 |
| As–C–Ge–O* | 6.2 | —(–) | 50.8 (33.4) | 10.2 | 37.1 | 13.9 | −16.1 | — | 33.3 (29.4) | −1.3 |
| As–C–Sn–O* | 1.8 | —(–) | 54.0 (34.4) | 7.4 | 40.6 | 8.5 | −18.2 | — | 33.2 (25.7) | −5.7 |
| P–C–Si–N | −16.8 | —(–) | 33.3 (21.0) | −13.2 | 29.7 | 5.4 | −25.2 | — | 26.5 (19.6) | −12.6 |
| P–C–Ge–N | −18.1 | —(–) | 27.5 (19.7) | −9.2 | 31.2 | 1.5 | −29.6 | — | 13.3 (9.4) | −21.3 |
| P–C–Sn–N | −19.6 | —(–) | 31.6 (21.1) | −8.6 | 16.9 | −4.9 | −28.9 | 1.6 | 14.9 (7.5) | −23.9 |
| As–C–Si–N* | −1.1 | —(–) | 48.8 (36.5) | 2.3 | 28.4 | 2.9 | −27.7 | — | 49.5 (42.7) | 10.4 |
| As–C–Ge–N* | −2.0 | —(–) | 43.4 (35.6) | 6.7 | 30.3 | −0.8 | −31.9 | — | 36.7 (32.8) | 2.0 |
| As–C–Sn–N | −16.0 | —(–) | 34.8 (24.3) | −5.4 | 22.6 | −19.7 | −43.7 | — | 25.7 (18.3) | −13.2 |

[a]All values are calculated relative to energies of open (betaine) forms except the compounds of second group (marked by asterisks) when open forms do not exist. In these cases energies are calculated relative to cyclic forms. Dash means that the state does not exist on PES. Values in brackets correspond to reactions in presence of one molecule of pyridine.

kinetically unstable and undergo exothermic cyclooligomerization to form dimers and trimers, which can compensate, to a great extent, for energy expenses of this decomposition.

The affinity of the $Me_2E^{14}$=X compounds to form stable complexes due to coordination at the $E^{14}$ atom of donor molecules or intramolecular coordination of donor groups is well known and has already been mentioned in Section 5. Therefore, we studied the model reaction of the retro-Wittig type decomposition of betaines of the first group in pyridine. Table XV contains the data on the geometric parameters of the $Me_2E^{14}$=X compounds and their complexes with pyridine and complex formation energies. The geometry of the $Me_2E^{14}$=X molecules changes markedly upon the formation of complexes with pyridine. The $E^{14}$=X and $E^{14}$–C bond lengths increase by 0.3–0.4 and $\sim 0.2$ Å, respectively. The configuration at the $E^{14}$ atom becomes pyramidal. It is of interest that the X–$E^{14}$–N angle in the complexes decreases in the series $E^{14}$ = Si, Ge, Sn from 103.8–107.7° for silicon to 92.5–101.1° for tin. The more electronegative is the X atom, the smaller is the X–$E^{14}$–N angle. This dependence is stipulated by the fact that with an increase in the atomic number of the $E^{14}$ atom the contribution increases of its $np_z$ orbital to LUMO of the $Me_2E^{14}$=X molecule with which a lone pair of the nitrogen atom of the pyridine molecule interacts.

As can be seen from the data presented, the high energies of complex formation decrease sharply the endothermicity of the retro-Wittig type decomposition and, moreover, fundamentally change the reaction mechanism. As has been shown for betaines $^{(-)}$X–$E^{14}Me_2$–$CH_2$–$E^{15(+)}Me_3$ (X = S, Se; $E^{14}$ = Si, Ge; $E^{14}$ = P, As), the reaction occurs as bimolecular nucleophilic substitution at the $E^{14}$ atom. For silicon betaines, the transition states TS-b-pyr with pentacoordinate silicon and nearby them no deep local minima corresponding to the C-b complexes can be localized in the reaction coordinate.

TS-b-pyr                         C-b

X = S, Se; $E^{14}$ = Si, Ge; $E^{15}$ = P, As.

The thermal effect of the retro-Wittig type decomposition of betaine $^-$S–$SiMe_2$–$CMe_2$–$P^+Ph_3$ (**20a**) in pyridine, which was observed experimentally (see Section 5), is much lower than that for the corresponding betaine $^-$S–$SiMe_2$–$CMe_2$–$P^+Me_3$ and amounts, according to the calculated data, 30.9 kcal/mol ignoring the cyclodimerization of S=$SiMe_2$ and −3.1 kcal/mol taking into account the latter. The values obtained agree satisfactorily with experiment.

## TABLE XV

Geometric Parameters of $Me_2E^{14}=X$ ($E^{14}$ = Si, Ge, Sn; X = S, Se, O, NMe, $CH_2$), their Complexes with Pyridine $Me_2E^{14}=X*Py$ (Values in Brackets) and Complex Formation Energies $\Delta E°$ (kcal/mol)

| $Me_2E^{14}=X$ | $E^{14}-X$ | $E^{14}-C$ | $X-E^{14}-C$ | $C-E^{14}-C$ | $E^{14}-N$ | $X-E^{14}-N$ | $\Delta E°$ |
|---|---|---|---|---|---|---|---|
| $Me_2Si=S$ | 1.978 (2.022) | 1.883 (1.899) | 124.5 (119.7) | 111.0 (109.2) | – (2.011) | – (106.3) | –21.3 |
| $Me_2Ge=S$ | 2.060 (2.103) | 1.968 (1.981) | 123.9 (121.0) | 112.1 (110.3) | – (2.173) | – (103.5) | –16.1 |
| $Me_2Sn=S$ | 2.260 (2.293) | 2.182 (2.190) | 124.6 (123.4) | 110.8 (108.9) | – (2.382) | – (99.9) | –18.2 |
| $Me_2Si=Se$ | 2.123 (2.164) | 1.884 (1.900) | 124.6 (119.4) | 110.8 (109.3) | – (2.010) | – (107.0) | –20.9 |
| $Me_2Ge=Se$ | 2.191 (2.240) | 1.969 (1.981) | 124.2 (120.9) | 111.7 (110.0) | – (2.170) | – (104.5) | –15.7 |
| $Me_2Sn=Se$ | 2.383 (2.420) | 2.183 (2.191) | 124.9 (123.5) | 110.3 (108.5) | – (2.385) | – (101.1) | –17.8 |
| $Me_2Si=O$ | 1.551 (1.572) | 1.879 (1.898) | 123.7 (120.6) | 112.5 (109.4) | – (2.013) | – (101.5) | –22.8 |
| $Me_2Ge=O$ | 1.661 (1.685) | 1.969 (1.981) | 122.9 (121.4) | 114.1 (111.2) | – (2.184) | – (97.8) | –17.4 |
| $Me_2Sn=O$ | 1.874 (1.896) | 2.186 (2.194) | 124.0 (123.7) | 112.0 (109.7) | – (2.384) | – (92.5) | –19.6 |
| $Me_2Si=NMe$ | 1.622 (1.638) | 1.875 (1.887) | 117.7 (115.4) | 117.3 (111.1) | – (2.073) | – (103.8) | –12.3 |
| $Me_2Ge=NMe$ | 1.729 (1.744) | 1.980 (1.963) | 127.7 (115.4) | 115.9 (113.9) | – (2.387) | – (100.1) | –7.8 |
| $Me_2Sn=NMe$ | 1.948 (1.963) | 2.174 (2.180) | 115.4 (114.7) | 115.0 (111.8) | – (2.505) | – (95.1) | –10.5 |
| $Me_2Si=CH_2$ | 1.722 (1.738) | 1.884 (1.896) | 122.7 (119.8) | 114.6 (113.1) | – (2.195) | – (107.7) | –6.9 |
| $Me_2Ge=CH_2$ | 1.789 (1.799) | 1.964 (1.967) | 122.6 (121.7) | 114.8 (114.2) | – (2.650) | – (103.9) | –3.9 |
| $Me_2Sn=CH_2$ | 1.999 (2.026) | 2.177 (2.187) | 123.1 (124.0) | 113.9 (111.8) | – (2.568) | – (95.9) | –7.5 |

### 3. Corey–Chaykovsky Type Decomposition (Direction **C**)

Betaine decomposition according to the Corey–Chaykovsky type (direction **C**) occurs as intramolecular $S_N$ substitution at the atom of the element and results in phosphine (arsine) $E^{15}Me_3$ elimination through the transition states TS-c. The energies of the transition states TS-c and thermal effects are presented in Table XIV. In the initial region of the coordinate of this reaction, the *gauche*-conformation of betaines is transformed into the *trans*-conformation. An increase in the solvent polarity facilitates this transition and favors an increase in the population of the *trans*-conformation. However, the influence of the solvent on the process rate can be estimated more exactly only in terms of more precise models because the polarity of the transition state TS 2 ($\mu \sim 7$ D) is lower than that of betaines in the *trans*-conformation. Elementaheterocyclopropanes formed as products are highly reactive and exothermically dimerize (Table XIV).

TS-c

The activation energy of this process increases in the Si–Ge–Sn series, except for the cases of X = NMe. On going from phosphorus to arsenic betaines, the activation energy decreases by 5–15 kcal/mol, and the stability of the reaction products increases similarly.

### 4. Wittig Type Decomposition of Betaines (Direction **D**)

The Wittig type decomposition of betaines to form $X=E^{15}Me_3$ (X = S, Se, O, NMe; $E^{15}$ = P, As) and elementaolefins $Me_2E^{14}=CH_2$ ($E^{14}$ = Si, Ge, Sn) passes through four-membered transition states TS-d. The reaction coordinates of betaine decomposition exhibit distinct local minima corresponding to post-reaction complexes of elementaolefins C-d in which the leaving $X=E^{15}Me_3$ molecules remain to be coordinated to the positively charged $E^{14}$ atoms by X atoms.

TS-d                              C-d

Elementaolefins $Me_2E^{14}=CH_2$ can further undergo exothermic cyclodimerization to form dielementacyclobutanes $(Me_2E^{14}CH_2)_2$. The activation barriers and thermal effects of the reactions are presented in Table XIV. The transition states TS-d are less polar ($\mu \sim 4$ D) than the transition states in the Corey–Chaykovsky type decompositions.

Compounds with the $E^{14}=C$ multiple bonds are polar and form complexes with pyridine. According to our estimations, the complex formation energies are 3.9–7.5 kcal/mol. The formation of such complexes allows an additional stabilization of products of their Wittig type decomposition.

According to the calculations, the activation barrier of the Corey–Chaykovsky type decomposition for silicon thiabetaines of the first group is lower by 5 kcal/mol than that for the Wittig type decomposition. According to this, photolysis and thermolysis of these betaines result only in the elimination of phosphine or arsine (see Section 5). However, the situation is opposite for germanium and tin betaines. In these cases, the Wittig type decomposition should become predominant because it is favored by the thermodynamic factor: high values of the total thermal effects of the reaction, taking into account elementaolefin cyclodimerization.

It is noteworthy that for betaines $X = O$, NMe $E^{14} = Ge$, Sn $E^{15} = P$ Wittig type decomposition becomes kinetically and thermodynamically the most favorable process. In the case of betaine $Me_3P^{(+)}–CH_2–GeMe_2–O^{(-)}$, the activation energy of germene $Me_2Ge=CH_2$ formation is only 5.5 kcal/mol. These conclusions need experimental checking.

### D. *Structures and Potential Energy Surfaces of Betaines* $^{(+)}Me_3E^{15}–CH_2–E^{14}Me_2{}^{(-)}$ ($E^{14}$ = Si, Ge, Sn; $E^{15}$ = P, As) II

We studied the structure and PES of betaines **II** for the model compounds $^{(+)}Me_3E^{15}–CH_2–E^{14}Me_2{}^{(-)}$ ($E^{14}$ = Si, Ge, Sn; $E^{15}$ = P, As). The most important calculated geometric parameters for these betaines are presented in Table XVI. In them the $E^{14}–C$ bonds in the $E^{14}–C–E^{15}$ fragment are much longer than the $E^{14}–Me$ bonds. This difference in lengths

TABLE XVI

THE MOST IMPORTANT GEOMETRIC PARAMETERS OF BETAINES $^{(+)}Me_3E^{15}–CH_2–E^{14}Me_2{}^{(-)}$ ($E^{14}$ = Si, Ge, Sn; $E^{15}$ = P, As) (INTERATOMIC DISTANCES IN Å, VALENCE ANGLES IN GRAD°)

| Betaine | $E^{14}–C$ | $E^{14}–C_{Me}$ | C–E | $E^{14} \cdots E$ | $E^{14}–C–E$ |
|---------|------------|------------------|-----|--------------------|---------------|
| P–C–Si | 2.057 | 1.952 | 1.777 | 3.103 | 107.8 |
| P–C–Ge | 2.192 | 2.045 | 1.763 | 3.209 | 107.9 |
| P–C–Sn | 2.410 | 2.248 | 1.756 | 3.445 | 110.6 |
| As–C–Si | 2.034 | 1.950 | 1.911 | 3.122 | 104.6 |
| As–C–Ge | 2.168 | 2.044 | 1.896 | 3.237 | 105.4 |
| As–C–Sn | 2.390 | 2.248 | 1.887 | 3.484 | 108.5 |

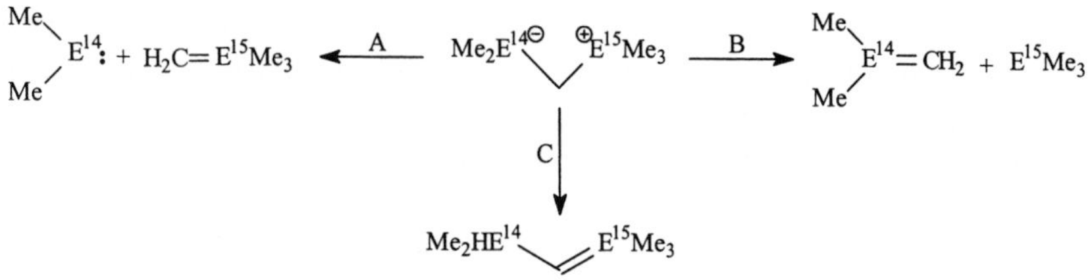

$$\text{Me}_2\text{E}^{14} \overset{\ominus}{\diagdown} \overset{\oplus}{\diagup} \text{E}^{15}\text{Me}_3$$

via reaction A: $\underset{\text{Me}}{\overset{\text{Me}}{\diagdown}} \text{E}^{14}\colon + \text{H}_2\text{C}=\text{E}^{15}\text{Me}_3 \overset{A}{\longleftarrow}$

via reaction B: $\overset{B}{\longrightarrow} \underset{\text{Me}}{\overset{\text{Me}}{\diagdown}} \text{E}^{14}=\text{CH}_2 + \text{E}^{15}\text{Me}_3$

via reaction C: $\overset{C}{\downarrow} \quad \text{Me}_2\text{HE}^{14} \diagdown \diagup \text{E}^{15}\text{Me}_3$

**SCHEME 39**

TABLE XVII

ENERGY PARAMETERS (kcal/mol) OF DECOMPOSITION OF BETAINES $^{(+)}\text{Me}_3\text{E}^{15}-\text{CH}_2-\text{E}^{14}\text{Me}_2^{(-)}$
($\text{E}^{14}=\text{Si, Ge, Sn; E}^{15}=\text{P, As}$)

| Betaine | Reaction A | Reaction B | | Reaction C | |
|---|---|---|---|---|---|
| | $\Delta E°$ | $E_a$ | $\Delta E°$ | $E_a$ | $\Delta E°$ |
| P–C–Si | 35.0 | 19.3 | −5.8 | 31.5 | −22.3 |
| P–C–Ge | 30.4 | 28.2 | 8.0 | 40.5 | −4.6 |
| P–C–Sn | 29.4 | 36.9 | 24.4 | 46.8 | 7.0 |
| As–C–Si | 38.4 | 8.1 | −20.5 | 34.5 | −18.5 |
| As–C–Ge | 33.0 | 15.1 | −7.5 | 41.6 | −1.6 |
| As–C–Sn | 31.5 | 22.5 | 8.4 | 48.4 | 9.4 |

is especially significant for tin betaines. The $\text{E}^{14}-\text{C}-\text{E}^{15}$ bond angles are somewhat smaller than the tetrahedral angle of 109.4°. The interatomic $\text{E}^{14}\cdots\text{E}^{15}$ distances in betaines are shorter by approximately 0.8 Å than the sums of van der Waals radii of these atoms (see Table XI).

We studied three main directions of chemical transformations of betaines **II** (Scheme 39):

- Decomposition with the elimination of arsenic or phosphorus ylide and formation of silylene (germylene, stannylene) $\text{Me}_2\text{E}^{14}$, which represents the process inverse to the synthesis of these betaines.
- Decomposition with the formation of elementaolefins $\text{Me}_2\text{E}^{14}=\text{CH}_2$ and phosphine (arsine) $\text{E}^{15}\text{Me}_3$.
- Isomerization to metallated ylides $\text{Me}_2\text{HE}^{14}-\text{CH}=\text{EMe}_3$.

The formation of betaines in the reaction of silylenes (germylenes, stannylenes) $\text{Me}_2\text{E}^{14}$ ($\text{E}^{14}=\text{Si, Ge, Sn}$) with ylides $\text{H}_2\text{C}=\text{E}^{15}\text{Me}_3$ ($\text{E}^{15}=\text{P, As}$) under gas-phase conditions occurs without a barrier as a strongly exothermic process. The thermal effects of the reactions are presented in Table XVII.

The elimination of phosphine (arsine) $\text{E}^{15}\text{Me}_3$ occurs through the transition states TS-1. The activation barriers of this process for betaines of the phosphorus series are higher than those for arsenic betaines. They increase and the thermal effect of the reaction decreases in the series

$E^{14} = Si$, Ge, Sn (Table XVII). This type of decomposition is thermo-dynamically favorable for betaines $Me_2E^{14-}-CH_2-As^+Me_3$ ($E^{14} = Si$, Ge), and global minima on PES of these betaines correspond to the formation of $Me_2E^{14}=CH_2$ ($E^{14} = Si$, Ge).

$$Me_2E^{14} \cdots E^{15}Me_3 \qquad\qquad Me_2E^{14} \diagdown\diagup EMe_3$$

$Me_2E^{14} \cdots E^{15}Me_3$      $Me_2E^{14} \underset{H}{\diagdown\diagup} EMe_3$

TS 1        TS 2

$E^{14} = Si$, Ge, Sn; $E^{15} = P$, As

For betaines $Me_2E^{14}-CH_2-P^+Me_3$ ($E^{14} = Si$, Ge), this decomposition is also kinetically most favorable but isomerization to ylides $Me_2HE^{14}-CH=PMe_3$ ($E^{14} = Si$, Ge) leads to global minima on PES. Intramolecular isomerization through the transition states TS-2 is associated with over-coming of the high activation barriers (31.5–48.4 kcal/mol, Table XVII). Therefore, the intermolecular route of isomerization can be more preferable from the kinetic point of view. In polar media a donating molecule of the solvent can act as a proton carrier. All three directions of chemical transformations are endothermic for tin betaines $^{(+)}Me_3E^{15}-CH_2-SnMe_2^{(-)}$, and global minima on PES correspond to them.

The thermodynamic data presented in Table XVI are calculated for the temperature $T = 0\,K$. Note that the entropy factor favors betaine decomposition via directions **A** and **B** at higher temperatures. The reactions of organoelement analogs of carbenes with phosphorus and arsenic ylides are yet poorly studied. The presented above results of calculations allow an optimistic prognosis about the possibility of developing a new method for the synthesis of elementaolefins $R_2E^{14}=CH_2$ ($E^{14} = Si$, Ge, Sn) on the basis of these reactions.

We theoretically studied the reactions of stable West' silylenes **32** and **73** with phosphorus ylide $H_2C=PMe_3$.[74] Similarly to the simplest analogs of carbenes, these compounds can form betaines in which the negative charge is localized on the silicon atom and the positive charge is localized on the phosphorus atom. These betaines can thermally decompose to form silenes (direction **A**, Scheme 39) or be isomerized to ylides via direction **B**.

t-Bu–N–Si–N–t-Bu     t-Bu–N–Si–N–t-Bu

**32**         **73**

Silenes formed in direction **A** can exothermically be dimerized to form dimers of two types (head-to-tail and head-to-head), which differ in thermal stability (Scheme 40). The energy parameters of these reactions are presented in Table XVIII.

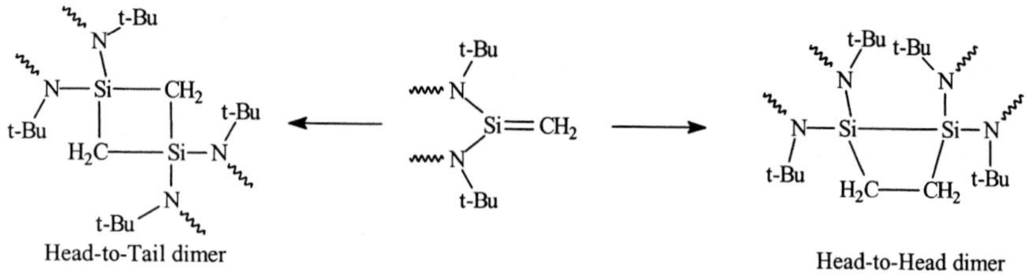

Scheme 40

## TABLE XVIII
Energies (kcal/mol) of Stationary Points of PES's for
Reactions of West's Silylenes 32 and 73 with $H_2C=PMe_3$

| ?? | $E^\circ$ | |
|---|---|---|
| | **32** | **73** |
| Sylilene + $H_2C=PMe_3$ | 0.0 | 0.0 |
| Betaine | −2.9 | −6.8 |
| TS-a | 8.5 | 2.9 |
| Silene + $PMe_3$ | −1.0 | −9.0 |
| $\frac{1}{2}$ H–T–T dimer + $PMe_3$ | −29.3 | −39.2 |
| $\frac{1}{2}$ H–T–H dimer + $PMe_3$ | −21.5 | −30.5 |
| TS-b | 35.6 | 29.3 |
| Ylide | −18.4 | −26.1 |

Analysis of the data in Table XVIII suggests that silene formation is kinetically the most favorable process. However, according to experiment, metallated silenes are formed. This is related to the fact that in polar solvents proton transfer from the carbon atom to silicon is intermolecular, which leads to a considerable decrease in the reaction barrier. We believe that when the migration of substituents from the carbon atom to silicon is suppressed, for example, by the introduction of two alkyl radicals, the elimination of phosphines resulting in silene formation becomes the most probable process.

# VII

# CONCLUSION

The data on the structure and reactivity of organoelement betaines **I** and **II** presented in the review shows that they lie at the "very vivid crossroads of chemical ways," which open alluring new challenges for mutual transitions between heavy analogs of carbenes (silylenes, germylenes, and stannylenes), on the one hand, and compounds with double bonds $E^{14}$=X and small cycles including Group 14 elements, on the other hand. The chemistry of these three types of compounds is being especially intensely developed. It becomes clear that it should be considered as a single area because the structure and properties of these substances are determined, to a great extent, by resembling general factors, common approaches are used in their synthesis and study, and several chemical reactions are known which allow their mutual transformations. The chemistry of betaines **I** and **II** is at the very beginning of its development. However, it is quite clear that the scope of their reactions is wide and diverse. In particular, the results of theoretical analysis give an optimistic prognosis about a possibility of the development of new general methods for the preparation or generation as short-lived intermediates of elementaolefins $R_2E^{14}$=$CH_2$ ($E^{14}$ = Si, Ge, Sn) on the basis of these reactions. Therefore, we can expect that betaines **I** and **II** will find various synthetic applications in future.

ACKNOWLEDGEMENTS

The authors are grateful to Dr. M.S. Nechaev and Dr. V.N. Khrustalev for help in the preparation of the text, to Professors R. West, C. Eaborn, D.J. Smith, J. Barrau, K. Izod, S. Berger, M. Veith, E. Vedejs, E. Nifant'ev, and O. Kolodyazhnyi and to Dr. D. Laikov for fruitful discussions. This work was financially supported by the Russian Foundation for Basic Research (Project Nos. 96-03-33188 and 00-03-32889).

REFERENCES

(1) Tacke, R.; Pulm, M.; Wagner, B. *Adv. Organomet. Chem.* **1999**, *44*, 221.

(2) Venne-Dunker, S.; Ahlers, W.; Erker, G.; Frohlich, R. *Eur. J. Inorg. Chem.* **2000**, 1671.

(3) Chauvin, R. *Eur. J. Inorg. Chem.* **2000**, 577.

(4) Alcaide, B.; Cassarrubios, L.; Deminguer, G.; Sierra, M. A. *Curr. Org. Chem.* **1998**, *2*, 551.

(5) Johnson, A. W. Ylides and Imines of Phosphorus, Wiley-Interscience, New York, 1993.

(6) Kolodiazhnyi, O. I. Phosphorus Ylides. Chemistry and Application in Organic Synthesis, Wiley-VCH, Weinheim-New York-Chichester, 1999.

(7) (a) Vedejs, E.; Peterson, M. J., (Sniekus, V. Ed.), *Advances in Carbanion Chemistry*, Vol. 2, JAI Press, Greenwich, CT, 1996. pp. 1–85. (b) Vedejs, E.; Peterson, M. J. *Top. Stereochem.*, Vol. 2119941.

(8) Vedejs, E.; Marth, C. F., (Quin, D. L., Verkade, J. G. Eds.) Phosphorus-31 NMR Spectral Properties in Compound Characterization and Structural Analysis, VCH, New York, 1994. pp. 297.

(9) Geletneky, C.; Forsterling, F. H.; Bock, W.; Berger, S. *Chem. Ber.* **1993**, *126*, 2397.

(10) Neumann, R. A.; Berger, S. *Eur. J. Org. Chem.* **1998**, 1085.

(11) Borisova, I. V.; Zemlyansky, N. N.; Shestakova, A. K.; Ustynyuk, Yu.A. *Mendeleev Commun.* **1996**, 90.

(12) Puke, C.; Erker, G.; Aust, N. C.; Wurtheim, E. U.; Frohlich, R. *J. Am. Chem. Soc.* **1998**, *120*, 4863.

(13) Puke, C.; Erker, G.; Wibbeling, B.; Frohlich, R. *Eur. J. Org. Chem.* **1999**, 1831.

(14) Borisova, I. V.; Zemlyansky, N. N.; Shestakova, A. K.; Ustynyuk, Yu.A. *Russ. Chem. Bull.* **1993**, *42*, 2053.

(15) Borisova, I. V.; Zemlyansky, N. N.; Ustynyuk, Yu.A.; Khrustalev, V. N.; Lindeman, S. V.; Struchkov, Yu.T. *Russ. Chem. Bull.* **1994**, *43*, 318.

(16) Seyferth, D.; Lim, T. F. O. *J. Am. Chem. Soc.* **1978**, *100*, 7074.

(17) (West, R., Stone, F. G. A. Eds.) Multiply Bonded Main group Metals and Metalloids, Academic Press, San Diego, 1996.

(18) (Auner, N., Weis, J. Eds.) Organosilicon Chemistry III. From Molecules to Materials, Wiley-VCH, Weinheim, Germany, 1998.

(19) Li, A. H.; Dai, L. X.; Aggarval, V. K. *Chem. Rev.* **1997**, *97*, 2341.

(20) Borisova, I. V.; Zemlyansky, N. N.; Ustynyuk, Yu.A.; Beletskaya, I. P.; Chernyshev, E. A. *Metalloorganich. Khimia* **1992**, *5*, 548.

(21) Ustynyuk, Yu.A.; Zakharov, P. I.; Azizov, A. A.; Shchembelov, G. A. *J. Organomet. Chem.* **1975**, *96*, 195.

(22) Nakadaira, Y.; Sakaba, H.; Sarurai, H. *Chem. Lett.* **1980**, 1071.

(23) Barton, T. G.; Burns, G. T.; Arnold, E. V.; Clardy, J. *Tetrahedron Lett.* **1981**, *22*, 287.

(24) Borisova, I. V.; Zemlyansky, N. N.; Shestakova, A. K.; Ustynyuk, Yu.A. *Mendeleev Commun.* **1996**, 229.

(25) Zemlyansky, N. N.; Borisova, I. V.; Shestakova, A. K.; Ustynyuk, Yu.A.; Chernyshev, E. A. *Russ. Chem. Bull.* **1998**, *47*, 469.

(26) Couret, C.; Escudié, J.; Satgé, J.; Lazraq, M. *J. Am. Chem. Soc.* **1987**, *109*, 4411.

(27) Lazraq, M.; Escudié, J.; Couret, C.; Satgé, J.; Dräger, M.; Dammel, R. *Angew. Chem., Int. Ed. Engl.* **1988**, *27*, 828.

(28) Lazraq, M.; Couret, C.; Escudié, J.; Satgé, J.; Soufiaoui, M. *Polyhedron* **1991**, *10*, 1153.

(29) Chaubon, M. A.; Escudié, J.; Ranaivonjatovo, H.; Satgé, J. *J. Chem. Soc., Dalton Trans.* **1996**, 893.

(30) Couret, C.; Escudié, J.; Delpon-Lacaze, G.; Satgé, J. *J. Organomet. Chem.* **1992**, *440*, 233.

(31) Anselme, G.; Ranaivonjatovo, H.; Escudié, J.; Couret, C.; Satgé, J. *Organometallics* **1992**, *11*, 2748.

(32) Anselme, G.; Couret, C.; Escudié, J.; Rishelme, S.; Satgé, J. *J. Organomet. Chem.* **1991**, *418*, 321.

(33) Borisova, I. V.; Zemlyansky, N. N.; Belsky, V. K.; Kolosova, N. D.; Sobolev, A. N.; Luzikov, Yu.N.; Ustynyuk, Yu.A.; Beletskaya, I. P. *J. Chem. Soc., Chem. Commun.* **1982**, 1090.

(34) Borisova, I. V.; Zemlyansky, N. N.; Luzikov, Yu.N.; Ustynyuk, Yu.A.; Belsky, V. K.; Kolosova, N. D.; Shtern, M. M.; Beletskaya, I. P. *Dokl. Akad. Nauk SSSR* **1983**, *269*, 90.

(35) Beletskaya, I. P.; Zemlyansky, N. N. (Reutov, O. A., Ed.), *Advances in Organometallic Chemistry*, Mir, Moscow, 1984, pp. 88–116.

(36) Ustynyuk Yu. A.; Nechaev M. S.; Laikov D. N.; Borisova I. V.; Zemlyansky N. N.; Khrustalev V. N.; Kuznetzova M. G.; Shestakova A. K. Presented in Part at the Lomonosov Scientific Conference (Moscow), Apr 2001.

(37) Zemlyansky, N. N.; Borisova, I. V.; Luzikov, Yu.N.; Kolosova, N. D.; Ustynyuk, Yu.A.; Beletskaya, I. P. *Izv. Akad. Nauk SSSR, Ser. Khim.* **1980**, 2668.

(38) Zemlyansky, N. N.; Borisova, I. V.; Luzikov, Yu.N.; Ustynyuk, Yu.A.; Kolosova, N. D.; Beletskaya, I. P. *J. Org. Chem. USSR* **1981**, *17*, 1323.

(39) Zemlyansky, N. N.; Borisova, I. V.; Kolosova, N. D.; Luzikov, Yu.N.; Bel'skii, V. K.; Ustynyuk, Yu.A.; Beletskaya, I. P. *Izv. Akad. Nauk SSSR, Ser. Khim.* **1981**, 2837.

(40) Bel'skii, V. K.; Zemlyansky, N. N.; Borisova, I. V.; Kolosova, N. D.; Beletskaya, I. P. *Izv. Akad. Nauk SSSR, Ser. Khim.* **1981**, 1184.

(41) Belsky, V. K.; Zemlyansky, N. N.; Borisova, I. V.; Kolosova, N. D.; Beletskaya, I. P. *Cryst. Struct. Commun.* **1982**, *11*, 497.

(42) Belsky, V. K.; Zemlyansky, N. N.; Borisova, I. V.; Kolosova, N. D.; Beletskaya, I. P. *Cryst. Struct. Commun.* **1982**, *11*, 881.

(43) Yarnykh, V. L.; Mstyslavsky, V. I.; Zemlyansky, N. N.; Borisova, I. V.; Roznyatovskii, V. A.; Ustynyuk, Yu.A. *Russ. Chem. Bull.* **1997**, *46*, 1228.

(44) Jones, P. R.; Rozell, J. M.; Campbell, B. M. *Organometallics* **1988**, *4*, 1321.

(45) Brook, A. G.; MacMillan, A. *J. Organomet. Chem.* **1988**, *341*, C9.

(46) Purrello, G.; Fiandaca, P. *J. Chem. Soc., Perkin I* **1976**, 693.

(47) Bombieri, G.; Forsellini, E.; Chiacchio, U.; Fiandaca, P.; Purrello, G.; Foresti, E.; Graziani, R. *J. Chem. Soc., Perkin II* **1976**, 1404.

(48) Schaumann, E.; Grabley, F.-F. *Liebigs Ann. Chem.* **1979**, 1702.

(49) Bestmann, H. J.; Engler, R.; Hartung, H.; Roth, K. *Chem. Ber.* **1979**, *112*, 28.

(50) Kunze, U.; Merkel, R.; Winter, W. *Chem. Ber.* **1982**, *115*, 3653.

(51) Erker, G.; Hock, R.; Wilker, S.; Laurent, C.; Puke, R.; Wurtheim, E.-U.; Frohlich, R. *Phosphorus, Sulfur, Silicon Relat. Elem.* **1999**, *153–154*, 79.

(52) Brook, A. G.; Brook, A. M. *Adv. Organomet. Chem.* **1996**, *39*, 71.

(53) Hemme, I.; Klingebiel, U. *Adv. Organomet. Chem.* **1996**, *39*, 159.

(54) Driess, M. *Adv. Organomet. Chem.* **1996**, *39*, 193.

(55) Okazaki, R.; West, R. *Adv. Organomet. Chem.* **1996**, *39*, 232.

(56) Escudie, J.; Ranaivonjatovo, H. *Adv. Organomet. Chem.* **1999**, *44*, 113.

(57) Zemlyansky, N. N.; Borisova, I. V.; Shestakova, A. K.; Ustynyuk, Yu.A. *Russ. Chem. Bull.* **1993**, *42*, 2056.

(58) Borisova, I. V.; Zemlyansky, N. N.; Shestakova, A. K.; Ustynyuk, Yu.A.; Chernyshev, E. A. *Russ. Chem. Bull.* **2000**, *49*, 920.

(59) Li, G. M.; Segi, M.; Kamogawa, T.; Nakajima, T. *Chem. Express* **1993**, *8*, 53.

(60) Borisova, I. V.; Zemlyansky, N. N.; Khrustalev, V. N.; Kuznetzova, M. G.; Ustynyuk, Yu.A.; Nechaev, M. S. *Russ. Chem. Bull.* **2001**, *50*, 1679.

(61) Borisova, I. V. D. Sci. (Chem.) Dissertation, M. V. Lomonosov Moscow State University, M., 2003.

(62) Kocheshkov, K. A.; Zemlyansky, N. N.; Sheverdina, N. I.; Panov, E. M. Methods of Organometallic Chemistry. Germanium. Tin. Lead, Nauka Publishers, Moscow, 1968.

(63) Lloyd, D.; Gosney, I.; Ormiston, R. A. *Chem. Soc. Rev.* **1987**, *16*, 45.

(64) Lloyd, D.; Gosney, I., (Patai, S. Ed.) The Chemistry of Organic Arsenic, Antimony, and Bismuth Compounds, Wiley, Chichester, 1994.

(65) Borisova, I. V.; Zemlyansky, N. N.; Khrustalev, V. N.; Kuznetzova, M. G.; Ustynyuk, Yu.A.; Nechaev, M. S. *Russ. Chem. Bull.* **2002**, *51*, 678.

(66) Oda, R.; Ito, Y.; Okano, M. *Tetrahedron Lett.* **1964**, 7.

(67) Ito, Y.; Okano, M.; Oda, R. *Tetrahedron* **1966**, *22*, 2615.

(68) Wheaton, G. A.; Burton, N. J. *Tetrahedron Lett.* **1976**, 895.

(69) Veith, M.; Huch, V. *J. Organomet. Chem.* **1985**, *293*, 161.

(70) Denk, M.; Lennon, R.; Hayashi, R.; West, R.; Belyakov, A. V.; Verne, H. P.; Haaland, A.; Wagner, M.; Metzler, N. *J. Am. Chem. Soc.* **1994**, *116*, 2691.

(71) Khrustalev, V. N.; Borisova, I. V.; Zemlyansky, N. N.; Ustynyuk, Yu.A.; Antipun, M.Yu. *Crystallogr. Rep.* **2002**, *47*, 670.

(72) Zemlyansky, N. N.; Borisova, I. V.; Khrustalev, V. N.; Ustynyuk, Yu. A.; Nechaev, M. S.; Lunin, V. V.; Barrau, J.; Rima, G. *Organometallics* **2003**, *22*, 1675.

(73) Barrau, J.; Rima, G.; El Amraoui, T. *Organometallics* **1998**, *17*, 607.

(74) Nechaev, M. S.; Ustynyuk, Yu. A.; Lunin, V. V.; Borisova, I. V.; Zemlyansky, N. N.; West, R. *Organometallics* **2003**, *22*, in press.

(75) Steiner, M.; Pritzkow, H.; Grützmacher, H. *Chem. Ber.* **1994**, *127*, 1177.

(76) Grützmacher, H.; Deck, W.; Pritzkow, H.; Sander, M. *Angew. Chem., Int. Ed. Engl.* **1994**, *33*, 456.

(77) Khrustalev, V. N.; Zemlyansky, N. N.; Borisova, I. V.; Ustynyuk, Yu.A.; Chernyshev, E. A. *Russ. Chem. Bull.* **2000**, *49*, 929.

(78) Allen, F. H.; Kennard, O.; Watson, D. G.; Orpen, A. G.; Brammer, L.; Taylor, R. *J. Chem. Soc., Perkin Trans. II* **1987**, S1.

(79) Suzuki, H.; Tokitoh, N.; Nagase, S.; Okazaki, R. *J. Am. Chem. Soc.* **1994**, *116*, 11,578.

(80) Cambridge Crystal Structure Database, Cambridge, Release 2000.

(81) Veith, M.; Huch, V. *J. Organometal. Chem.* **1986**, *308*, 263.

(82) Dreihäupt, K.-H.; Angermaier, K.; Riede, J.; Schmidbaur, H. *Chem. Ber.* **1994**, *127*, 1599.

(83) Schmidbaur, H.; Pichl, R.; Muller, G. *Chem. Ber.* **1987**, *120*, 789.

(84) Borisova, I. V.; Zemlyansky, N. N.; Shestakova, A. K.; Khrustalev, V. N.; Ustynyuk, Yu.A.; Chernyshev, E. A. *Russ. Chem. Bull.* **2000**, *49*, 933.

(85) Albright, T. A.; Freeman, W. J.; Schweizer, E. E. *J. Am. Chem. Soc.* **1975**, *97*, 2542.

(86) Albright, T. A.; Gordon, M.; Freeman, W. J.; Schweizer, E. E. *J. Am. Chem. Soc.* **1976**, *98*, 6249.

(87) Schmidbaur, H.; Tronich, W. *Chem. Ber.* **1968**, *101*, 595.

(88) Maryanoff, B. E.; Reitz, A. B. *Chem. Rev.* **1989**, *89*, 863.

(89) Borisova, I. V.; Zemlyansky, N. N.; Shestakova, A. K.; Khrustalev, V. N.; Ustynyuk, Yu.A.; Chernyshev, E. A. *Russ. Chem. Bull.* **2000**, *49*, 1583.

(90) Tsumuraya, T.; Sato, S.; Ando, W. *Organometallics* **1989**, *8*, 161.

(91) Jutzi, P.; Mohrke, A. *Angew. Chem., Int. Ed. Engl.* **1989**, *28*, 762.

(92) Barrau, J.; Rima, G.; El Amin, M.; Satge, J. *J. Organomet. Chem.* **1988**, *345*, 39.

(93) Barrau, J.; Rima, G.; Satge, J. *J. Organomet. Chem.* **1983**, *252*, C73.

(94) Gornowicz, G. A.; West, R. *J. Am. Chem. Soc.* **1968**, *90*, 4478.

(95) Peterson, D. J. *J. Organomet. Chem.* **1967**, *9*, 373.

(96) Borisova I. V.; Zemlyansky N. N.; Shestakova A. K.; Khrustalev V. N.; Ustynyuk Yu. A.; Chernyshev E. A. XVIIIth Int. Conference on Organometallic Chemistry (XVIIIth ICOMC), Munich, Germany, August 21–22, 1998, Book of abstracts, Part 1, p. B19.

(97) Borisova, I. V.; Zemlyansky, N. N.; Shestakova, A. K.; Khrustalev, V. N.; Ustynyuk, Yu.A. *Mendeleev Commun.* **1997**, 10.

(98) Scheithauer, S.; Mayer, R. *Chem. Ber.* **1967**, *100*, 1413.

(99) Schmidbaur, H. *Acc. Chem. Res.* **1975**, *8*, 62.

(100) Zemlyansky, N. N.; Borisova, I. V.; Shestakova, A. K.; Ustynyuk, Yu.A.; Chernyshev, E. A. *Russ. Chem. Bull.* **1994**, *43*, 2126.

(101) Voronkov, M. G.; Basenko, S. V. *J. Organomet. Chem.* **1995**, *500*, 325.

(102) Kapp, J.; Remko, M.; Schleyer, P.v.R. *J. Am. Chem. Soc.* **1996**, *118*, 5745.

(103) Schmidbaur, H.; Tronich, W. *Chem. Ber.* **1967**, *100*, 1023.

(104) Starzewski, K. A. O.; Dieck, H. T. *Phosphorus* **1976**, *6*, 177.

(105) Schmidbaur, H.; Eberlein, J.; Richter, W. *Chem. Ber.* **1977**, *110*, 677.

(106) Schmidbaur, H.; Stuhler, H.; Buchner, W. *Chem. Ber.* **1973**, *106*, 1238.

(107) Seyferth, D.; Duncan, D. P.; Schmidbaur, H.; Holl, P. *J. Organomet. Chem.* **1978**, *159*, 137.

(108) Schmidbaur, H.; Wolf, W. *Chem. Ber.* **1975**, *108*, 2834.

(109) Schmidbaur, H.; Richter, W.; Wolf, W.; Kohler, F. H. *Chem. Ber.* **1975**, *108*, 2642.

(110) Gilman, H.; Tomasi, R. A. *J. Org. Chem.* **1962**, *27*, 3647.

(111) (Schleyer, P.v.R., Allinger, N. L., Clark, T., Kollman, P. A., Schafer, H. F., III, Scheiner, P. R. Eds.), Encyclopedia of Computational Chemistry1–5 1998, Wiley-VCH, Chichester.

(112) Nechaev, M. S.; Borisova, I. V.; Zemlyansky, N. N.; Laikov, D. N.; Ustynyuk, Yu.A. *Russ. Chem. Bull.* **2000**, *49*, 1823.

(113) Nechaev M. S.; Borisova I. V.; Zemlyansky N. N.; Laikov D. N.; Ustynyuk Yu. A. 2nd V. A. Fock School (conference) on Quantum and Computational Chemistry, Novgorod, January 31–February 4, 2000, Book of abstracts, p. 43, 44.

(114) Nechaev M. S.; Borisova I. V.; Zemlyansky N. N.; Laikov D. N.; Ustynyuk Yu. A. 3rd V. A. Fock Conference for Quantum and Computational Chemistry, Novgorod, May 21– 22, 2001, Book of abstracts, p. 100.

(115) Perdew, J. P.; Burke, K.; Ernzerhof, M. *Phys. Rev. Lett.* **1996**, *77*, 3865.

(116) Laikov, D. N. *Chem. Phys. Lett.* **1997**, *281*, 151.

(117) Laikov D. N. Ph.D. Dissertation, Moscow State University (Moscow), 2000.

# Coordination Compounds with Organoantimony and Sb$_n$ Ligands

## HANS JOACHIM BREUNIG and IOAN GHESNER

*Institut für Anorganische und Physikalische Chemie (Fb 2),*
*Universität Bremen, D-28334 Bremen, Germany*

# I

# INTRODUCTION

The chemistry of antimony ligands has usually been reviewed[1–12] together with analogous compounds of the other group 15 elements (pnicogens). This perspective is well justified by the strong relations between analogous pnicogen derivatives and also reflects the conceptualization of many research groups which study compounds of the heavier elements in extension of their work with phosphorus.

In this review, however, aspects of the coordination chemistry of organoantimony ligands and antimony ligands free of substituents are discussed without including the other pnicogens. We attempt to give a timely overview on the different types of ligands and coordination patterns with emphasis on recent developments. The ligand systems discussed here include tertiary stibines, R$_3$Sb and R$_2$Sb, RSb or Sb units in different environments and with various coordination patterns. The majority of the complexes involve coordination on transition metals but complexes with main group (including Sb) acceptors are also discussed.

General features of organoantimony ligands compared with their lighter congeners are related to the low stability of Sb–C or Sb–Sb bonds, to the relative weak Lewis basicity and to a larger tendency for higher coordination. Some comparative studies reveal a higher reactivity and flexibility of antimony ligand systems compared to analogous P or As compounds. However, the differences between related antimony and bismuth ligands

ADVANCES IN ORGANOMETALLIC CHEMISTRY
VOLUME 49 ISSN 0065-3055/DOI 10.1016/S0065-3055(03)49003-5

©2003 Elsevier Science (USA)
All rights reserved.

appear to be considerably larger. Many tertiary stibines have donor properties similar to phosphines or arsines and they are easily coordinated to a large variety of metal complex fragments, whereas bismuthines are often weak donors.

## II

## COMPLEXES WITH $R_3Sb$, $R_2SbX$, AND $RSbX_2$ (R = ALKYL, ARYL, $SiMe_3$; X = HALOGEN, OR, SR) LIGANDS

Tertiary stibines, $R_3Sb$, are the most frequently used antimony ligands and the coordination chemistry of these donors to transition metals of the groups 3–12 and also to main group 13 and 15 elements has been reviewed several times, often together with other pnicogen ligands.[1,2,6–8]

From a structural point of view the complexes with tertiary stibine ligands are very uniform. The predominant type is $\eta^1$-coordination (type **1**), with electron pair donation of the Sb atom to the metal centre.[1,2,6,8] However, complexes with *bridging* stibine ligands (type **2**) are also known.[13,14] (Scheme 1).

The majority of complexes with antimony ligands contain triphenyl-stibine, which is a versatile donor, easy to handle and readily available. Recent developments include examples with coordination of $Ph_3Sb$ on Co,[15] Os,[16,17] Rh,[18–20] Ru,[21,22] Mn,[23] Re,[23] Pd,[24] Pt,[25,26] Ag[27–32] or Cu.[31,33–37] Trialkylstibines have been used only rarely as ligands. However, complexes of *i*-$Pr_3Sb$ coordinated to Ru,[38–41] Rh,[42–44] or Ir[45] have recently received considerable attention because of a specific high reactivity of these compounds with olefins and other organic $\pi$-systems. Other reported examples for the use of trialkylstibine or mixed alkylarylstibine ligands include complexes where $Et_3Sb$ is bonded to Rh[42] or $PhMe_2Sb$ to Ru.[46]

Recently, also the coordination of a tertiary stibine to antimony acceptors with formation of a coordinative Sb–Sb bond was investigated. Examples are the complexes of $Me_3Sb$ with $SbMeI_2$,[47] $SbI_3$[48] or $Me_2Sb^+$.[49]

In $[Me_3Sb–SbMeI_2]$[47] the trimethylstibine ligand is coordinated to the antimony atom of a T-shaped $SbMeI_2$ moiety. The structure is depicted in Fig. 1a.

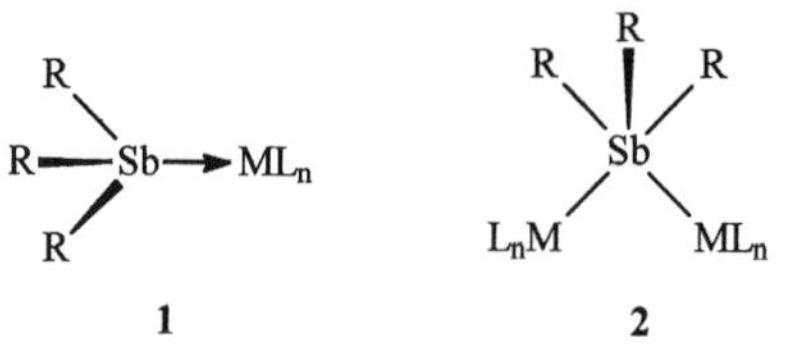

SCHEME 1

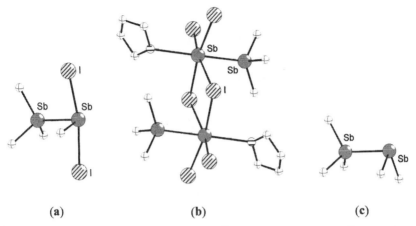

(a)                          (b)                          (c)

FIG. 1. Molecular structure of [Me$_3$Sb–SbMeI$_2$][47] (a), [(THF)(Me$_3$Sb)SbI$_3$]$_2$[48] (b), and the cation in [Me$_3$Sb–SbMe$_2$]$_2$[MeSbBr$_3$]$_2$[49] (c).

The length of the coordinative bond (2.86 Å) corresponds to a normal Sb–Sb single bond. The coordination geometries of the donor or acceptor antimony atoms are distorted tetrahedral for the former and pseudo trigonal bipyramidal for the latter with the iodine atoms in axial (I–Sb–I 169.71°) and the lone pair, the Me$_3$Sb and the methyl group in equatorial positions.

Coordination of Me$_3$Sb to SbI$_3$ is achieved in [(THF)(Me$_3$Sb)SbI$_3$]$_2$[48] where two T-shaped SbI$_3$ moieties are connected through iodine bridges. The Me$_3$Sb and THF ligands are coordinated *trans* to each other above and below the plane of the SbI$_3$ units and an overall geometry of edge-sharing octahedra results (Fig. 1b).

An ionic complex of the trimethylstibine ligand coordinated to an Sb acceptor is the cation of [Me$_3$Sb–SbMe$_2$]$_2$[MeSbBr$_3$]$_2$, a complex salt formed by a scrambling reaction of Me$_2$SbBr.[49] The structure (Fig. 1c) consists of pyramidal Me$_3$Sb bound to bent Me$_2$Sb units through short (2.82 Å) Sb–Sb bonds.

Related to tertiary stibines is the (Me$_3$Si)$_3$Sb ligand. Recent studies of this ligand comprise coordination on transition metal and main group element centers. An example for a transition metal complex with a known crystal structure is [(CO)$_5$CrSb(SiMe$_3$)$_3$],[50] the structure of which is depicted in Fig. 2.

The coordination of (Me$_3$Si)$_3$Sb leads to an increase of the mean Si–Sb–Si angles from 99.2° in the free ligand to 103.6° in the complex. This structural change corresponds to the use of orbitals with predominant p character for the bonding in the free ligand and sp$^3$ orbitals in the complex. 1:1 complexes of (Me$_3$Si)$_3$Sb with main group acceptors such as X$_3$B (X = Cl, Br, I),[51] R$_3$Al (R = Me, Et, t-Bu),[52] R$_2$AlCl (R = Et, t-Bu),[52]

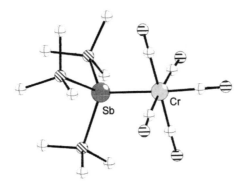

FIG. 2. Molecular structure of $[(CO)_5CrSb(SiMe_3)_3]$.[50]

$R_3Ga$ ($R = Et$,[53] $t$-$Bu$[54]), $R_3In$ ($R = Me_3SiCH_2$),[53] which have been the subject of a recent review[55] and will not be discussed here in detail, show similar trends. The Si–Sb–Si angles in these main group complexes range between 100.2 and 106.7°.

A novel type of ligands are tertiary stibines containing heterocyclic aromatic groups ($Y$-2-$C_4H_3)_3Sb$ ($Y = S$, O, or NMe) which with silver nitrate form polymeric 2:1 complexes, $[\{(Y$-2-$C_4H_3)_3Sb\}_2AgNO_3]_x$.[56]

Bridging coordination of a tertiary stibine ligand (type **2**) occurs in binuclear complexes of rhodium (Equation 1).[13,14]

$$2 \; Cl\!-\!\underset{i\text{-}Pr_3Sb}{\overset{i\text{-}Pr_3Sb}{Rh}}\!\!=\!\!C\!\!\overset{R}{\underset{R}{<}} \quad \xrightarrow{-\,3\; i\text{-}Pr_3Sb} \quad \underset{R\;\;R}{\overset{i\text{-}Pr}{\underset{Cl}{Rh}}} \tag{1}$$

Mixed organo antimony ligands with one or two electronegative groups have been only rarely used in transition metal chemistry. The resulting complexes belong to type **3** or **4** with $^1\eta$-Sb coordination (Scheme 2).

Complexes with diorganoantimony halide ligands are the pentacarbonylchromium complexes of $Me_2SbBr$,[57] $(t$-$Bu)_2SbCl$[58] or $Ph_2SbCl$,[58] $[(CO)_2\{P(OPh)_3\}_2Fe(Ph_2SbCl)]$,[59] $[(CO)_2(MeC_5H_4)Mn(Ph_2SbCl)]$,[59] and $[(CO)_4M(Ph_2SbCl)_2]$ ($M = Cr$, Mo)[60]. Complexes with diphenylalkoxides and thiolates include $[(CO)_3M(Ph_2SbY)_3]$ ($M = Cr$, $Y = OEt$, SEt, SPh, $SCHMe_2$, $SCH_2Ph$; $M = Mo$, $Y = SPh$, $SCHMe_2$, $SCH_2Ph$).[61]

An X-ray crystal structure analysis of $[(CO)_3Mo\{Ph_2Sb(SPh)\}_3]$ revealed the $^1\eta$-coordination of the $Ph_2Sb(SPh)$ ligand (Mo–Sb 2.742–2.751 Å) (Fig. 3a).

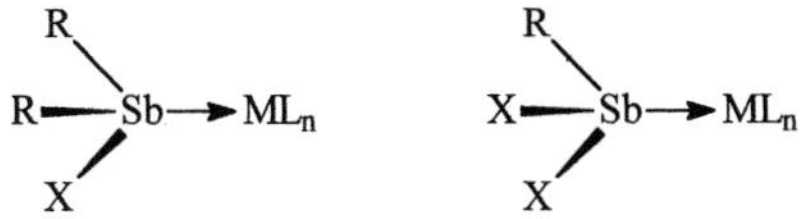

$$R\text{---}Sb\rightarrow ML_n \qquad X\text{---}Sb\rightarrow ML_n$$

**3**                              **4**

SCHEME 2

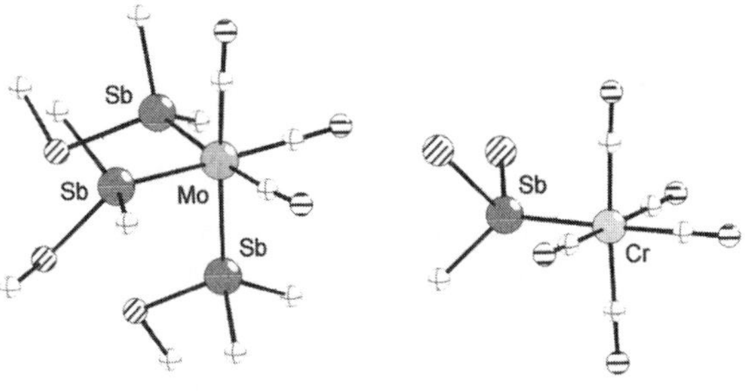

**(a)**                   **(b)**

FIG. 3. Molecular structure of $[(CO)_3Mo\{SbPh_2(SPh)\}_3]$[61] (a) and $[(CO)_5Cr(MeSbBr_2)]$[62] (b).

A complex of the type $[ML_n(RSbY_2)]$ with a known crystal structure is $[(CO)_5Cr(MeSbBr_2)]$.[62] The $MeSbBr_2$ unit is coordinated through the antimony atom to the Cr atom (Cr–Sb 2.556 Å) (Fig. 3b). The environment of the Sb atom is distorted tetrahedral. Other complexes of the type $[ML_n(RSbY_2)]$ are $[(CO)_5Cr(RSbCl_2)]$ ($R = t$-Bu, Ph;[58] $R = Me$[63]) and $[(CO)_5Cr(PhSbI_2)]$.[64]

The reaction of racemic Sb-chiral 1-phenyl-2-trimethylsilylstibindole with the optically active ortho-palladated benzylamine derivative, di-μ-chlorobis{(S)-2-[1-dimethylamino)ethyl]phenyl-C,N}dipalladium, leads to diastereomeric complexes which were used for the separation of the enantiomers of the stibindoles.[65] The molecular structures of the diastereomeric palladium complexes are depicted in Fig. 4.

# III

## COMPLEXES WITH $R_2Sb$–$SbR_2$ AND $R_2Sb$ (R = ALKYL, ARYL) LIGANDS

Preceding reviews on distibines, $R_2Sb$–$SbR_2$[5,7,66–68] feature mainly the structural aspects in relation to the thermochromic properties of some of these compounds, but also overviews on the earlier work of the coordination chemistry of distibine ligands have been reported.[7,68] All the distibines that

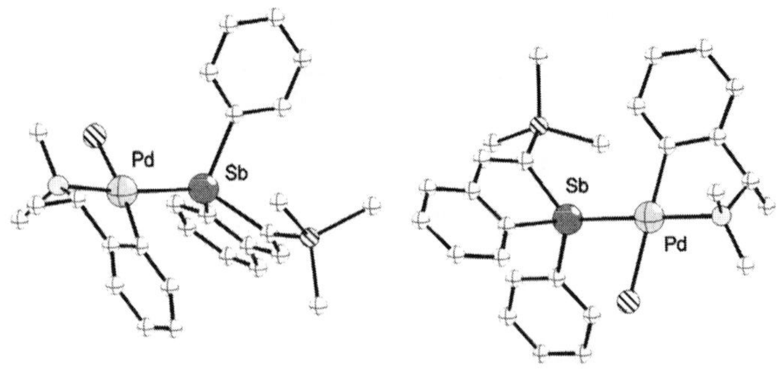

**(a)**          **(b)**

FIG. 4. Molecular structures of diasteromeric complexes [chloro(1-phenyl-2-trimethylsilylsti-bindole){(S)-2-[1-dimethylamino)ethyl]phenyl-C,N}palladium] **(a)** $S(C)$–$R(Sb)$, **(b)** $S(C)$–$S(Sb)$.[65]

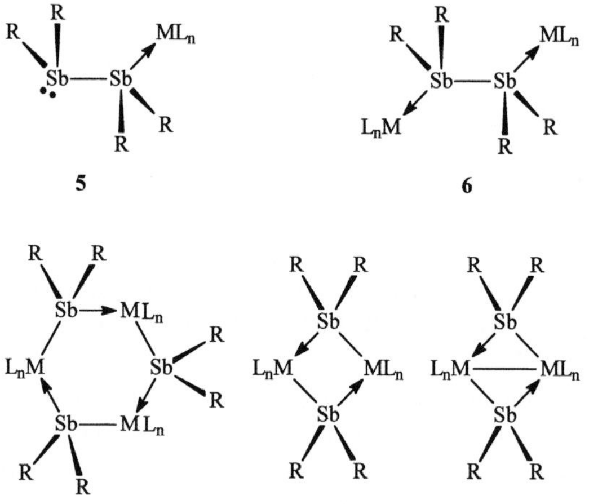

SCHEME 3

have been characterized by single crystal X-ray diffraction adopt the *anti* conformation in the solid state. In fluid phases however, *gauche* conformers were also detected and free rotation around the Sb–Sb bond is likely to occur even at low temperature in solution.

Distibines react as *monodentate* (type **5**) or *bridging bidentate* (type **6**) ligands through donation of the lone pairs of electrons to main group or transition metal centers. Fission of the Sb–Sb bond leads to complexes with *bridging* $R_2Sb$ ligands (type **7**, **8** and **9**) (Scheme 3).

In order to describe the different conformations in distibines or in distibine complexes, the torsion angles Ep–Sb–Sb–Ep ($\varphi$) for the free ligand

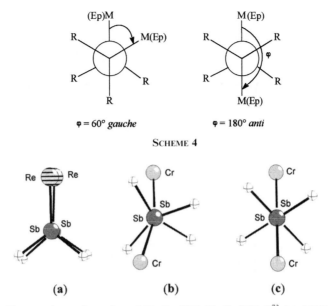

$\varphi = 60°$ *gauche*                    $\varphi = 180°$ *anti*

SCHEME 4

(a)                    (b)                    (c)

FIG. 5. View on the conformation of [Re$_2$Br$_2$(CO)$_6$(Ph$_2$Sb–SbPh$_2$)][73] (a), [(CO)$_5$Cr(Me$_2$Sb–SbMe$_2$)Cr(CO)$_5$][71] (b), and [(CO)$_5$Cr(Ph$_2$Sb–SbPh$_2$)Cr(CO)$_5$][69] (c).

or M–Sb–Sb–M ($\varphi$) for *bridging bidentate* coordination, where Ep is the assumed direction of the lone pair at antimony and M stands for the assumed direction of the Lewis acid group, is used. The molecular conformations are *gauche* and *anti* for $\varphi$ angles of 60° and 180° (Scheme 4).

*Monodentate* coordination of distibines (type **5**) has been observed exclusively in transition metal carbonyl complexes of the type [(CO)$_n$M(R$_2$Sb–SbR$_2$)] (R = Me, Et, Ph; M = Cr, W, $n$ = 5; M = Fe, $n$ = 4),[67,68] but crystal structures were not determined. The first examples for complexes with *bridging* distibine ligands (type **6**) were the metal carbonyl derivatives, [(CO)$_n$M(R$_2$Sb–SbR$_2$)M(CO)$_n$] (R = Me, Et, *t*-Bu, Ph; M = Cr, Mo, W, $n$ = 5; M = Ni, $n$ = 3).[67,68] Crystal structure analyses on [(CO)$_5$M(Ph$_2$Sb–SbPh$_2$)M(CO)$_5$] (M = Cr,[69] W[70]) revealed that the complexes with the tetraphenyldistibine ligand adopt the *anti* conformation ($\varphi$ = 180°) in the solid state (Fig. 5c). In [(CO)$_5$Cr(Me$_2$Sb–SbMe$_2$)Cr(CO)$_5$][71] (Fig. 5b) with $\varphi$ = 160° and in [CdI$_2$(Et$_2$Sb–SbEt$_2$)]$_n$[72] with $\varphi$ = 153° there is a significant deviation from the ideal *anti* conformation.

[Re$_2$Br$_2$(CO)$_6$(Ph$_2$Sb–SbPh$_2$)][73] (**10**) and [Rh$_2$(cod)$_2$(Ph$_2$Sb–SbPh$_2$)(SbPh$_2$)$_2$][74] (**11**) (cod = cyclooctadiene) are the only distibine complexes with the distibine ligand adopting a *syn* structure (Fig. 5a), which is a consequence of the positions of the coordination sites in the bridged transition metal fragments (Scheme 5).

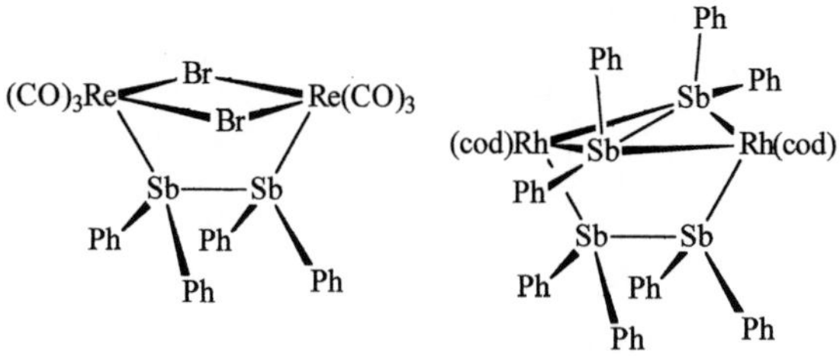

**Scheme 5**

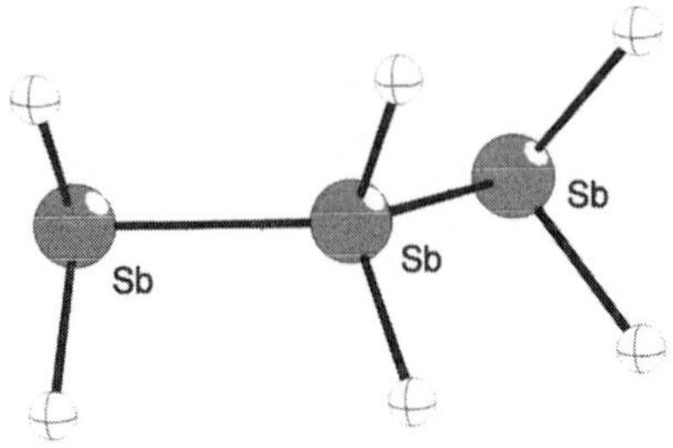

FIG. 6. Molecular structure of the cation of $[Me_2Sb-Me_2Sb-SbMe_2)][Me_2SbBr_2]$.[75]

Complexes with distibine donor ligands and main group acceptors have been investigated recently. Monodentate coordination of $Me_2Sb-SbMe_2$ to $Me_2Sb^+$ occurs in the cation of the salt $[Me_2Sb-Me_2Sb-SbMe_2)][Me_2SbBr_2]$ which is stable only in the solid state (Fig. 6).[75]

Bidentate coordination leading to bis-adducts is observed in reactions of distibines with aluminium or gallium trialkyls in absence of solvent at low temperatures (Equation 2).[76]

$$R_2Sb-SbR_2 + 2\ R'_3M \longrightarrow R'_3M-SbR_2-SbR_2-MR'_3$$

$$R = Me;\ R' = t\text{-Bu};\ M = Al,\ Ga$$ (2)

$$R = Et;\ R' = Me,\ Et;\ M = Al,\ Ga$$

The bis-adducts are stable in the solid state at low temperatures. They dissociate in solution at ambient temperature. In the case of $Me_2Sb-SbMe_2$ a stable complex was formed only with the sterically hindered and less electrophilic $t\text{-Bu}_3Al$. X-ray crystal structure analyses for $[(t\text{-Bu})_3M(R_2Sb-SbR_2)M(t\text{-Bu})_3]$ (R = Me, Et; M = Al, Ga) revealed that the distibine ligands adopt the *anti* conformation ($\varphi = 180°$). Structural parameters of distibine complexes are summarized in Table I.

A general reaction of binuclear distibine complexes is the fission of the Sb–Sb bond and transformation of the $R_2Sb-SbR_2$ into a $R_2Sb$ groups.

## TABLE I
### Geometric Parameters of Crystal Structures for Complexes with Bis(Stibino) Donors

| | Sb–Sb (Å) | Sb–M (Å) | C–Sb–C (°) | C–Sb–M (°) | $\varphi^a$ (°) | Ref. |
|---|---|---|---|---|---|---|
| Me₂Sb–SbMe₂ | 2.863 | | 92.2; 95.2 | | 180 | 81 |
| Ph₂Sb–SbPh₂ | 2.837 | | 94.4 | | 180 | 93 |
| [(t-Bu)₃Al(Me₂Sb–SbMe₂)Al(t-Bu)₃] | 2.811 | 2.919 | 97.8 | 112.2; 117.2 | 180 | 76 |
| [(t-Bu)₃Ga(Me₂Sb–SbMe₂)Ga(t-Bu)₃] | 2.814 | 2.919 | 96.6 | 113.5; 114.8 | 180 | 76 |
| [(CO)₅Cr(Me₂Sb–SbMe₂)Cr(CO)₅] | 2.810 | 2.621; 2.628 | 99.2; 100.3 | 112.4–118.9 | 160 | 71 |
| [Me₂Sb–Me₂Sb–SbMe₂]][Me₂SbBr₂] | 2.820 | | 92.4; 98.0 | 110.3 | 2.4 | 75 |
| [(t-Bu)₃Al(Et₂Sb–SbEt₂)Al(t-Bu)₃] | 2.838 | 3.001 | 96.8 | 115.7; 117.4 | 180 | 76 |
| [(t-Bu)₃Ga(Et₂Sb–SbEt₂)Ga(t-Bu)₃] | 2.839 | 3.022 | 96.1 | 115.9; 117.7 | 180 | 76 |
| [I₂Cd(Et₂Sb–SbEt₂)]ₙ | 2.784 | 2.822; 2.821 | 102.3; 103.9 | 106.0–119.3 | 153 | 72 |
| [(CO)₅Cr(Ph₂Sb–SbPh₂)Cr(CO)₅] | 2.866 | 2.626 | 101.0 | 110.6; 122.5 | 180 | 69 |
| [(CO)₅W(Ph₂Sb–SbPh₂)W(CO)₅] | 2.861 | 2.749 | 100.7 | 111.5; 122.2 | 180 | 70 |
| [Rh₂(cod)₂(Ph₂Sb–SbPh₂)(SbPh₂)₂] | 2.926; 3.155 | 2.584–2.678 | 93.6; 97.0 | 116.1; 124.7 | 7 | 74 |
| [Br₂Re₂(CO)₆(Ph₂Sb–SbPh₂)] | 2.826 | 2.726 | 101.2; 104.5 | 116.2–125.0 | 0.3 | 73 |

ª $\varphi$ = (Ep)M–Sb–Sb–M(Ep).

103

This type of reaction was observed when $[(CO)_5W(Ph_2Sb–SbPh_2)W(CO)_5]$ was dissolved in refluxing diglyme and $[(CO)_4W(SbPh_2)_2W(CO)_4]$, a heterocyclic complex (type **8**) with the *bridging* diphenylantimonido ligand was formed.[57]

Analogous reactions of $[(CO)_4Fe(SbR_2–SbR_2)Fe(CO)_4]$ leading to $[(CO)_3FeSbR_2]_2$[77] (R = Et, Ph) or the formation of $[(CO)_4MoSbR_2]_2$[57] (R = Me, Et, Ph) (type **8**) from the corresponding distibines and $[(CO)_4Mo(THF)_2]$ (THF = tetrahydrofuran) occur at ambient temperature. The crystal structures of these antimonido complexes are however not known. The structures of $[(Me_3P)_2CuSbMes_2]_2$[78] (type **8**), $[(CO)_2(\eta^5-C_5H_5)MoSbPh_2]_2$[79] (type **8**) or $[(CO)_2NiSb(t-Bu)_2]_2$[80] (type **9**) consist of $Sb_2M_2$ (M = Mo, Ni, Cu) cores with the $R_2Sb$ (R = Ph, Mes, $t$-Bu) fragments bridging two $ML_n$ moieties, which in the case of type **9** complexes are connected through a metal metal bond. The molecular structure of the Ni derivative is depicted in Fig. 7. The Sb–Sb distances in the Ni and Mo complexes (3.099 Å in $[(CO)_2NiSb(t-Bu)_2]_2$[80] and 3.05 Å in $[(CO)_2(\eta^5-C_5H_5)MoSbPh_2]_2$[79]) are smaller than the sum of the van der Waals radii for two Sb atoms (4.40 Å). The $Sb_2M_2$ rings are planar only in the case of $[(CO)_2NiSb(t-Bu)_2]_2$. In $[(CO)_2(\eta^5-C_5H_5)MoSbPh_2]_2$ the dihedral angles are 34.1° between the $MoSb_2$ planes and 47.1° between the $SbMo_2$ planes.[79]

Fission of the Sb–Sb bond is even more facile in the distibine-group 13 element complexes and with migration of an alkyl group four- and six-membered heterocycles (type **7** and **8**) are formed.[76,82]

$$[R'_3M(SbR_2-SbR_2)MR'_3] \longrightarrow [R_2Sb-MR'_2]_n + 2\ R'R_2Sb + R'_3M$$

$$R = Me;\ R' = t\text{-Bu};\ M = Ga;\ n = 3\ [76]$$
$$R = Et;\ R' = t\text{-Bu};\ M = Ga;\ n = 2\ [76] \qquad\qquad (3)$$
$$R = Me;\ R' = Me_3SiCH_2;\ n = 3\ [82]$$

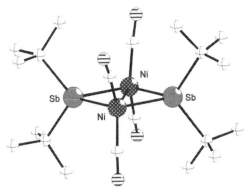

FIG. 7. Molecular structure of $[(CO)_2NiSb(t-Bu)_2]_2$.[80]

Recent developments are ring-cleavage reactions of the heterocycles $[R'_2M{-}SbR_2]_n$ with 4-(dimethylamino)pyridine leading to base-stabilized monomers, $[L - R'_2M{-}SbR_2]$, (R = Me, SiMe$_3$; R′ = Me, Et, i-Bu; M = Al, Ga).[83,84] Reaction of [L-Al(Me$_2$)–Sb(SiMe$_3$)$_2$] [L = 4-(dimethylamino)pyridine] with [Ni(CO)$_4$] leads to the corresponding tricarbonyl nickel complex (Equation 4).[85]

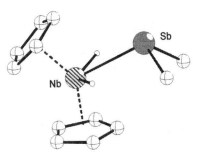

$$\text{1/n [Me}_2\text{Al-Sb(SiMe}_3)_2]_n \xrightarrow{\ +\text{L}\ } \text{Me}_2\text{Al-Sb(SiMe}_3)_2$$

L = 4-(dimethylamino)pyridine      − CO | + Ni(CO)$_4$

Ni(CO)$_3$
↑
Me$_2$Al-Sb(SiMe$_3$)$_2$
↑
L

(4)

In this Ni complex the ligand is amphoteric with a "hard" acidic site at Al and a "soft" basic site on Sb.

Other synthetic pathways for the preparation of group-13-element antimony compounds include dehydro- and dehalosilylation reactions of R$_2$SbSiMe$_3$ or Sb(SiMe$_3$)$_3$ with R′$_2$MX or MX$_3$ (R = SiMe$_3$, t-Bu; R′ = Me, Et, i-Bu; M = Ga, In; X = H, Cl).[53] A review of the chemistry of these potential precursors for 13–15 semiconductors was recently published.[55]

Compounds where a R$_2$Sb moiety is coordinated to one or two 17-e⁻ or 16-e⁻ complex fragments (type **12–15**) are also known. The synthetic and structural aspects in the chemistry of complexes of the type, [L$_n$M–SbR$_2$] (ML$_n$ = 17e⁻ fragment), (type **12**) and [L$_n$M = SbR$_2$] (type **14**) compounds were already reviewed.[3] A type **12** compound characterized by crystallography is [Cp$_2$(H)$_2$Nb–SbPh$_2$].[86] The structure of this niobocene derivative is depicted in Fig. 8. Other monometalla stibines, [L$_n$M–SbR$_2$] (type **12**) with

FIG. 8. Molecular structure of [Cp$_2$(H)$_2$Nb–SbPh$_2$] (only the *ipso* C atoms of the phenyl groups are represented).[86]

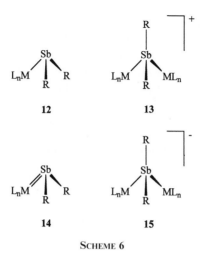

**SCHEME 6**

known crystal structures comprise $[(\eta^5\text{-}C_5Me_5)(CO)_3W\text{-}Sb(CF_3)_2]$[87] and $[(\eta^5\text{-}C_5Me_5)(CO)_2M\text{-}SbR_2]$[88] [M = Fe or Ru, R = SiMe_3; M = Ru, R = C(O)Ph]. In all these complexes the geometry around the antimony atom is pyramidal with C–Sb–C angles near to 90° (C–Sb–C 89.1° in $[(\eta^5\text{-}C_5Me_5)(CO)_3W\text{-}Sb(CF_3)_2]$;[87] 93.7° in $[(\eta^5\text{-}C_5Me_5)(CO)_2Ru\text{-}Sb\{C(O)Ph\}_2]$[88] and 94.3° in $[Cp_2(H)_2Nb\text{-}SbPh_2]$[86]) (Scheme 6).

The type **13** complexes can be considered as analogues of stibonium ions, $R_4Sb^+$ (R = organic group). Among the few examples of these complexes are $[\{Cp(CO)_2Fe\}_2\text{-}SbPh_2]X$ with X = Cl[89,90] and FeBr_4.[91]

Four coordinate tetrahedral antimony exists also in the anions of the ammonium salts $[NEt_4][\{(CO)_4Fe\}_2SbR_2]$ (R = Me, Et).[92] The molecular structure of the $[\{(CO)_4Fe\}_2SbMe_2]^-$ anion is depicted in Fig. 9.

## IV

## COMPLEXES WITH $R_2Sb\text{-}Y\text{-}SbR_2$ (R = ALKYL, ARYL;

## Y = $CH_2$, O, S) LIGANDS

The coordination chemistry of the well known bis(stibino)methane ligands, $R_2Sb\text{-}CH_2\text{-}SbR_2$ (R = Me, Ph) has been studied for 30 years and the results have been reviewed several times.[2,6] The related oxygen or sulfur compounds, $R_2Sb\text{-}X\text{-}SbR_2$ (R = Me, Ph; X = O, S) or ligands of the type $Ph_2Sb\text{-}EMe_2\text{-}SbPh_2$ (E = C, Si, Ge, Sn) have received less attention. Selena or tellura derivatives or bis(stibino) ligands with other mono atomic spacers are unknown.

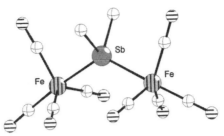

FIG. 9. Molecular structure of the anion of [NEt$_4$][{(CO)$_4$Fe}$_2$SbMe$_2$].[92]

The coordination modes achieved in complexes with bis(stibino) ligands having a monoatomic interdonor linkage include 1:1 complexes with *monodentate* (type **16**) or *chelating* (type **17**) coordination, or 1:2 complexes with *bridging bidentate* (type **18, 19** and **20**) coordination.[2,6] *Chelating* coordination is disfavored through the strain in the four-membered ring. Like their P or As analogues, R$_2$Sb–X–SbR$_2$ ligands are especially useful for *bridging bidentate* coordination (Scheme 7).

In order to describe the different conformations in type **16–20** molecules, the torsion angles (Ep)M–Sb–Y–Sb (Ep = assumed direction of the lone pair at antimony; M = transition metal; Y = CH$_2$, O, S) were used. The Ep–Sb–Y–Sb torsion angles, which are considered in complexes with *monodentate* (type **16**) coordination of the Sb ligand, were calculated from the corresponding pairs of torsion angles C–Sb–Y–Sb (Scheme 8).

The extreme molecular conformations are *syn–syn* with $\varphi_{1,2} = 0°$ (in *cis*, *trans*-[Br$_4$Pd$_2$($\mu$-Ph$_2$Sb–CH$_2$–SbPh$_2$)$_2$] · 1.1CH$_2$Cl$_2$[94] $\varphi_1 = 12.23°$, $\varphi_2 = 17.05°$), *gauche-gauche* with $\varphi_{1,2} = 60°$ (in [(CO)$_5$Fe($\eta^1$-Ph$_2$Sb–S–SbPh$_2$)][60] $\varphi_1 = 57.5°$, $\varphi_2 = 58.02°$) and *syn-anti* with $\varphi_1 = 0°$ and $\varphi_2 = 180°$ (in Me$_2$Sb–O–SbMe$_2$[95] $\varphi_1 = 15.7°$, $\varphi_2 = 177.9°$). The torsion angles $\varphi$ for complexes with R$_2$Sb–Y–SbR$_2$ ligands (R = alkyl, aryl; Y = CH$_2$, O, S) are given in Table II.

Recent examples of complexes with *monodentate* coordination of the R$_2$Sb–Y–SbR$_2$ (R = Me, Ph; Y = CH$_2$, O, S) ligand are listed in Table III. The identification of these complexes with *monodentate* coordination of the antimony ligand results mainly from their IR and NMR spectra. For [(CO)$_4$Fe($\eta^1$-Ph$_2$Sb–CH$_2$–SbPh$_2$)],[96] [(CO)$_5$W($\eta^1$-Ph$_2$Sb–CH$_2$–SbPh$_2$)][97], [I$_2$Ru($\eta^1$-Ph$_2$Sb–CH$_2$–SbPh$_2$)$_2$($\eta^2$-Ph$_2$Sb–CH$_2$–SbPh$_2$)][94] and [(CO)$_5$Cr($\eta^1$-Ph$_2$Sb–S–SbPh$_2$)][60] the *monodentate* coordination was confirmed by single crystal X-ray studies, which offer also the possibility to inspect directly the geometric consequences of the coordination by comparing the environment of the two antimony atoms.

The structures of [(CO)$_4$Fe($\eta^1$-Ph$_2$Sb–CH$_2$–SbPh$_2$)] (Fig. 10a), [(CO)$_5$W($\eta^1$-Ph$_2$Sb–CH$_2$–SbPh$_2$)] and [(CO)$_5$Cr($\eta^1$-Ph$_2$Sb–S–SbPh$_2$)][60]

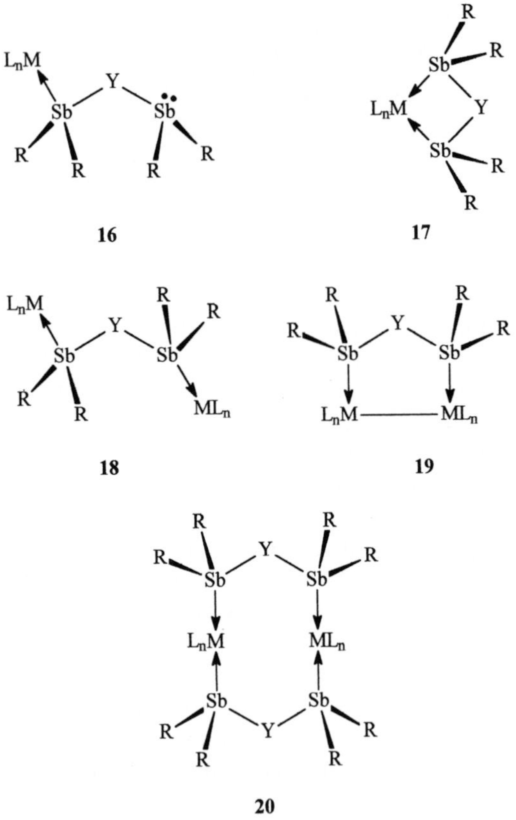

**16**

**17**

**18**

**19**

**20**

SCHEME 7

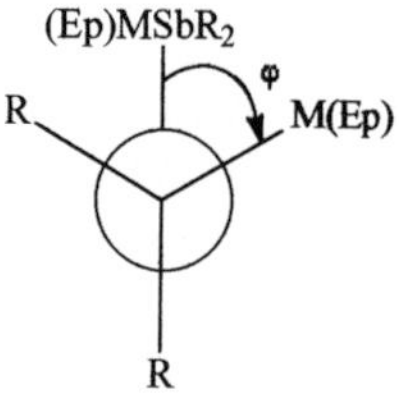

SCHEME 8

(Fig. 10b) are similar. In all these complexes the antimony ligand is coordinated to the transition metal center in the apical position. The angles around the four coordinated antimony atom (100.8–106.8° in $[(CO)_4Fe(\eta^1\text{-}Ph_2Sb\text{--}CH_2\text{--}SbPh_2)]$, 99.4–104.9° in $[(CO)_5W(\eta^1\text{-}Ph_2Sb\text{--}CH_2\text{--}SbPh_2)]$, and

## TABLE II

GEOMETRIC PARAMETERS OF CRYSTAL STRUCTURES FOR THE FREE LIGANDS AND COMPLEXES
WITH R$_2$Sb–Y–SbR$_2$ (R = Me, Ph; Y = CH$_2$, O, S) LIGANDS

| Compound | Sb–M$^a$ (Å) | Sb–Y–Sb (°) | $\varphi_1^g$ (°) | $\varphi_2^g$ (°) | Ref. |
|---|---|---|---|---|---|
| Ph$_2$Sb–CH$_2$–SbPh$_2$ | | 117.3 | 90.4 | 99.8 | 96 |
| Ph$_2$Sb–O–SbPh$_2$ | | 122.1 | 6.8 | 43.3 | 106 |
| Me$_2$Sb–O–SbMe$_2$ | | 123.0 | 15.7 | 177.9 | 95 |
| Me$_2$Sb–S–SbMe$_2$ | | 92.3 | 41.5 | 41.5 | 95 |
| [(CO)$_5$W(Ph$_2$Sb–CH$_2$–SbPh$_2$)] | 2.743 | 106.8 | 31.2 | 55.8 | 97 |
| [(CO)$_4$Fe(Ph$_2$Sb–CH$_2$–SbPh$_2$)] | 2.491 | 107.9 | 31.8 | 55.3 | 96 |
| [(CO)$_5$Cr(Ph$_2$Sb–S–SbPh$_2$)] | 2.598 | 96.7 | 57.5 | 58.0 | 60 |
| [I$_2$Ru(Ph$_2$Sb–CH$_2$–SbPh$_2$)$_3$] | 2.598; 2.623$^b$<br>2.578; 2.584$^d$ | 93.5$^b$<br>118.9; 122.2$^d$ | 19.5$^b$<br>20.9; 27.6$^d$ | 19.7$^b$<br>77.5; 85.7$^d$ | 94 |
| [(OC)$_5$W(Ph$_2$Sb–CH$_2$–SbPh$_2$)W(CO)$_5$] | 2.753; 2.756 | 122.0 | 90.6 | 103.8 | 97 |
| [(CO)$_4$W(Ph$_2$Sb–CH$_2$–SbPh$_2$)$_2$W(CO)$_4$]$^c$ | 2.742; 2.757 | 125.4 | 76.3 | 139.2 | 97 |
| [W(CO)$_4$(Me$_2$Sb–CH$_2$–SbMe$_2$)$_2$W(CO)$_4$]$^c$ | 2.752; 2.754 | 119.7 | 72.2 | 130.2 | 97 |
| [(CO)$_4$Cr(Ph$_2$Sb–O–SbPh$_2$)$_2$Cr(CO)$_4$] | 2.581; 2.587 | 136.9 | 100.8 | 141.1 | 60 |
| [(CO)$_4$Cr(Me$_2$Sb–O–SbMe$_2$)$_2$Cr(CO)$_4$] | 2.566; 2.573 | 137.8 | 87.5 | 153.0 | 99 |
| [(CO)$_4$Cr(Me$_2$Sb–S–SbMe$_2$)$_2$Cr(CO)$_4$] | 2.584; 2.598 | 114.5 | 77.3 | 135.5 | 99 |
| [(CO)$_2$Ni(Ph$_2$Sb–O–SbPh$_2$)$_2$Ni(CO)$_2$] | 2.447; 2.450 | 130.7 | 30.8 | 95.7 | 102 |
| [(CO)$_6$Mn$_2$(Ph$_2$Sb–CH$_2$–SbPh$_2$)$_2$] | 2.500–2.487 | 103.0;<br>106.7 | 20.3;<br>22.5 | 47.2;<br>52.1 | 98 |
| [(CO)$_6$Co$_2$(Me$_2$Sb–CH$_2$–SbMe$_2$)] | 3.508; 3.529 | 114.5 | 20.6 | 21.5 | 96 |
| [Cl$_4$Pt$_2$(Ph$_2$Sb–CH$_2$–SbPh$_2$)$_2$] · Me$_2$CO | 2.479; 2.494$^e$<br><br>2.554; 2.572$^f$ | 112.5;<br>114.5 | 18.6;<br>19.4 | 22.5;<br>43.4 | 94 |
| [Br$_4$Pd$_2$(Ph$_2$Sb–CH$_2$–SbPh$_2$)$_2$] · xCH$_2$Cl$_2$ | 2.507; 2.510$^e$ | 111.3;<br>115.0 | 12.2;<br>16.8 | 17.0;<br>18.9 | 94 |
| (x = 1.1) | 2.556; 2.567$^f$ | | | | |
| [Cl$_2$Ph$_2$Pd$_2$(Ph$_2$Sb–CH$_2$–SbPh$_2$)$_2$]$^c$ | 2.530; 2.560 | 115.0 | 47.1 | 55.1 | 105 |

$^a$M refers to transition metal.
$^b$chelating coordination.
$^c$centrosymmetric molecule.
$^d\eta^1$-coordination.
$^e$cis coordination.
$^f$trans coordination.
$^g\varphi_{1,2}$ = Ep(M)–Sb–Y–Sb.

99.4–119.9° in [(CO)$_5$Cr($\eta^1$-Ph$_2$Sb–S–SbPh$_2$)]) are wider than in the case of
three coordinated antimony atoms (96.5–97.7° in [(CO)$_4$Fe($\eta^1$-Ph$_2$Sb–CH$_2$–
SbPh$_2$)], 96.4–97.4° in [(CO)$_5$W($\eta^1$-Ph$_2$Sb–CH$_2$–SbPh$_2$)] and 94.1–98.1° in
[(CO)$_5$Cr($\eta^1$-Ph$_2$Sb–S–SbPh$_2$)]). The widening of the angles with the increase
of the coordination number corresponds to a change from an approximate
p$^3$ configuration to sp$^3$ hybridization. The Sb–C–Sb angles in [(CO)$_4$Fe($\eta^1$-
Ph$_2$Sb–CH$_2$–SbPh$_2$)] (107.9°) and [(CO)$_5$W($\eta^1$-Ph$_2$Sb–CH$_2$–SbPh$_2$)] (106.8°)

TABLE III

RECENT EXAMPLES OF COMPLEXES WITH *MONODENTATE* $R_2Sb-Y-SbR_2$ ($R = Me$, $Ph$; $Y = CH_2$, O, S) LIGANDS (TYPE 16)

| Compound | M/Y/X/R | Ref. |
|---|---|---|
| $[(CO)_5M(Ph_2Sb-CH_2-SbPh_2)]$ | $M = Cr$, Mo, W | 97 |
| $[(CO)_3M(Ph_2Sb-CH_2-SbPh_2)_3]$ | $M = Cr$, Mo, W | 97 |
| $[(MeC_5H_4)(CO)_2Mn(Ph_2Sb-CH_2-SbPh_2)]$ | | 59 |
| $[(CO)_2\{P(OPh)_3\}_2Fe(Ph_2Sb-CH_2-SbPh_2)]$ | | 59 |
| $[X(CO)_4Mn(Ph_2Sb-CH_2-SbPh_2)]$ | $X = Cl$, Br, I | 96 |
| $[(CO)_4Fe(Ph_2Sb-CH_2-SbPh_2)]$ | | 96 |
| $[(CO)_3Ni(Ph_2Sb-CH_2-SbPh_2)]$ | | 96 |
| $[X_2Ru(Ph_2Sb-CH_2-SbPh_2)_4]$ | $X = Cl$, Br | 94 |
| $[X_2M(Ph_2Sb-CH_2-SbPh_2)_3]$ | $M = Ru$, $X = I$; $M = Os$, $X = Br$ | 94 |
| $[X_3Rh(Ph_2Sb-CH_2-SbPh_2)_2]$ | $X = Cl$, Br; I | 94 |
| $[(CO)_9Re_2(R_2Sb-CH_2-SbR_2)]$ | $R = Me$, Ph | 98 |
| $[(MeC_5H_4)(CO)_2Mn(Ph_2Sb-Y-SbPh_2)]$ | $Y = O$, S | 59 |
| $[(CO)_2\{P(OPh)_3\}_2Fe(Ph_2Sb-Y-SbPh_2)]$ | $Y = O$, S | 59 |
| $[(CO)_5M(Ph_2Sb-Y-SbPh_2)]$ | $M = Cr$, Mo; $Y = O$, S | 60 |
| $[(CO)_4Cr(Me_2Sb-S-SbMe_2)_2]$ | | 99 |

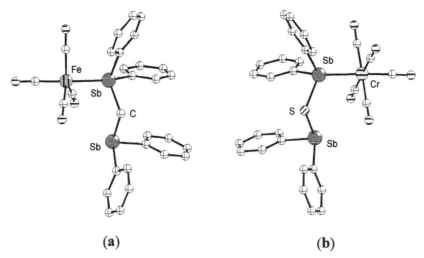

**(a)**                    **(b)**

FIG. 10. Molecular structures of $[(CO)_4Fe(\eta^1-Ph_2Sb-CH_2-SbPh_2)]^{96}$ **(a)** and $[(CO)_5Cr(\eta^1-Ph_2Sb-S-SbPh_2)]^{60}$ **(b)**.

are smaller than in the free $Ph_2Sb-CH_2-SbPh_2$ ligand ($117.3°$)[96]. However, in $[I_2Ru(\eta^1-Ph_2Sb-CH_2-SbPh_2)_2(\eta^2-Ph_2Sb-CH_2-SbPh_2)]$ (Fig. 11) where steric factors interfere, larger Sb–C–Sb angles were found for the $\eta^1$-coordinated bis(diphenylstibino)methane ligands ($122.2$ and $118.9°$).[94]

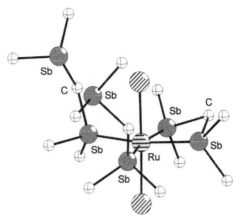

FIG. 11. Molecular structure of $[I_2Ru(\eta^1\text{-}Ph_2Sb\text{–}CH_2\text{–}SbPh_2)_2(\eta^2\text{-}Ph_2Sb\text{–}CH_2\text{–}SbPh_2)]$ (only the *ipso* carbon atoms are represented).[94]

In $[(CO)_5Cr(\eta^1\text{-}Ph_2Sb\text{–}S\text{–}SbPh_2)]$ (Fig. 10b) the coordination to the chromium center occurs through one of the antimony atoms of the ligand and not through sulfur. In $[(CO)_4Fe(\eta^1\text{-}Ph_2Sb\text{–}CH_2\text{–}SbPh_2)]$ and $[(CO)_5W(\eta^1\text{-}Ph_2Sb\text{–}CH_2\text{–}SbPh_2)]$ the conformation of the Sb ligand can be described as being near to *syn-syn* ($\varphi_1 = 31.8$, $\varphi_2 = 55.3°$ and $\varphi_1 = 31.2$; $\varphi_2 = 55.8°$, respectively), whereas in $[(CO)_5Cr(\eta^1\text{-}Ph_2Sb\text{–}S\text{–}SbPh_2)]$ the conformation of the Sb ligand is near to *gauche-gauche* ($\varphi_1 = 57.5$ and $\varphi_2 = 58.0°$).

The first crystallographically characterized complex with a chelating bis(stibino) ligand, $[I_2Ru(\eta^1\text{-}Ph_2Sb\text{–}CH_2\text{–}SbPh_2)_2(\eta^2\text{-}Ph_2Sb\text{–}CH_2\text{–}SbPh_2)]$ was recently described.[94] It was obtained by reacting $[Ru(dmf)_6][CF_3SO_3]_3$ (dmf = dimethylformamide) with four equivalents of the antimony ligand and an excess of LiI.[94] Other complexes where the evidence for *chelating* coordination of the bis(stibino)methane ligand was based on UV/VIS and $^1H$ and $^{13}C$ NMR spectra are *mer*-$[X_3Rh(\eta^1\text{-}Ph_2Sb\text{–}CH_2\text{–}SbPh_2)(\eta^2\text{-}Ph_2Sb\text{–}CH_2\text{–}SbPh_2)]$[94] (X = Cl, Br, I) obtained by reacting $RhX_3$ with two equivalents of the antimony ligand, and *trans*-$[Br_2Os(\eta^1\text{-}Ph_2Sb\text{–}CH_2\text{–}SbPh_2)_2(\eta^2\text{-}Ph_2Sb\text{–}CH_2\text{–}SbPh_2)]$[94] obtained from *trans*-$[Br_2Os(dmso)_4]$ (dmso = dimethylsulfoxide) and the antimony ligand.

The chelating coordination of bis(diphenylstibino)methane leads to considerable distortion of the ligand. In the structure of $[I_2Ru(\eta^1\text{-}Ph_2Sb\text{–}CH_2\text{–}SbPh_2)_2(\eta^2\text{-}Ph_2Sb\text{–}CH_2\text{–}SbPh_2)]$ (Fig. 11) the Sb–C–Sb angle of the $\eta^2$-coordinated ligand is 93.5°, but 118.9 and 122.2° for the $\eta^1$-ligands. Also the chelate Sb–Ru–Sb angle (74.3°) is smaller than the other Sb–Ru–Sb angles (91.6–101.3°).

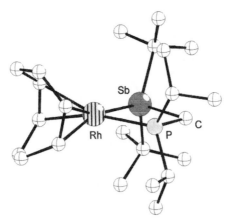

FIG. 12. Molecular structure of the cation of $[(\eta^4\text{-cod})Rh\{\eta^2\text{-}(i\text{-Pr})_2P\text{-}CH_2\text{-}Sb(t\text{-Bu})_2\}]PF_6$.[100]

Related to bis(diorganostibino)methane are the (diorganophosphino) (diorganostibino)-methane, $R_2P\text{-}CH_2\text{-}SbR'_2$ (R, R' = alkyl or cycloalkyl) ligands. The chelating coordination of this mixed-donor atoms ligand was observed in the complex cations of $[(\eta^4\text{-cod})Rh(\eta^2\text{-}R_2P\text{-}CH_2\text{-}SbR'_2)]Y$ (R = i-Pr, R' = i-Pr or t-Bu; Y = PF$_6$ or BPh$_4$; cod = cyclooctadiene).[100]

The formation of these complexes is remarkable since related chelate complexes of the type $[(\eta^4\text{-cod})Rh(\eta^2\text{-}R_2P\text{-}CH_2\text{-}PR_2)]Y$ do not exist. The chelate coordination of the phosphino(stibino)methane ligand in the case of $[(\eta^4\text{-cod})Rh\{\eta^2\text{-}(i\text{-Pr})_2P\text{-}CH_2\text{-}Sb(t\text{-Bu})_2\}]PF_6$ was proven by X-ray crystallography. In the structure of this complex (Fig. 12) the four-membered RhSbCP ring is planar with a P–C–Sb angle of 96.6°. In contrast, in the structure of $[I_2Ru(\eta^1\text{-}Ph_2Sb\text{-}CH_2\text{-}SbPh_2)_2(\eta^2\text{-}Ph_2Sb\text{-}CH_2\text{-}SbPh_2)]$[94] (Fig. 11) the four-membered chelate ring is non planar with a dihedral angle between the Ru–Sb–Sb and Sb–Sb–C planes of 23.1°.

Attempts to substitute the cyclooctadiene ligand from $[(\eta^4\text{-cod})Rh(\eta^2\text{-}R_2P\text{-}CH_2\text{-}SbR'_2)][BPh_4]$ (R = i-Pr, R' = i-Pr or t-Bu) by H$_2$ led to neutral half-sandwich compounds where one of the phenyl rings of the BPh$_4$ anion is coordinated to rhodium (**21**) (Scheme 9).[100]

Another structurally well characterized rhodium complex containing phosphino(stibino)methane ligand as a chelate is $[H_3Rh_2(\eta^2\text{-}O_2CCF_3)_2\{\eta^2\text{-}(i\text{-Pr})_2P\text{-}CH_2\text{-}Sb(t\text{-Bu})_2\}_2]PF_6$.[101]

In the dinuclear cation of this PF$_6$ salt (Fig. 13) the P–Rh–Sb (75.7 and 75.9°) and Sb–C–P (94.5 and 95.4°) angles are similar with those in the cation of $[(\eta^4\text{-cod})Rh\{\eta^2\text{-}(i\text{-Pr})_2P\text{-}CH_2\text{-}Sb(t\text{-Bu})_2\}]PF_6$[100] (Fig. 12) (P–Rh–Sb 74.9°;Sb–C–P 96.6°). The four-membered ring RhSbCP however is not planar (dihedral angle between the Rh–Sb–P and Sb–P–C planes of 7.7 and 2.1°).

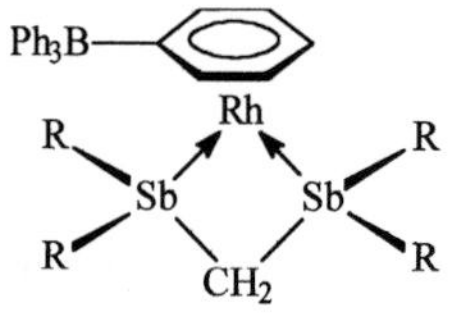

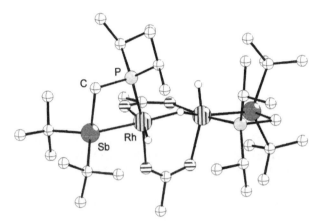

**21**

SCHEME 9

FIG. 13. Molecular structure of the cation of $[H_3Rh_2(\eta^2-O_2CCF_3)_2\{\eta^2-(i-Pr)_2P-CH_2-Sb(t-Bu)_2\}_2]PF_6$.[101]

Binuclear complexes with *bridging bidentate* bis(dimethylstibino)-methane ligands are the group 6 metal compounds $[(OC)_5M(\mu - Me_2Sb-CH_2-SbMe_2)M(CO)_5]$ (M = Cr, Mo, W)[97] which were obtained by reaction of $M(CO)_6$ with the ligand at elevated temperatures in high-boiling solvents. The analogous complexes of Cr and W with the bis(diphenyl-stibino)methane ligand are best obtained from $[(CO)_5M(\eta^1-Ph_2Sb-CH_2-SbPh_2)]$ and $(THF)M(CO)_5$.[97]

The structure of $[(OC)_5W(\mu-Ph_2Sb-CH_2-SbPh_2)W(CO)_5]$ as determined by single crystal X-ray diffraction is depicted in Fig. 14.

As a result of the coordination of both antimony atoms in $[(OC)_5W(\mu-Ph_2Sb-CH_2-SbPh_2)W(CO)_5]$ the Sb–C–Sb angle is about 15° larger than in the case of $[W(CO)_5(\eta^1-Ph_2Sb-CH_2-SbPh_2)]$ (Sb–C–Sb 106.8).[97]

Other recent examples of complexes with acyclic M–Sb–Y–Sb–M (M = transition metal; Y = $CH_2$, O, S) frames are $[Br(CO)_4Mn(\mu-Me_2Sb-CH_2-SbMe_2)Mn(CO)_4Br]$,[96]  $[(MeC_5H_4)(CO)_2Mn(\mu-Ph_2Sb-Y-SbPh_2)Mn$

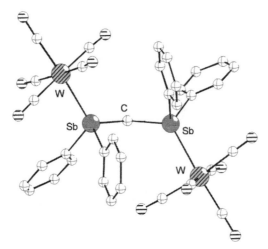

FIG. 14.  Molecular structure of $[(OC)_5W(\mu\text{-}Ph_2Sb\text{-}CH_2\text{-}SbPh_2)W(CO)_5]$.[97]

$(CO)_2(MeC_5H_4)]^{59}$ (Y = CH$_2$, O, S), $[(CO)_2\{P(OPh)_3\}_2Fe(\mu\text{-}Ph_2Sb\text{-}Y\text{-}Sb$ Ph$_2$)Fe(CO)$_2\{P(OPh)_3\}_2]^{59}$ (Y = CH$_2$, O, S), $[(CO)_3Ni(\mu\text{-}R_2Sb\text{-}CH_2\text{-}SbR_2)$ Ni(CO)$_3]^{96}$ (R = Me or Ph) and $[(CO)_5M(\mu\text{-}Ph_2Sb\text{-}Y\text{-}SbPh_2)M(CO)_5]^{60}$ (M = Cr, Mo or W; Y = O or S).

Bis(stibino) ligands are suitable for *bridging bidentate* coordination on M–M (M = transition metal) units (type **19**). Examples for this type of coordination are $[(CO)_8Mn_2(\mu\text{-}R_2Sb\text{-}CH_2\text{-}SbR_2)]$ (R = Me or Ph), which are synthesized by thermal reactions of Mn$_2$(CO)$_{10}$ with the corresponding Sb ligands.[96] The bridging bidentate coordination was confirmed by IR and $^1$H, $^{13}$C and $^{55}$Mn NMR data and by X-ray structure determination for $[(CO)_6Mn_2(\mu\text{-}Ph_2Sb\text{-}CH_2\text{-}SbPh_2)_2]$.[98]

An example of a complex of type **19** with a known crystal structure is $[(CO)_6Co_2(\mu\text{-}Me_2Sb\text{-}CH_2\text{-}SbMe_2)]$.[96,103] The structure is shown in Fig. 15. The Sb–C–Sb angle of 114.5° indicates no strain in the Co$_2$Sb$_2$C ring.

Photolysis of Mn$_2$(CO)$_{10}$ in the presence of bis(diphenylstibino)methane gave $[(CO)_6Mn_2(\mu\text{-}Ph_2Sb\text{-}CH_2\text{-}SbPh_2)_2]$.[98] This unique doubly bridged complex was characterized by X-ray diffractometry. The molecular structure is shown in Fig. 16. The *bridging* ligands are in *trans* positions to each other with Sb–C–Sb angles (103.0 and 106.7°) smaller than in the uncomplexed ligand (117.3°).[96] The Mn–Mn bond (3.098 Å) in the complex is longer than in Mn$_2$(CO)$_{10}$[104] (2.895 Å).

Complexes of type **20** with two bidentate ligands bridging metal centers without M–M bonds are the Pd or Pt compounds $[X_2M(\mu\text{-}R_2Sb\text{-}CH_2\text{-}SbR_2)_2MX_2]$ (M = Pd or Pt; R = Me or Ph; X = Cl, Br or I)[6,94,105] which exist as *cis,trans* or *trans,trans* isomers.

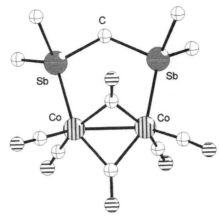

FIG. 15. Molecular structure of [(CO)$_6$Co$_2$($\mu$-Me$_2$Sb–CH$_2$–SbMe$_2$)].[96]

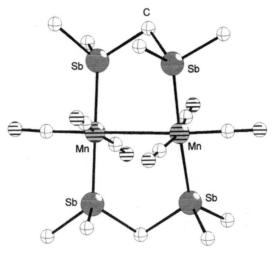

FIG. 16. Molecular structure of [(CO)$_6$Mn$_2$($\mu$-Ph$_2$Sb–CH$_2$–SbPh$_2$)$_2$] (only *ipso* C atoms of the phenyl groups are shown).[98]

The molecular structure of *trans,trans*-[Cl$_2$Ph$_2$Pd$_2$($\mu$-Ph$_2$Sb–CH$_2$–SbPh$_2$)$_2$][105] is shown in Fig. 17a. This complex was formed by photolysis of solutions of [Cl$_2$Pd($\mu$-Ph$_2$Sb–CH$_2$–SbPh$_2$)$_2$] in CH$_2$Cl$_2$ which resulted in the migration of phenyl groups from the Sb ligand to Pd. In *trans,trans*-[Cl$_2$Ph$_2$Pd$_2$($\mu$-Ph$_2$Sb–CH$_2$–SbPh$_2$)$_2$] two bis(diphenylstibino)methane ligands in *trans* positions bridge two parallel square planar palladium units with Sb–C–Sb angles of 115.0°, which are similar as in the free

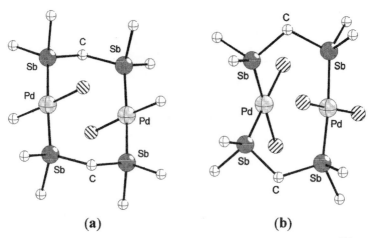

**(a)**                                    **(b)**

FIG. 17. Molecular structure of *trans,trans*-[Cl$_2$Ph$_2$Pd$_2$(μ-Ph$_2$Sb–CH$_2$–SbPh$_2$)$_2$][105] (a) and *cis,trans*-[Cl$_4$Pt$_2$(μ-Ph$_2$Sb–CH$_2$–SbPh$_2$)$_2$][94] (b).

ligand (117.3°),[96] but wider than in *trans*-[(CO)$_6$Mn$_2$(μ-Ph$_2$Sb–CH$_2$–SbPh$_2$)$_2$][98] (106.7 and 103.0°).

The bridging *cis,trans* coordination mode of the bis(diphenylstibino)-methane units was proven by X-ray crystal structure analysis for [Cl$_4$Pt$_2$(μ-Ph$_2$Sb–CH$_2$–SbPh$_2$)$_2$] and [Br$_4$Pd$_2$(μ-Ph$_2$Sb–CH$_2$–SbPh$_2$)$_2$].[94] The structures are similar with Sb–C–Sb angles in the range 111.3–115.0°. The structure of the Pt complex is depicted in Fig. 17b.

Two bis(stibino)methane ligands coordinate to group 6 metals in the dinuclear *cis* tetracarbonyl complexes [(OC)$_4$M(μ-R$_2$Sb–CH$_2$–SbR$_2$)$_2$ M(CO)$_5$] (M = Cr,Mo,W; R = MeorPh) (type **20**) which are obtained in low yields by reacting [(nbd)M(CO)$_4$] (M = Cr, Mo; nbd = norbornadiene) or [(CO)$_4$W(Me$_2$N(CH$_2$)$_3$NMe$_2$)] with the antimony ligands.[97] The structures of the tungsten complexes were determined by single crystal X-ray studies. The folding of the eight membered heterocycles in both complexes is close to the chair conformation.

The molecular structures of [(CO)$_4$Cr(μ-Ph$_2$Sb–O–SbPh$_2$)$_2$Cr(CO)$_4$][60] and [(CO)$_4$Cr(μ-Me$_2$Sb–Y–SbMe$_2$)$_2$Cr(CO)$_4$][99] (Y = O, S) are very similar to that of [(OC)$_4$Cr(μ-R$_2$Sb–CH$_2$–SbR$_2$)$_2$Cr(CO)$_4$] (R = Me, Ph), consisting of eight membered heterocycles in close to a chair conformation (Fig. 18). The Sb–Y–Sb–M (Y = CH$_2$, O or S; M = Cr, W) dihedral angles $\varphi_1$ = 72.2–100.8° and $\varphi_2$ = 130.2–153.0° (Table 2) indicate a far from *syn-anti* conformation for the antimony ligands. This conformation was found also in [(OC)$_2$Ni(μ-Ph$_2$Sb–O–SbPh$_2$)$_2$Ni(CO)$_2$][102] however, the dihedral angles $\varphi_1$ = 30.8° and $\varphi_2$ = 95.7° differ considerably.

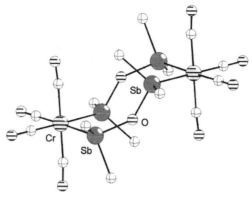

FIG. 18.  Molecular structure of $[(CO)_4Cr(\mu\text{-}Me_2Sb\text{-}O\text{-}SbMe_2)_2Cr(CO)_4]$.[60]

## V

## COMPLEXES WITH $R_2Sb\text{-}Y_n\text{-}SbR_2$ (R = ALKYL, ARYL; $Y_n$ = $(SbR)_2$, $o\text{-}C_6H_4$, $(CH_2)_3$, S(MeSb)S) LIGANDS

Only very few complexes with bis(stibino) ligands with two or three spacer atoms between the antimony donors are known. Newcomers in this class of ligands are *catena*-tetrastibines $R_2Sb(SbR')_2SbR_2$ (R = Me, Ph; R' = Ph, $Me_3SiCH_2$) and the antimony–sulfur chain $Me_2Sb\text{-}S\text{-}SbMe\text{-}S\text{-}SbMe_2$. Bidentate Sb ligands with $o\text{-}C_6H_4$ or $(CH_2)_3$ spacers are well established. All complexes known feature *chelating* coordination with formation of five- (type **22**) or six-membered (type **23**) heterocycles (Scheme 10).

Although, *catena*-stibines have been under investigation for a long time[7,9,107,108] well defined examples are rare and the isolation of oligomers of the type $R_2Sb(SbR)_nSbR_2$ (n = 1, 2) has not yet been achieved. The existence of *catena*-tri- and tetrastibines in ring-chain equilibria is however well established (Equation 5).[109,110]

$$R_2Sb\text{-}SbR_2 + cyclo\text{-}(RSb)_n \rightleftharpoons R_2Sb(SbR)_nSbR_2 \tag{5}$$
$$R = \text{organic group}$$

In these mixtures the *catena*-tristibines are by far the most abundant chain species and even under favorable conditions [excess of $cyclo\text{-}(RSb)_n$] the *catena*-tetrastibines form only as minor components. Nevertheless, recently the selective extraction and stabilization of these tetrastibines in the coordination sphere of a transition metal carbonyl complex was achieved by reacting mixtures of distibines and cyclostibines with

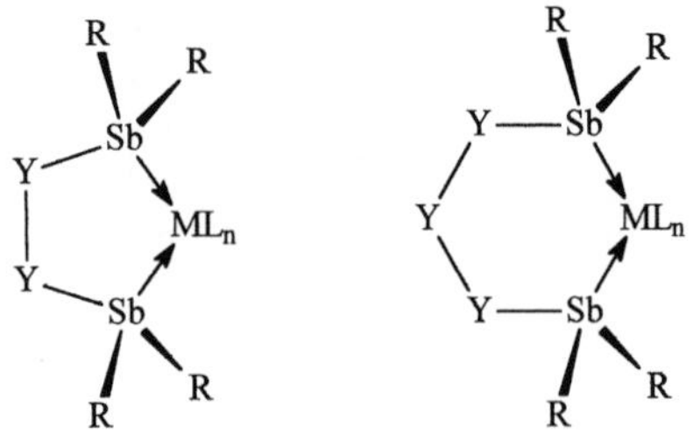

**22**          **23**

SCHEME 10

$Cr(CO)_4(nbd)$ (nbd = norbornadiene).[109] The *catena*-stibine complexes *cyclo*-$[Cr(CO)_4(R'_2Sb–SbR–SbR–SbR'_2)]$ (R' = Me, Ph; R = Me$_3$SiCH$_2$), were obtained in good yields. The high selectivity of this trapping reaction is certainly due to the favorable chelate ring size and the good fit of the ligand bite in these complexes where the *catena*-tetrastibines act as bidentate ligands through the terminal antimony atoms. Apparently the coordination of distibines or tristibines on Cr(CO)$_4$ fragments is less favorable.

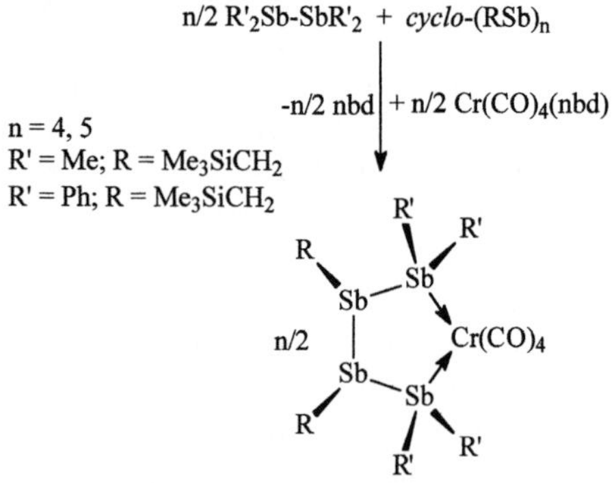

$$n/2\ R'_2Sb-SbR'_2\ +\ cyclo\text{-}(RSb)_n$$

$$n = 4, 5$$
$$R' = Me;\ R = Me_3SiCH_2$$
$$R' = Ph;\ R = Me_3SiCH_2$$

$$-n/2\ nbd \mid +\ n/2\ Cr(CO)_4(nbd)$$

(6)

The influence of the organic substituents of tetrastibine ligands on the donor strengths towards Cr(CO)$_4$(nbd) is reflected in the different reactivities of derivatives with methyl or phenyl groups in terminal positions. The former react already at room temperature, whereas for the latter high temperatures with reflux of the solvent are required. Under these conditions also migration of the organic groups is possible and *cyclo*-$[Cr(CO)_4(Ph_2Sb–SbPh–SbR–SbPh_2)]$ (R = Me$_3$SiCH$_2$),[111] a complex where a Me$_3$SiCH$_2$ was replaced by a phenyl group on a central antimony atom of the *catena*-stibine chain, forms as side product.

Further coordination of the potentially tetradentate *catena*-tetrastibine ligands is achieved by reaction[109] with $W(CO)_5(THF)$.

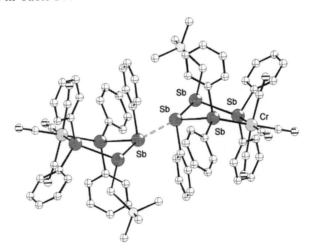

$$R = Me$$
$$R' = Me_3SiCH_2$$

$$(7)$$

X-ray diffraction studies on these complexes reveal the chelating bidentate coordination of the organo antimony chains as four-electron donors via lone pair donation through the terminal antimony atoms. The resulting five-membered $CrSb_4$ rings are nonplanar (Figs. 19 and 20), the $Sb(2)–Sb(3)$ unit being twisted out of the $Sb(1)–Cr(1)–Sb(4)$ plane with twist angles between 22.4 and 26.3°. The organic groups bonded to the central antimony atoms occupy *trans* positions. Of the possible isomeric forms of a *catena*-tetrastibine, *meso* and *d,l*, only the latter acts as ligand in these complexes.

The crystal structures consist of discrete (Fig. 20) or pair wise associated (Fig. 19) molecules.

Selected geometric parameters for *catena*-stibine and related complexes are listed in Table IV.

FIG. 19. Molecular structure of dimers of $d,l\text{-}cyclo\text{-}[Cr(CO)_4(Ph_2Sb–SbPh–SbR–SbPh_2)]$ ($R = Me_3SiCH_2$); $Sb \cdots Sb$ 3.636 Å.[111]

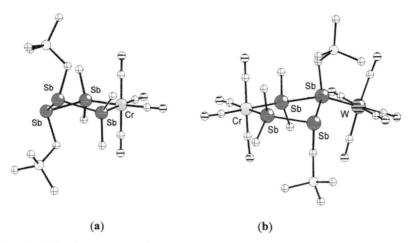

(a)                                    (b)

Fig. 20. Molecular structures of *cyclo*-[Cr(CO)$_4$(Me$_2$Sb–SbR–SbR–SbMe$_2$)] (a) and *cyclo*-[Cr(CO)$_4$(Me$_2$Sb–SbR–SbR–SbMe$_2$)W(CO)$_5$] (R = Me$_3$SiCH$_2$) (b).[109]

TABLE IV

GEOMETRIC PARAMETERS OF CRYSTAL STRUCTURES FOR COMPLEXES WITH *CATENA*-STIBINE AND
RELATED PNICOGEN LIGANDS

| Compound | E–M (Å) | E–M–E (°) | Twist of the central RE–ER bond (°) | Ref. |
|---|---|---|---|---|
| [Cr(CO)$_4$(R$_2$Sb–SbR′–SbR′-SbR$_2$)] | M = Cr 2.596 | 92.1 | 26.3 | 109 |
| R = Me; R′ = Me$_3$SiCH$_2$ | E = Sb 2.588 | | | |
| [Cr(CO)$_4$(R$_2$Sb–SbR′–SbR′-SbR$_2$)W(CO)$_5$] | M = Cr 2.593 | 94.2 | 22.4 | 109 |
| R = Me; R′ = Me$_3$SiCH$_2$ | E = Sb 2.575 | | | |
| | M = W 2.791 | | | |
| | E = Sb | | | |
| [Cr(CO)$_4$(R$_2$Sb–SbR′–SbR-SbR$_2$)] | M = Cr 2.606 | 91.6 | 23.3 | 111 |
| R = Ph; R′ = Me$_3$SiCH$_2$ | E = Sb 2.611 | | | |
| [Cr(CO)$_4${(R)ClAs-As(R)-As(R)-AsCl(R)}][a] | M = Cr 2.418 | 92.4 | 1.8 | 112 |
| R = *t*-Bu | E = As 2.422 | 92.4 | 0.6 | |
| | 2.420 | | | |
| | 2.426 | | | |
| [Mo(CO)$_4$(R$_2$P–AsR–AsR–PR$_2$)][b] | M = Mo 2.453 | 89.2 | 16 | 113 |
| R = Me | E = P | | | |
| [Mo(CO)$_4$(R$_2$P–PR–PR–PR$_2$)][b] | M = Mo 2.489 | 86.8 | 16.5 | 114 |
| R = Me | E = P | | | |

[a]Two molecules per asymmetric unit.
[b]The molecules posses crystallographic $C_2$ symmetry.

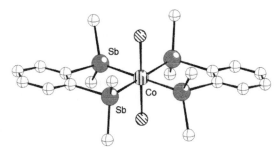

FIG. 21. Structure of the cation of *trans*-[Cl$_2$Co{*o*-C$_6$H$_4$(SbMe$_2$)$_2$}$_2$]$_2$[CoCl$_4$].[115]

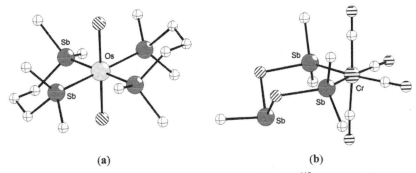

**(a)**                                  **(b)**

FIG. 22. Molecular structures of *trans*-[OsCl$_2$(Ph$_2$Sb(CH$_2$)$_3$SbPh$_2$)$_2$][117] **(a)** (only the *ipso*-carbon atoms are represented) and *cyclo*-[Cr(CO)$_4$(Me$_2$Sb–S–SbMe–S–SbMe$_2$)]$^{99}$ **(b)**.

Closely related to *catena*-stibines are the organo bridged bis(stibino) ligands R$_2$Sb(X)SbR$_2$ [R = Me, Ph; X = (CH$_2$)$_3$ or *o*-C$_6$H$_4$] which also show preferences for chelating coordination in transition metal complexes.$^{2,6}$

*trans*-[Cl$_2$Co{*o*-C$_6$H$_4$(SbMe$_2$)$_2$}$_2$]$_2$[CoCl$_4$] is the only crystallographically characterized transition metal complex of type 22.[115] In the cations there are two *o*-C$_6$H$_4$(SbMe$_2$)$_2$ ligands coordinating to a cobalt center (Fig. 21). The chelating coordination is a result of the placement in *ortho* positions of the two donating Me$_2$Sb units in the free ligand. The Sb–Co–Sb angles are close to 90° [Sb–Co–Sb(endocyclic) 88.3° and Sb–Co–Sb(exocyclic) 91.7°].

NMR ($^1$H, $^{103}$Rh) spectroscopic evidence for chelating bidentate coordination was found in the rhodium complexes, [X$_2$Rh{o − C$_6$H$_4$ (SbMe$_2$)$_2$}$_2$]BF$_4$ (X = Cl, Br, or I), which were obtained from, Rh X$_3$·$n$H$_2$O, (X = Cl, Br, or I) and the antimony ligand after addition of HBF$_4$.[116]

A complex with a ligand where the Sb donors are separated by a three-atomic spacer is *trans*-[OsCl$_2$(Ph$_2$Sb(CH$_2$)$_3$SbPh$_2$)$_2$], obtained by reacting *trans*-[OsCl$_2$(dmso)$_4$] (dmso = dimethylsulfoxide) with the Sb ligand.[117] The structure (Fig. 22a) was determined by X-ray crystallography. Both 1,3-bis(diphenylstibino)propane ligands are coordinated as chelates on Os.

The analogous Ru complexes $trans$-$[Br_2Ru(Ph_2Sb(CH_2)_3SbPh_2)_2]$[118] and the cation of $trans$-$[Cl_2Rh(Ph_2Sb(CH_2)_3SbPh_2)_2][ClO_4]$[116] have closely related structures.

A complex with a ligand with two sulfur atoms and one SbR group separating the Sb donor atoms is $cyclo$-$[Cr(CO)_4(Me_2Sb$–$S$–$SbMe$–$S$–$SbMe_2)]$ obtained as side product on reacting $Me_2Sb$–$S$–$SbMe_2$ with $(nbd)Cr(CO)_4$ (nbd = norbornadiene) in 1:1 molar ratio (Equation 8).[99] The tristibadisulfur ligand, $Me_2Sb$–$S$–$SbMe$–$S$–$SbMe_2$ is a possible decomposition product of $Me_2Sb$–$S$–$SbMe_2$.

$$2\ Me_2Sb\text{-}S\text{-}SbMe_2 \longrightarrow Me_2Sb\text{-}S\text{-}SbMe\text{-}S\text{-}SbMe_2 + Me_3Sb$$

(8)

The structure (Fig. 22b) consists of six-membered $Sb(SSb)_2Cr$ ring in a near to chair conformation (dihedral angle between the Sb–Sb–S–S and S–S–Sb planes 100.2° and between the Sb–Sb–S–S and Sb–Sb–Cr planes 9.2°). The same conformation was found also for the six-membered $MSb_2C_3$ (M = Ru, Os, Rh) heterocycles in $[Br_2Ru(Ph_2Sb(CH_2)_3SbPh_2)_2]$[118], $[Cl_2Rh(Ph_2Sb(CH_2)_3SbPh_2)_2][ClO_4]$[116], $[Cl_2Os(Ph_2Sb(CH_2)_3SbPh_2)_2]$[117], but with different dihedral angles (dihedral angle between the Sb–Sb–C–C and C–C–C planes 58° in $[Cl_2Rh(Ph_2Sb(CH_2)_3SbPh_2)_2][ClO_4]$, 62.4° in $[Cl_2Os(Ph_2Sb(CH_2)_3SbPh_2)_2]$ and 60.9° in $[Br_2Ru(Ph_2Sb(CH_2)_3SbPh_2)_2]$; dihedral angle between the Sb–Sb–C–C and Sb–Sb–M planes 25.2° in $[Cl_2Rh(Ph_2Sb(CH_2)_3SbPh_2)_2][ClO_4]$, 37.9° in $[Cl_2Os(Ph_2Sb(CH_2)_3SbPh_2)_2]$ and 27.1° in $[Br_2Ru(Ph_2Sb(CH_2)_3SbPh_2)_2]$).

# VI

## COMPLEXES WITH (RSb)$_n$ (R = ALKYL, ARYL) AND Sb$_n$ LIGANDS

The coordination chemistry of RSb, $(RSb)_2$ or "naked" $Sb_1$ ligands is well established and was summarized in reviews several times.[4,7,10–12] More recent developments include syntheses of complexes with antimony rings or chains $(RSb)_n$, and reactions of $(t\text{-}BuSb)_4$ leading to complexes with "naked" $Sb_2$, cyclo-$Sb_3$ or cyclo-$Sb_5$ ligands.[9]

Among transition metal complexes with RSb groups different classes of compounds may be distinguished. When the RSb moiety is coordinated to two 16-e⁻ transition metal complex fragments [$W(CO)_5$, $Fe(CO)_4$, $CpMn(CO)_2$] typical (open) stibinidene (type **24**) complexes result, which are deeply colored and where the antimony center is in a trigonal planar environment. A representative example is the deep blue complex [$(Me_3Si)_2CHSb(W(CO)_5)_2$][119], where the lone pair of electrons is delocalized in the $W_2Sb$ π-system. In "closed" stibinidene complexes (type **25**), the coordination of the antimony center is pyramidal and through lone pair donation from antimony a third 16-e⁻ fragment may be coordinated (type **26**). Examples for the latter types are [$(Me_3Si)_2CHSb$][$Fe(CO)_4$]$_2$[120] (type **25**) and [$t\text{-}BuSb(W(CO)_5)$]$_3$[121] (type **26**) (Scheme 11).

Taking into account the isolobal analogy between RSb and 16-e⁻ $ML_n$ fragments metallocycles, the complexes of type **25** may be considered as analogs of cyclo-tristibine systems. Type **26** is related to complexes with cyclo-tristibine ligands.

When a RSb moiety is coordinated to two or three 17-e⁻ complex fragments the resulting compounds compare well with tertiary stibines,

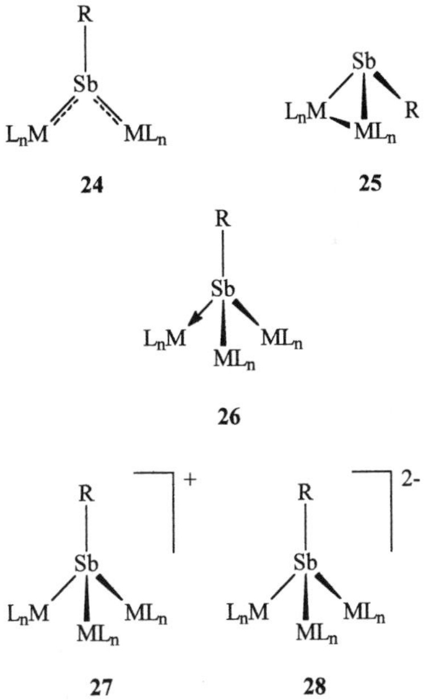

**24**         **25**

**26**

**27**         **28**

SCHEME 11

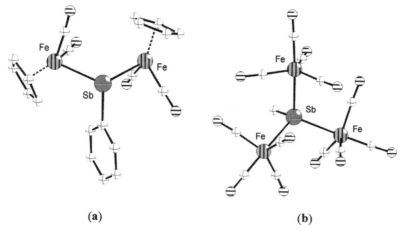

(a)                                          (b)

FIG. 23. Molecular structures of $[PhSb\{Fe(CO)_2Cp\}_2]^{97}$ (a) and of the anion of $[NEt_4]_2$ $[MeSb\{Fe(CO)_4\}_3].^{92}$ (b).

$R_3Sb$ or stibonium, $R_4Sb^+$ compounds. Example for a dimetalla stibine is $[PhSb(Fe(CO)_2Cp)_2]^{97}$ (Fig. 23a) and for trimetalla stibonium cations (type **27**) $[PhSb\{Fe(CO)_2Cp\}_3]_2[FeCl_4]^{91}$.

Anions of type **28** where a $RSb^{2-}$ moiety is coordinated to three 16-e$^-$ complex fragments are also known. An example for type **28** is $[NEt_4]_2[MeSb\{Fe(CO)_4\}_3]^{92}$ (Fig. 23b).

Reactions of $RSbCl_2$ with $Na_2W_2(CO)_{10}$ or $Na_2Fe(CO)_4$ lead to complexes of type **29** or **30** with $RSb = SbR$ (distibene) ligands. Examples are $[(RSb)_2\{W(CO)_5\}_3]$ (R = Ph$^{122}$, $t$-Bu$^{121}$) (type **29**). A complex of type **30** with a distibene coordinated to a 14-e$^-$ complex fragment is $[(RSb)_2Pt(PEt_3)_2]$ $[R = t$-Bu(O)C].$^{123}$

With Sb–Sb distances corresponding to bond orders 1.5 these complexes can be considered either as $\eta^2$-coordinated distibenes $RSb = SbR$ or as metallacycles (Scheme 12).

Considering the isolobal relations between $W(CO)_5$ or $RSb$, the complexes of type **29** and **30** may be considered as analogs of $cyclo$-$R_3Sb_3$ or of complexes with a $cyclo$-$R_3Sb_3$ ligand, which are still unknown. Closely related are the complexes $[MeC(CH_2Sb)_3M(CO)_5]$ (M = Cr, Mo, W)$^{124}$ where a polycyclic tristibine is coordinated to metal carbonyl units.

The reaction of $cyclo$-$[(Me_3Si)_2CHSb]_3$ with $Fe_2(CO)_9$ leads to insertion of a $Fe(CO)_4$ fragment into the $cis$-RSb–SbR unit with formation of the heterocycle $[\{(Me_3Si)_2CHSb\}_3Fe(CO)_4].^{125}$ The structure of this complex (Fig. 24) consists of a chelating RSb–SbR–SbR chain bonded to Fe through the terminal antimony atoms. The resulting four membered heterocycle is folded with dihedral angles $Sb_3/FeSb_2$ 131.4° and $FeSb_2/FeSb_2$ 125.8°.

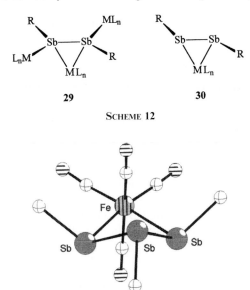

SCHEME 12

FIG. 24. Molecular structure of [{(Me$_3$Si)$_2$CHSb}$_3$Fe(CO)$_4$] (without SiMe$_3$ groups).[125]

TABLE V

GEOMETRIC PARAMETERS OF CRYSTAL STRUCTURES FOR COMPLEXES WITH (RSb)$_n$ (R = ALKYL, ARYL) LIGANDS

| Compound | Sb–M (Å) | Sb–C (Å) | C–Sb–M (°) | Sb–Sb–M (°) | Ref. |
|---|---|---|---|---|---|
| [(Me$_3$SiCH$_2$Sb)$_5${W(CO)$_5$}$_2$] | 2.80; 2.807 | 2.172–2.191 | 93.9–105.1 | 112.4–118.3 | 125 |
| [{(Me$_3$Si)$_2$CHSb}$_3$Fe(CO)$_4$] | 2.706; 2.709 | 2.222–2.235 | 100.4; 102.7 | 90.4; 90.8 | 125 |
| [(PhSb)$_2${W(CO)$_5$}$_3$] | 2.756–3.092 | 2.135; 2.183 | 114.3–121.2 | 128.3; 130.6 | 122 |
| [(t-BuSb)$_2${Cr(CO)$_5$}$_3$] | 2.687–2.924 | 2.25; 2.26 | 112.6; 113.5 | 130.2; 130.4 | 121 |
| [(Me$_3$Si)$_2$CHSb{Fe(CO)$_4$}$_2$] | 2.641; 2.663 | 2.208 | 110.8; 112.1 | | 120 |
| [PhSb{FeCp(CO)$_2$}$_2$] | 2.634; 2.639 | 2.177 | 98.7; 102.2 | | 97 |
| [t-BuSb{W(CO)$_5$}$_3$] | 2.805–2.854 | 2.250 | 109.0–115.0 | | 121 |
| [MeSb{Fe(CO)$_4$}$_3$][NEt$_4$] | 2.604–2.618 | 2.161 | 102.4–105.9 | | 92 |
| [PhSb{Fe(CO)$_2$Cp}$_3$]$_2$[FeCl$_4$] | 2.575–2.581 | 2.172 | 100.0–108.0 | | 91 |

The coordination of the iron atom is distorted octahedral, with the CO groups in *cis*-positions to the antimony atoms being inclined towards the center of the heterocycle.

Selected geometric parameters of crystal structures for complexes with (RSb)$_n$ (R = alkyl, aryl) ligands are listed in Table V.

Examples for complexes with (RSb)$_4$ ligands are the *t*-Bu compounds *cyclo*-[*t*-Bu$_4$Sb$_4$–Mo(CO)$_5$], *cyclo*-[*t*-Bu$_4$Sb$_4$–Fe(CO)$_4$] and *cyclo*-[1,3-(*t*-Bu$_4$Sb$_4$)–{W(CO)$_5$}$_2$]. The structures of these complexes consist of the

folded cyclotetrastibine ring with the four *t*-Bu groups in all-*trans* positions
and of one or two 16-e⁻ metal carbonyl fragments coordinated through lone
pair donation of Sb. The structural chemistry of these complexes was
summarized in previous review.[9]

The only known complex with a $(RSb)_5$ ligand is *cyclo*-[μ-
$(Me_3SiCH_2Sb)_5–Sb^1,Sb^3–\{W(CO)_5\}_2]$.[125] The molecular structure is depicted
in Fig. 25. The antimony ring adopts a slightly distorted envelope conformation with a $Sb_4/Sb_3$ dihedral angle of 129.1°. The metalcarbonyl units are
in 1–3 positions *trans* to each other.

Recently not only the coordination of *cyclo*- or *catena*-stibine moieties,
but with *t*-$Bu_4Sb_4$ for example also, the fragmentation leading ultimately to
complexes with "naked" $Sb_n$ ligands was investigated.

Thus reactions of *t*-$Bu_4Sb_4$ with $[Cp^*Mo(CO)_3]_2$ at elevated temperatures
in toluene or decalin lead to the substitution of alkyl groups with formation

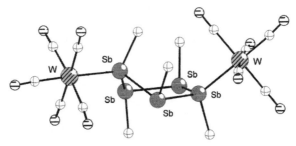

FIG. 25. Molecular structure of $[μ-(Me_3SiCH_2Sb)_5-Sb^1,Sb^3-\{W(CO)_5\}_2]$.[125]

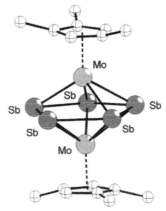

FIG. 26. Molecular structure of $[(η^5-1,2,4-tBu_3C_5H_2)Mo(μ-η^5-Sb_5)Mo(η^5-1,4-tBu_2-2-MeC_5H_2)]$
(without the Me groups from *t*-Bu substituents).[126]

of [$t$-Bu$_3$Sb$_4$- Mo(CO)$_3$Cp*] (Cp* = η$^5$-C$_5$Me$_5$), a complex with the R$_3$Sb$_4$-ligand or complexes with 'naked' antimony ligands [Sb$_2${Mo(CO)$_2$Cp*}], or [Mo(Sb$_3$)(CO)$_2$Cp$^*$]. The chemistry of these novel Sb$_n$ complexes was summarized recently.[9] Important developments since then are the syntheses and structural analyses of complexes with the *cyclo*–Sb$_5$ (pentastibacyclopentadienyl) ligand in the triple-decker sandwich complexes [{(η$^5$-1,2,4-$t$-Bu$_3$C$_5$H$_2$)Mo}$_2$(μ-η$^5$-Sb$_5$)] and [(η$^5$-1,2,4-$t$-Bu$_3$C$_5$H$_2$)Mo(μ-η$^5$-Sb$_5$)Mo(η$^5$-1,4-$t$Bu$_2$-2-MeC$_5$H$_2$)].[126] The molecular structure of the latter complex is depicted in Fig. 26.

The Sb–Sb bond lengths range between 2.759 and 2.850 Å. The Sb$_5$ ring is almost planar, one of the Sb atoms lying 0.2 Å above the plane of the other four Sb atoms.

# VII

# CONCLUDING REMARKS

Knowledge in the field of coordination compounds with antimony ligands at present is still very limited compared to more widely used donors like phosphorus ligands and there is little doubt that some of the ligand systems discussed in this work will not lose their exotic status in the near future. On the other hand it is emerging that some stibine ligands may play a specific role in complexes used for organic syntheses or other purposes. Further studies in this field will certainly lead to new useful applications. One of the fields of increasing importance for antimony ligand systems is the study of precursors for semiconducting materials where the MOCVD techniques are involved. With respect to fundamental research an impetus for the study of antimony compounds results from the progress of analytical techniques. Especially X-ray diffractometry will continue to be a very useful tool for the study of antimony ligands, including labile systems which are protected from rapid decomposition only in the sphere of complex.

ACKNOWLEDGEMENTS

We thank Professor Dr. C. Silvestru from the Babes-Bolyai University (Cluj-Napoca, Romania) for helpful discussions and the University of Bremen for financial support.

REFERENCES

(1) McAuliffe, C. A. *Transition Metal Complexes of Phosphorus, Arsenic and Antimony Ligands.* Macmillan Press, London, 1973, pp. 207–255.

(2) McAuliffe, C. A.; Levason, W. *Phosphine, Arsine and Stibine Complexes of the Transition Elements*. Vol. 1, Elsevier, North-Holland, 1979.

(3) Malisch, W.; Hanak, H.; Lorz, P.; Lother, S.; Schemm, R.; Reich, W.; Meyer, A., Krebs, B. *Unkonventionelle Wechselwirkungen in der Chemie metallischer Elemente*. VCH Publishers, Weinheim, 1992. pp. 245–255.

(4) (a) Scherer, O. J. *Angew. Chem. Int. Ed. Engl.* **1990**, *29*, 1104. (b) Scherer, O. J. *Angew. Chem.* **1990**, *102*, 1137.

(5) Ashe III, A. J. *Adv. Organomet. Chem.* **1990**, *30*, 77.

(6) Champness, N. R.; Levason, W. *Coord. Chem. Rev.* **1994**, *133*, 115.

(7) Breunig, H. J.; Rösler, R. *Coord. Chem. Rev.* **1997**, *163*, 33.

(8) Godfrey, S. M.; McAuliffe, C. A.; Mackie, A. G.; Pritchard, R. G., Norman, N. C. *Chemistry of Arsenic, Antimony and Bismuth*. Blackie Academic and Professional, London, 1998. pp. 159–205.

(9) Breunig, H. J.; Rösler, R. *Chem. Soc. Rev.* **2000**, *29*, 403.

(10) Whitmire, K., Norman, N. C. *Chemistry of Arsenic, Antimony and Bismuth*. Blackie Academic and Professional, London, 1998. pp. 345–402.

(11) Whitmire, K. *Adv. Organomet. Chem.* **1998**, *42*, 1.

(12) Jones, C. *Coord. Chem. Rev.* **2001**, *215*, 151.

(13) (a) Schwab, P.; Mahr, N.; Wolf, J.; Werner, H. *Angew. Chem., Int. Ed. Engl.* **1994**, *33*, 97. (b) Schwab, P.Mahr, N.Wolf, J.Werner, H. *Angew. Chem.* **1994**, *106*, 82.

(14) (a) Herber, U.; Weberndörfer, B.; Werner, H. *Angew. Chem., Int. Ed. Engl.* **1999**, *38*, 1609. (b) Herber, U.Weberndörfer, B.Werner, H. *Angew. Chem.* **1999**, *111*, 1707.

(15) Godfrey, S. M.; McAuliffe, C. A.; Pritchard, R. G. *J. Chem. Soc., Chem. Commun.* **1994**, 45.

(16) Liu, Y.; Leong, W. K.; Pomeroy, R. K. *Organometallics* **1998**, *17*, 3387.

(17) Dallmann, K.; Preetz, W. *Z. Anorg. Allg. Chem.* **1998**, *624*, 267.

(18) Cini, R.; Cavaglioni, A.; Tizzi, E. *Polyhedron* **1999**, *18*, 669.

(19) Cavaglioni, A.; Cini, R. *J. Chem. Soc., Dalton Trans.* **1997**, 1149.

(20) Cavaglioni, A.; Cini, R. *Polyhedron* **1997**, *16*, 4045.

(21) Chand, S.; Coll, R. K.; Scott McIndoe, J. *Polyhedron* **1998**, *17*, 507.

(22) Wache, S.; Herrmann, W. A.; Artus, G.; Nuyken, O.; Wolf, D. *J. Organomet. Chem.* **1995**, *491*, 181.

(23) Holmes, N. J.; Levason, W.; Webster, M. *J. Organomet. Chem.* **1998**, *568*, 213.

(24) Mentes, A.; Kemmitt, R. D. W.; Fawcett, J.; Russell, D. R. *J. Organomet. Chem.* **1997**, *528*, 59.

(25) Wendt, O. F.; Elding, L. I. *J. Chem. Soc., Dalton Trans.* **1997**, 4725.

(26) Wendt, O. F.; Scodinu, A.; Elding, L. I. *Inorg. Chim. Acta* **1998**, *277*, 237.

(27) Hill, A. M.; Levason, W.; Webster, M. *Inorg. Chem.* **1996**, *35*, 3428.

(28) Domasevitch, K. V.; Petkova, E. G.; Nazarenko, A. Y.; Ponamareva, V. V.; Sieler, J.; Dalley, N. K.; Rusanov, E. B. *Z. Naturforsch.* **1999**, *54b*, 904.

(29) Bowmaker, G.A.; Effendy; Reid, J.C; Rickard, C.E.F.; Skelton, B.W. *J. Chem. Soc. Dalton Trans.* **1998**, *0*, 2139.

(30) Effendy; Kildea, J.D.; White, A.H. *Aust. J. Chem.* **1997**, *50*, 587.

(31) Bowmaker, G.A.; Effendy; Hart, R.D.; Kildea, J.D.; de Silva, E.N.; Skelton, B.W.; White, A.H. *Aust. J. Chem.* **1997**, *50*, 539.

(32) Park, Y.-W.; Kim, J.; Do, Y. *Inorg. Chem.* **1994**, *33*, 1.

(33) Black, J. R.; Levason, W.; Spicer, M. D.; Webster, M. J. *J. Chem. Soc., Dalton Trans.* **1993**, 3129.

(34) Bowmaker, G. A.; Hart, R. D.; de Silva, E. N.; Skelton, B. W.; White, A. H. *Aust. J. Chem.* **1997**, *50*, 621.

(35) Bowmaker, G. A.; Hart, R. D.; White, A. H. *Aust. J. Chem.* **1997**, *50*, 567.

(36) Bowmaker, G.A.; Effendy; de Silva, E. N.; White, A.H. *Aust. J. Chem.* **1997**, *50*, 641.

(37) Effendy; Kildea, J.D.; White, A.H. *Aust. J. Chem.* **1997**, *50*, 671.

(38) Werner, H.; Grünwald, C.; Steinert, P.; Gevert, O.; Wolf, J. *J. Organomet. Chem.* **1998**, *565*, 231.

(39) Braun, T.; Laubender, M.; Gevert, O.; Werner, H. *Chem. Ber. Recueil* **1997**, *130*, 559.

(40) Werner, H.; Grünwald, C.; Laubender, M.; Gevert, O. *Chem. Ber.* **1996**, *129*, 1191.

(41) Grünwald, C.; Laubender, M.; Wolf, J.; Werner, H. *J. Chem. Soc., Dalton Trans.* **1998**, 833.

(42) Werner, H.; Schwab, P.; Bleuel, E.; Mahr, N.; Steinert, P.; Wolf, J. *Chem. Eur. J.* **1997**, *3*, 1375.

(43) Werner, H.; Schwab, P.; Heinemann, A.; Steinert, P. *J. Organomet. Chem.* **1995**, *496*, 207.

(44) Werner, H.; Heinemann, A.; Windmüller, B.; Steinert, P. *Chem. Ber.* **1996**, *129*, 903.

(45) Werner, H.; Ortmann, D. A.; Gevert, O. *Chem. Ber.* **1996**, *129*, 411.

(46) Holmes, N. J.; Levason, W.; Webster, M. *J. Chem. Soc., Dalton Trans.* **1997**, 4223.

(47) Breunig, H. J.; Denker, M.; Ebert, K. H. *J. Chem. Soc., Chem. Commun.* **1994**, 875.

(48) Breunig, H. J.; Denker, M.; Schulz, R. E.; Lork, E. *Z. Anorg. Allg. Chem.* **1998**, *624*, 81.

(49) Althaus, H.; Breunig, H. J.; Lork, E. *J. Chem. Soc., Chem. Commun.* **1999**, 1971.

(50) Breunig, H. J.; Jönsson, M.; Rösler, R.; Lork, E. *J. Organomet. Chem.* **2000**, *608*, 60.

(51) Lube, M. S.; Wells, R. L.; White, P. S. *J. Chem. Soc., Dalton Trans.* **1997**, 285.

(52) Schulz, S.; Nieger, M. *Organometallics* **1999**, *18*, 315.

(53) Baldwin, R. A.; Foos, E. E.; Wells, R. L.; White, P. S.; Rheingold, A. L.; Yap, G. P. A. *Organometallics* **1996**, *15*, 5035.

(54) Wells, R. L.; Foos, E. E.; White, P. S.; Rheingold, A. L.; Liable-Sands, L. M. *Organometallics* **1997**, *16*, 4771.

(55) Schulz, S. *Coord. Chem. Rev.* **2001**, *215*, 1.

(56) Vela, J.; Sharma, P.; Cabrera, A.; Alvarez, C.; Rosas, N.; Hernandez, S.; Toscano, A. *J. Organomet. Chem.* **2001**, *634*, 5.

(57) Breunig, H. J.; Fichtner, W. *Z. Anorg. Allg. Chem.* **1981**, *477*, 119.

(58) Herberhold, M.; Schamel, K. *Z. Naturforsch.* **1988**, *43b*, 1274.

(59) Graf, N.; Wieber, M. *Z. Anorg. Allg. Chem.* **1993**, *619*, 2061.

(60) Wieber, M.; Graf, N. *Z. Anorg. Allg. Chem.* **1993**, *619*, 1991.

(61) Wieber, M.; Höhl, H.; Burschka, Ch. *Z. Anorg. Allg. Chem.* **1990**, *583*, 113.

(62) Breunig, H. J.; Denker, M.; Ebert, K. H. *J. Organomet. Chem.* **1994**, *470*, 87.

(63) Lang, H.; Huttner, G. *Z. Naturforsch.* **1986**, *41b*, 191.

(64) Seyerl, J.v.; Scheidsteiger, O.; Berke, H.; Huttner, G. *J. Organomet. Chem.* **1986**, *311*, 85.

(65) Kurita, J.; Usuda, F.; Yasuike, S.; Tsuchiya, T.; Tsuda, Y.; Kiuchi, F.; Hosoi, S. *J. Chem. Soc., Chem. Commun.* **2000**, 191.

(66) Breunig, H. J., Patai, S. *The Chemistry of Organic Arsenic, Antimony, and Bismuth Compounds.* Wiley J. & Sons, Chichester, 1994. pp. 441–456.

(67) Haiduc, I.; Edelmann, F. T. Supramolecular Organometallic Chemistry. Weinhem, Wiley-VCH, 1999.

(68) Sharma, P.; Cabrera, A.; Jha, N. K.; Rosas, N.; Le Lagadec, R.; Sharma, M.; Arias, J. L. *Main Group Metal Chemistry* **1997**, *20*, 697.

(69) Seyer, J.v.; Huttner, G. *Cryst. Struct. Commun.* **1980**, *9*, 1099.

(70) Breunig, H. J.; Pawlik, J. *Z. Anorg. Allg. Chem.* **1995**, *621*, 817.

(71) Breunig, H.J.; Ghesner I. *Unpublished results.*

(72) Dickson, R. S.; Heazle, K. D.; Pain, G. N.; Deacon, G. B.; West, O. B.; Fallon, G. D.; Rowe, R. S.; Leech, P. W.; Faith, M. *J. Organomet. Chem.* **1993**, *449*, 131.

(73) Bernal, I.; Korp, J. D.; Calderazzo, F.; Poli, R.; Vitali, D. *J. Chem. Soc., Dalton Trans.* **1984**, 1945.

(74) Sharma, P.; Rosas, N.; Hernandez, S.; Cabrera, A. *J. Chem. Soc., Chem. Commun.* **1995**, 1325.

(75) (a) Breunig, H. J.; Denker, M.; Lork, E. *Angew. Chem., Int. EdEngl.* **1996**, *35*, 1005. (b) Breunig, H. J.Denker, M.Lork, E. *Angew. Chem.* **1996**, *108*, 1081.

(76) Kuczkowski, A.; Schulz, S.; Nieger, M.; Saarenketo, P. *Organometallics* **2001**, *20*, 2000.

(77) Breunig, H. J.; Fichtner, W.; Knobloch, T. P. *Z. Anorg. Allg. Chem.* **1981**, *477*, 126.

(78) (a) Cowley, A. H.; Jones, R. A.; Nunn, C. M.; Westmoreland, D. L. *Angew. Chem., Int. Ed. Engl.* **1989**, *28*, 1018. (b) Cowley, A. H.Jones, R. A.Nunn, C. M.Westmoreland, D. L. *Angew. Chem.* **1989**, *101*, 1089.

(79) Gibbson, M. N.; Sowerby, D. B. *J. Organomet. Chem.* **1998**, *571*, 289.

(80) Mlynek, P. D.; Dahl, L. F. *Organometallics* **1997**, *16*, 1641.

(81) Ashe III, A. J.; Ludwig, E. G.; Oleksyszyn, J.; Huffman, J. C. *Organometallics* **1984**, *3*, 337.

(82) Breunig, H. J.; Stanciu, M.; Rösler, R.; Lork, E. *Z. Anorg. Allg. Chem.* **1998**, *624*, 1965.

(83) Schulz, S.; Nieger, M. *Organometallics* **2000**, *19*, 2640.

(84) Thomas, F.; Schulz, S.; Nieger, M. *Eur. J. Inorg. Chem.* **2001**, 161.

(85) Thomas, F.; Schulz, S.; Nieger, M. *Organometallics* **2001**, *20*, 2405.

(86) Nikonov, G. I.; Kuzmina, L. G.; Howard, J. A. K. *Organometallics* **1997**, *16*, 3723.

(87) Grobe, J.; Golla, W.; Van, D. L.; Krebs, B.; Läge, M. *Organometallics* **1998**, *17*, 5717.

(88) Weber, L.; Mast, C. A.; Scheffer, M. H.; Schumann, H.; Uthmann, S.; Boese, R.; Bläser, D.; Stammler, H.-G.; Stammler, A. *Z. Anorg. Allg. Chem.* **2000**, *626*, 421.

(89) Matsumura, Y.; Harakawa, M.; Okawara, R. *J. Organomet. Chem.* **1974**, *71*, 403.

(90) Cullen, W. R.; Patmore, D. J.; Sams, J. R.; Scott, J. C. *Inorg. Chem.* **1974**, *13*, 649.

(91) Lorenz, I.-P.; Schneider, R.; Nöth, H.; Polborn, K.; Breunig, H. J. *Z. Naturforsch.* **2001**, *56b*, 671.

(92) Shieh, M.; Sheu, C.; Fang, L.; Cherng, J.-J.; Jang, L.-F.; Ueng, C.-H.; Peng, S.-M.; Lee, G.-H. *Inorg. Chem.* **1996**, *35*, 5504.

(93) Deuten, K. V.; Rehder, D. *Cryst. Struct. Commun.* **1980**, *9*, 167.

(94) Even, T. E.; Genge, A. R. J.; Hill, A. M.; Holmes, N. J.; Levason, W.; Webster, M. *J. Chem. Soc., Dalton Trans.* **2000**, 655.

(95) Breunig, H. J.; Lork, E.; Rösler, R.; Becker, G.; Mundt, O.; Schwarz, W. *Z. Anorg. Allg. Chem.* **2000**, *626*, 1595.

(96) Hill, A. M.; Levason, W.; Webster, M.; Albers, I. *Organometallics* **1997**, *16*, 5641.

(97) Hill, A. M.; Holmes, N. J.; Genge, A. R. J.; Levason, W.; Webster, M.; Rutschow, S. *J. Chem. Soc., Dalton Trans.* **1998**, 825.

(98) Genge, A. R. J.; Holmes, N. J.; Levason, W.; Webster, M. *Polyhedron* **1999**, *18*, 2673.

(99) Breunig, H. J.; Jönsson, M.; Rösler, R.; Lork, E. *Z. Anorg. Allg. Chem.* **1999**, *625*, 2120.

(100) Manger, M.; Wolf, J.; Laubender, M.; Teichert, M.; Stalke, D.; Werner, H. *Chem. Eur. J.* **1997**, *3*, 1442.

(101) Manger, M.; Gevert, O.; Werner, H. *Chem. Ber. Recueil* **1997**, *130*, 1529.

(102) DesEnfants II, R. E.; Gavney, J. A., Jr.; Hayashi, R. K.; Rae, A. D.; Dahl, L. F.; Bjarnason, A. *J. Organomet. Chem.* **1990**, *383*, 543.

(103) Fukumoto, T.; Matsumura, Y.; Okawara, R. *J. Organomet. Chem.* **1974**, *69*, 437.

(104) Martin, N.; Rees, B.; Mitschler, A. *Acta Crystallogr.* **1982**, *38B*, 6.

(105) Chiffey, A.F; Evans, J.; Levason, W.; Webster, M. *Organometallics* **1995**, *14*, 1522.

(106) Bordner, J.; Andrews, B. C.; Long, G. G. *Cryst. Struct. Commun.* **1974**, *3*, 53.

(107) Schmidt, H. *Liebigs Ann. Chem.* **1920**, *421*, 174.

(108) Wieber, M. *"Gmelin Handbook of Inorganic Chemistry,"* Sb Organoantimony Compounds; Berlin: Springer Verlag; 1981; part 2.

(109) Breunig, H. J.; Ghesner, I.; Lork, E. *Organometallics* **2001**, *20*, 1360.

(110) Ates, M.; Breunig, H. J.; Ebert, K.; Gülec, S.; Kaller, R.; Dräger, M. *Organometallics* **1992**, *11*, 145.
(111) Breunig, H. J.; Ghesner, I. *Unpublished results.*
(112) Jones, R. A.; Whittlesey, B. R. *Organometallics* **1984**, *3*, 469.
(113) Sheldrick, W. S. *Acta Crystallogr.* **1975**, *B31*, 1789.
(114) Sheldrick, W. S. *Chem. Ber.* **1975**, *108*, 2242.
(115) Jewiss, H. C.; Levason, W.; Spicer, M. D.; Webster, M. *Inorg. Chem.* **1987**, *26*, 2102.
(116) Hill, A. M.; Levason, W.; Webster, M. *Inorg. Chim. Acta* **1998**, *271*, 203.
(117) Barton, A. J.; Levason, W.; Reid, G.; Tolhurst, V.-A. *Polyhedron* **2000**, *19*, 235.
(118) Holmes, N. J.; Levason, W.; Webster, M. *J. Chem. Soc., Dalton Trans.* **1998**, 3457.
(119) Arif, A. M.; Cowley, A. H.; Norman, N. C.; Pakulski, M. *Inorg. Chem.* **1986**, *25*, 4836.
(120) Cowley, A. H.; Norman, N. C.; Pakulski, M. *J. Am. Chem. Soc.* **1984**, *106*, 6844.
(121) Weber, U.; Huttner, G.; Scheidsteger, O.; Zsolnai, L. *J. Organomet. Chem.* **1985**, *289*, 357.
(122) (a) Huttner, G.; Weber, U.; Sigwarth, B.; Scheidsteger, O. *Angew. Chem., Int. Ed. Engl.* **1982**, *21*, 215. (b) Huttner, G.Weber, U.Sigwarth, B.Scheidsteger, O. *Angew. Chem.* **1982**, *94*, 210.
(123) Black, S. J.; Hibbs, D. E.; Hursthouse, M. B.; Jones, C.; Steed, J. W. *J. Chem. Soc., Chem. Commun.* **1998**, 2199.
(124) Ellerman, J.; Veit, A. *J. Organomet. Chem.* **1985**, *290*, 307.
(125) Balazs, G.; Breunig, H. J.; Lork, E. *Z. Anorg. Allg. Chem.* **2001**, *627*, 2666.
(126) (a) Breunig, H. J.; Burford, N.; Rösler, R. *Angew. Chem. Int. Ed. Engl.* **2000**, *39*, 4150. (b) Breunig, H. J.Burford, N.Rösler, R. *Angew. Chem.* **2000**, *112*, 4320.

# Ladder Polysilanes

## SOICHIRO KYUSHIN and HIDEYUKI MATSUMOTO

*Department of Nano-Material Systems,*
*Graduate School of Engineering, Gunma University,*
*Kiryu, Gunma 376-8515, Japan*

# I

# INTRODUCTION

## A. *Ladder Polysilanes: Unique Polycyclopolysilanes*

Construction of fused polycyclic frameworks with silicon atoms has attracted much attention.[1] In 1972, West and his co-worker obtained fused polycyclopolysilanes in the coupling reaction of trichloromethylsilane and dichlorodimethylsilane with sodium–potassium alloy.[2] The bicyclo[2.2.1]-heptasilane, bicyclo[2.2.2]octasilane, bicyclo[3.3.1]nonasilane, and bicyclo[4.4.0]decasilane derivatives were assigned by [1]H NMR spectra. The structures of bicyclo[3.3.1]nonasilane[3] and bicyclo[4.4.0]decasilane[4] derivatives were confirmed later by X-ray crystallography. Recently, much effort has been concentrated on the synthesis of highly strained polycyclopolysilane systems. For example, octamethylspiropentasilane was synthesized by Boudjouk and his co-worker and was trapped by $LiAlH_4$, $MeMgBr$, or $PCl_5$.[5] Masamune and co-workers reported the synthesis of bicyclo[1.1.0]-tetrasilane,[6] bicyclo[1.1.1]pentasilane,[7] tricyclo[2.2.0.0$^{2,5}$]hexasilane,[8] and tetracyclo[3.3.0.0$^{2,7}$.0$^{3,6}$]octasilane[8] derivatives. In 1988, Matsumoto, Nagai, and co-workers accomplished the synthesis of octasilacubane.[9] The X-ray structures of octasilacubanes were reported independently by three groups

**ADVANCES IN ORGANOMETALLIC CHEMISTRY**
**VOLUME 49 ISSN 0065-3055/DOI 10.1016/S0065-3055(03)49004-7**

©2003 Elsevier Science (USA)
All rights reserved.

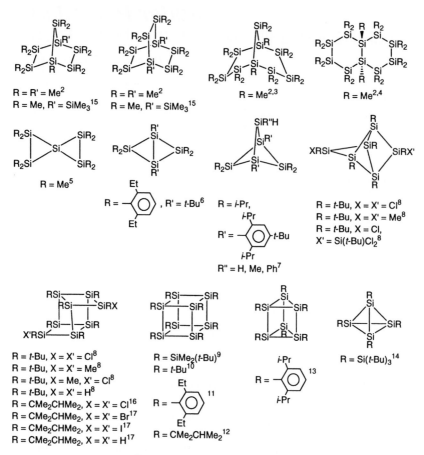

FIG. 1.  Fused polycyclopolysilanes.

in 1992.[10–12] In 1993, the synthesis of hexasilaprismane was reported by Sekiguchi, Sakurai, and co-workers,[13] and the synthesis of tetrasilatetrahedrane was reported by Wiberg and co-workers.[14] The fused polycyclopolysilanes so far reported are summarized in Fig. 1.

In this review, we explain the synthesis, structures, and properties of ladder polysilanes.[18] The ladder polysilanes have the unique structure consisting of catenated and fused cyclotetrasilane rings. This system has remarkable features:

1) The systematic catenation of cyclotetrasilane rings gives rise to a double helix structure of two polysilane main chains.

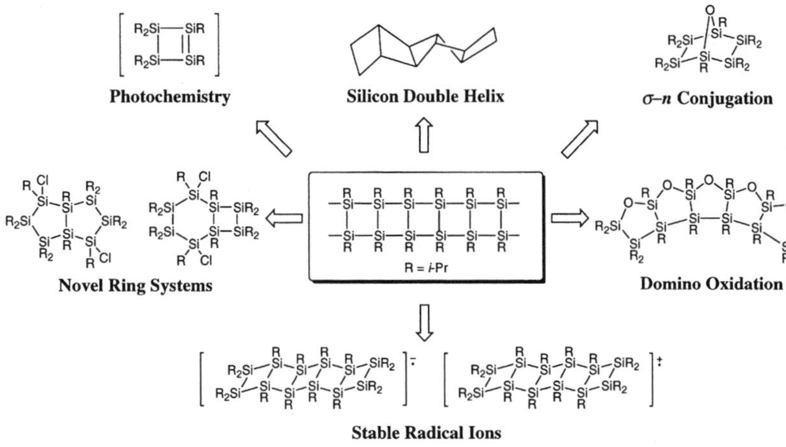

FIG. 2. Chemistry of ladder polysilanes.

2) The ladder polysilanes are highly $\sigma$-conjugated systems, and they are easily oxidized and reduced to give unique oxidation products and persistent radical anions.

3) As the ladder frameworks are highly strained, the Si–Si bonds are easily cleaved by thermolysis, photolysis, and the action of electrophiles and transition metal complexes.

An outline of the chemistry of the ladder polysilanes is illustrated in Fig. 2.

## B. *Ladder Compounds of Other Group 14 Elements*

Before the explanation of ladder polysilanes is started, ladder compounds of other Group 14 elements are briefly mentioned. Hydrocarbons with a ladder-shaped carbon framework are known as ladderanes. The study on ladderanes goes back to 1927, when bicyclo[2.2.0]hexane ([2]ladderane) was synthesized by the reduction of *cis*-1,4-dibromocyclohexane with sodium.[19] Ladderanes so far reported are tricyclo[4.2.0.0$^{2,5}$]octane ([3]ladderane),[20] tetracyclo[4.4.0.0$^{2,5}$0$^{7,10}$]decane[21] ([4]ladderane), and a number of their derivatives (Fig. 3).

The structures of [2]ladderane and [3]ladderane were determined by electron diffraction.[22,23] Each cyclobutane ring of bicyclo[2.2.0]hexane has a folded structure with fold angle 11.5°, and the molecule has $C_2$ symmetry.[22] *Anti*- and *syn*-tricyclo[4.2.0.0$^{2,5}$]octanes also have folded cyclobutane rings with fold angles of 8.0 and 9.0°, respectively.[23] MM2 calculations on

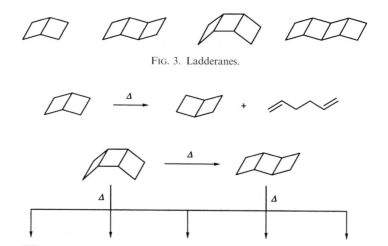

FIG. 3. Ladderanes.

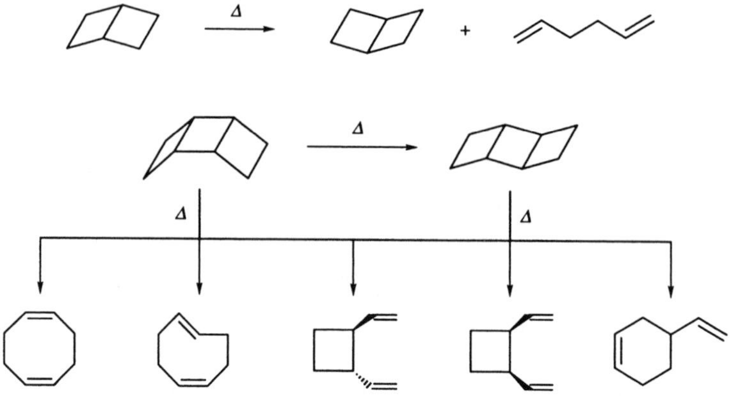

SCHEME 1

bicyclo[2.2.0]hexane show that the $C_2$ structure is more stable by 2.1 kJ mol$^{-1}$ than the $C_{2v}$ structure containing planar cyclobutane rings.[24] Similarly, the twisted $C_2$ structure of [10]ladderane was calculated by the MM2 method to be slightly (by 0.3 kcal mol$^{-1}$) more stabilized than the $C_{2v}$ structure containing planar cyclobutane rings.[25]

The thermolysis of ladderanes has been studied in detail (Scheme 1). On heating, bicyclo[2.2.0]hexane and its derivatives exhibit skeletal inversion and cleavage to 1,5-hexadiene derivatives.[26] The thermolysis of *anti*- and *syn*-tricyclo[4.2.0.0$^{2,5}$]octanes and their derivatives gives *cis,cis*- and *cis, trans*-1,5-cyclooctadienes, *cis*- and *trans*-1,2-divinylcyclobutanes, and 4-vinylcyclohexene as ring-opening products.[27–29] Furthermore, *syn*-tricyclo-[4.2.0.0$^{2,5}$]octane isomerizes to *anti*-tricyclo[4.2.0.0$^{2,5}$]octane.[29c,d] The thermodynamic parameters and the reaction mechanisms for these thermal reactions have been discussed.

Recently, synthetic routes to longer ladderanes with C–C double bonds in the terminal rungs have been improved: the [3]-, [5]-, and [7]ladderane derivatives were synthesized by the repeated cycloaddition of cyclobutadiene derivatives,[30] and the [*n*]ladderane derivatives ($n = 3,4,5,6,7,9$) were synthesized by the successive and alternate cycloaddition with cyclobutadiene and dimethyl acetylenedicarboxylate.[31] X-ray crystallographic analysis of the [5]ladderane derivative[31] shows a corrugated backbone without twist, in contrast with saturated and unsubstituted ladderanes.[22,23]

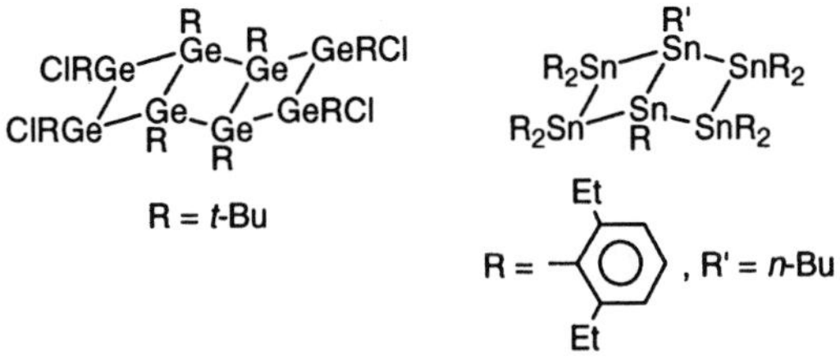

FIG. 4. Ladder polygermane and ladder polystannane.

As the ladderanes have rigid rodlike structures consisting of C–C $\sigma$ bonds, they have been used as spacer for electron-transfer reactions[32,33] and as templates for controlling oligomerization.[34]

A ladder polygermane and a ladder polystannane have also been reported (Fig. 4). 1,2,3,4,5,6,7,8-Octa-*tert*-butyl-3,4,7,8-tetrachlorotricyclo[4.2.0.0$^{2,5}$]-octagermane was isolated in the reaction of *all-trans*-1,2,3,4-tetra-*tert*-butyl-1,2,3,4-tetrachlorocyclotetragermane with sodium, and the structure was determined by X-ray crystallography.[35] The neighboring chlorine atoms have trans configuration. The central cyclotetragermane ring has a planar structure, while the other cyclotetragermane rings are folded with a fold angle of 27°. 1-Butyl-2,2,3,3,4,5,5,6,6-nonakis(2,6-diethylphenyl)bicyclo-[2.2.0]hexastannane was synthesized by the reaction of 2,6-diethylphenyl-lithium with tin(II) chloride.[36] X-ray crystallographic analysis showed that each of the four-membered rings is puckered rather than planar. This compound shows dramatic reversible thermochromic behavior, being pale yellow at $-196\ ^\circ$C and orange-red at room temperature.

## II

## SYNTHESIS

Ladder polysilanes were synthesized for the first time by the co-condensation of Cl($i$-Pr)$_2$SiSi($i$-Pr)$_2$Cl and Cl$_2$($i$-Pr)SiSi($i$-Pr)Cl$_2$ with lithium (Scheme 2).[37] The reaction gave a mixture of ladder polysilanes with various numbers of cyclotetrasilane rings as shown in Fig. 5. The mixture was separated by recycle-type HPLC to give decaisopropylbicyclo[2.2.0]-hexasilane (**1**), *anti*- and *syn*-dodecaisopropyltricyclo[4.2.0.0$^{2,5}$]octasilanes (*anti*-**2** and *syn*-**2**), *anti,anti*- and *anti,syn*-tetradecaisopropyltetracyclo-[4.4.0.0$^{2,5}$.0$^{7,10}$]decasilanes (*anti,anti*-**3** and *anti,syn*-**3**), and *anti,anti,anti*-hexadecaisopropylpentacyclo[6.4.0.0$^{2,7}$.0$^{3,6}$.0$^{9,12}$]dodecasilane (*anti,anti,anti*-**4**). These compounds were obtained as colorless (**1** and *anti*-**2**) and yellow (*syn*-**2**, *anti,anti*-**3**, *anti,syn*-**3**, and *anti,anti,anti*-**4**) crystals which

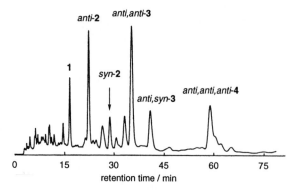

FIG. 5. HPLC analysis of ladder polysilanes (ODS, CH₃OH–THF (7:3), UV detection at 250 nm).

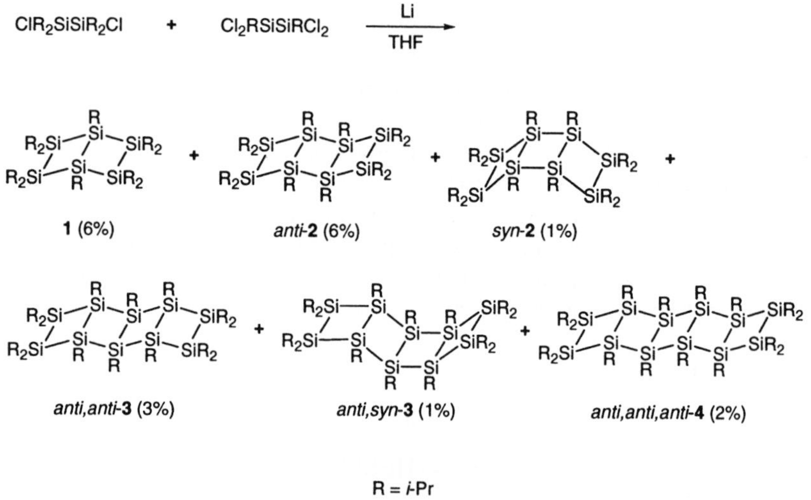

SCHEME 2

are stable in the air and can be handled without protection of an inert atmosphere.

Selective synthetic routes to a specific ladder polysilane have also been developed. *Anti*-1,2,5,6-tetra-*tert*-butyl-3,3,4,4,7,7,8,8-octaisopropyltricyclo[4.2.0.0$^{2,5}$]octasilane (**6**) was synthesized by the coupling reaction of *all-trans*-[(*t*-Bu)ClSi]₄ (**5**) and Cl(*i*-Pr)₂SiSi(*i*-Pr)₂Cl with lithium in 40% yield (Scheme 3).[38] The syn isomer was not formed in this reaction. This reaction is interesting because the all-trans structure of **5** is not reflected in the stereochemistry of the anti (i.e., cis-trans-cis) structure of **6**, indicating

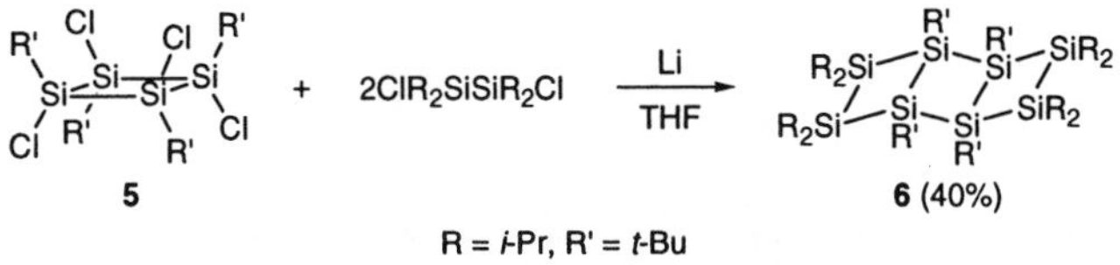

R = i-Pr, R' = t-Bu

SCHEME 3

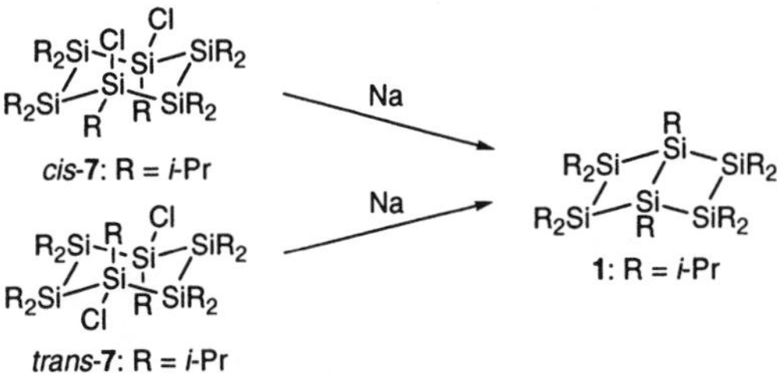

SCHEME 4

that inversion of configuration around the silicon atoms occurred during the course of the reaction.

The transannular coupling reactions of *cis*- and *trans*-1,4-dichloro-1,2,2,3,3,4,5,5,6,6-decaisopropylcyclohexasilanes (*cis*-7, *trans*-7) with sodium are an effective route to the bicyclic ladder polysilane **1** (Scheme 4).[39] The stereochemistry of the starting dichlorocyclohexasilanes does not affect the course of the reactions. These reactions gave **1** quantitatively without producing intermolecular coupling products. These results are in contrast with the coupling reaction of dichlorodecamethylcyclohexasilanes with sodium, in which only intermolecular coupling occurs (Scheme 4).[40]

# III

## STRUCTURE

The structures of ladder polysilanes were determined by X-ray crystallography.[41,42] In Fig. 6, molecular structures of the ladder polysilanes with the anti structure are shown. A remarkable feature of the anti ladder polysilanes is a helical structure. These molecules can be regarded as silicon double helices, in which two polysilane main chains twist in one way and are bridged by Si–Si bonds. The twist angles between the terminal rungs are

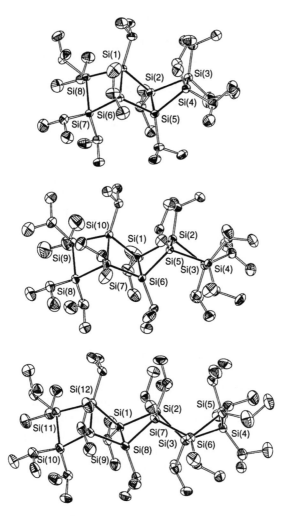

FIG. 6. Molecular structures of *anti*-**2** (top), *anti,anti*-**3** (middle), and *anti,anti,anti*-**4** (bottom). Thermal ellipsoids are drawn at the 30% probability level.

44.0° (*anti*-**2**), 63.0° (*anti,anti*-**3**), and 80.3° (*anti,anti,anti*-**4**). As shown in Fig. 7, left-hand skewed and right-hand skewed molecules of *anti,anti,anti*-**4** are paired in a unit cell since the crystals of *anti,anti,anti*-**4** have a centrosymmetric space group. Therefore, these crystals are racemates of chiral helical molecules. The helical structures of the ladder polysilanes arise from the systematic catenation of folded cyclotetrasilane rings.[43] The folding of catenated cyclotetrasilane rings all upward or all downward along

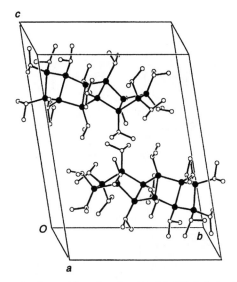

FIG. 7. View of the molecular packing of *anti,anti,anti*-**4**. Filled circles denote silicon atoms.

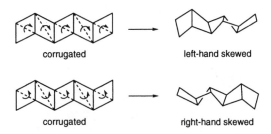

corrugated                           left-hand skewed

corrugated                           right-hand skewed

SCHEME 5

diagonals leads to left-hand skewed or right-hand skewed helical structures (Scheme 5).

The bond lengths of the bridgehead Si–Si bonds are comparable with those of the peripheral Si–Si bonds. The Si–Si bond lengths of **1** (2.385(2)–2.426(2) Å, average 2.400 Å), *anti*-**2** (2.346(3)–2.405(4) Å, average 2.388 Å), *anti,anti*-**3** (2.356(2)–2.408(2) Å, average 2.388 Å), and *anti,anti,anti*-**4** (2.357(4)–2.411(4) Å, average 2.390 Å) are similar. The fold angles of cyclotetrasilane rings are 21.4–21.8° (**1**), 22.6–25.2° (*anti*-**2**), 24.5–27.1° (*anti,anti*-**3**), and 23.8–28.8° (*anti,anti,anti*-**4**).

On the other hand, the ladder polysilanes containing a syn moiety have more strained structures (Fig. 8). The Si–Si bond lengths of *syn*-**2**

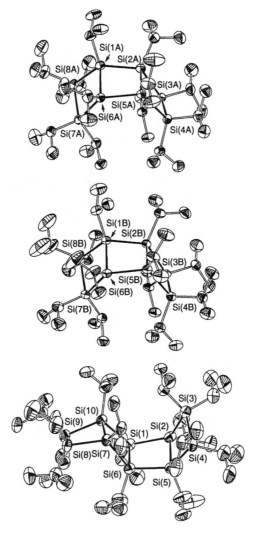

FIG. 8. Molecular structures of *syn*-**2** (top and middle) and *anti,syn*-**3** (bottom). Thermal ellipsoids are drawn at the 30% probability level.

(2.369(4)–2.436(4) Å, average 2.402 Å) and *anti,syn*-**3** (2.361(5)–2.477(4) Å, average 2.398 Å) are longer than those of the corresponding anti isomers. The fold angles of the Si(1)–Si(6)–Si(7)–Si(8), Si(1)–Si(2)–Si(5)–Si(6), and Si(2)–Si(3)–Si(4)–Si(5) rings of *syn*-**2** are 12.7, 18.8, and 14.6° (molecule A) and 12.8, 21.0, and 10.1° (molecule B), respectively. The fold angles of the Si(7)–Si(8)–Si(9)–Si(10), Si(1)–Si(6)–Si(7)–Si(10), Si(1)–Si(2)–Si(5)–Si(6), and Si(2)–Si(3)–Si(4)–Si(5) rings of *anti,syn*-**3** are 22.2, 12.8, 17.9, and 14.8°, respectively. Apparently, the cyclotetrasilane rings facing each other in the syn moiety have relatively planar structures. These structural features of the syn ladder polysilanes seem favorable for reducing the steric repulsion between the facing isopropyl groups.

## IV

### ELECTRONIC PROPERTIES

The highly strained structures of the ladder polysilanes lead to unique electronic properties.[42] Although the ladder polysilanes contain only $\sigma$ bonds, they show relatively strong absorption in the UV–visible region (Fig. 9). In the series of the anti ladder polysilanes, the lowest energy absorption maximum shifts to longer wavelength as the number of cyclotetrasilane rings progressively increases (**1**: 310 nm, *anti*-**2**: 345 nm, *anti,anti*-**3**: 380 nm, *anti,anti,anti*-**4**: 414 nm). A similar bathochromic shift has been reported in permethylpolysilanes as shown in Fig. 10,[44] and the shift is explained by the extension of $\sigma$ conjugation of Si–Si bonds.[45] On the other hand, the absorption of the syn ladder polysilanes extends to the longer wavelength region than that of the corresponding anti isomers: the absorption maxima with the longest wavelength exist at 398 nm (*syn*-**2**) and 400 nm (*anti,syn*-**3**) (Fig. 9). These results are explained by destabilization of the highest occupied molecular orbital (HOMO) of the syn ladder

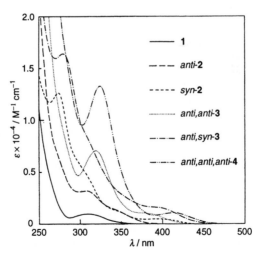

FIG. 9. UV–visible spectra of ladder polysilanes in hexane at room temperature.

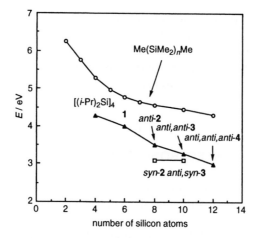

FIG. 10. Lowest transition energies for ladder and linear polysilanes. Values for linear
polysilanes are cited from reference 44.

TABLE I
OXIDATION POTENTIALS OF LADDER POLYSILANES[a]

| Compound | $E_p^{ox}$/V vs. SCE |
|---|---|
| [(*i*-Pr)$_2$Si]$_4$ | 1.24[b] |
| **1** | 0.87 |
| *anti*-**2** | 0.85 |
| *anti,anti*-**3** | 0.82 |
| *anti,anti,anti*-**4** | 0.74 |
| *syn*-**2** | 0.51 |
| *anti,syn*-**3** | 0.51 |

[a]In dichloromethane.
[b]Reference 46.

polysilanes which have more strained polysilane skeletons than those of the
anti isomers.

The ladder polysilanes exhibit a highly electron-donating property. The
oxidation potentials measured by cyclic voltammetry show several
interesting features (Table I). The oxidation potentials of the anti ladder
polysilanes decrease slightly as the number of cyclotetrasilane rings
increases. Since oxidation potentials are mainly related to the energy levels
of the HOMO, the decrease of the oxidation potentials is explained by the
destabilization of the HOMO with extension of $\sigma$-conjugation. In addition,
the syn ladder polysilanes show far lower oxidation potentials than the
corresponding anti isomers. The low oxidation potentials of the syn isomers

are attributable to the further destabilization of the HOMO by the highly strained syn structures. These results are in good accord with the tendency observed in the UV–visible spectra.

From these results, the electronic properties of the ladder polysilanes are found to be highly affected by their stereochemistry.

# V

# REACTIONS

## A. *Oxidation*

Oxidation of Si–Si bonds with peracids is one of the fundamental reactions of polysilanes.[47] The ladder polysilanes have several unequivalent Si–Si bonds, and it seems interesting to study the selectivity of their oxidation positions.[48,49]

The oxidation of **1** with a slightly deficient amount (0.7 equiv.) of *m*-chloroperbenzoic acid (MCPBA) gave the monooxidation products **8** (32%) and **9** (18%) and the dioxidation product **10** (39%) (Scheme 6). The oxidation product which contains an oxygen atom in a terminal Si–Si rung was not formed under these conditions. This result suggests that **9** is quite easily oxidized to **10** because a significant amount of **10** was formed in spite of a deficient amount of MCPBA. When **1** was oxidized with 2 equivalents of MCPBA to transform **9** into **10** completely, **10** and **8** were obtained as the final products.

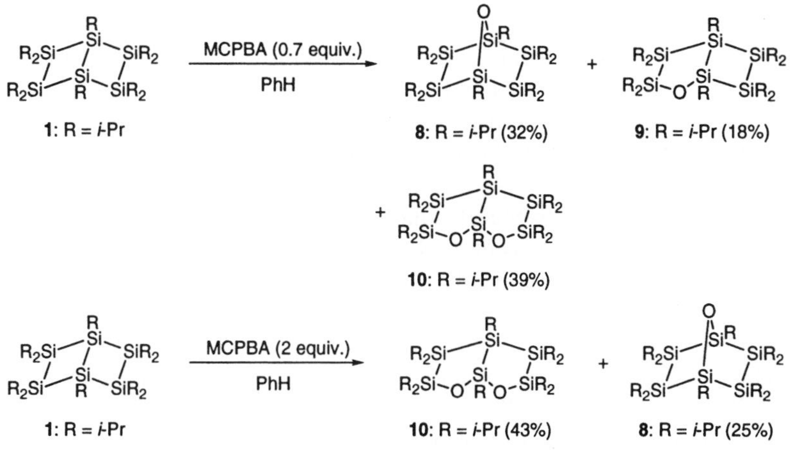

SCHEME 6

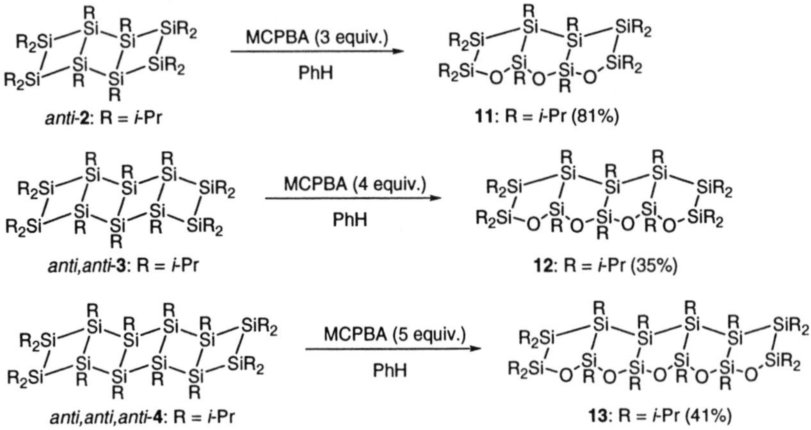

SCHEME 7

When *anti-2* was oxidized with 3 equivalents of MCPBA, the trioxidation product **11** was obtained in 81% yield (Scheme 7). Similarly, the oxidation of *anti,anti-3* and *anti,anti,anti-4* with 4 and 5 equivalents of MCPBA gave the tetraoxidation product **12** and the pentaoxidation product **13**, respectively, in moderate yields. Therefore, these ladder polysilanes were found to be oxidized in a unique manner; one of two polysilane main chains was oxidized selectively, and novel ladder compounds consisting of polysiloxane and polysilane chains were formed.

There are several features in this polyoxidation. Only small or negligible amounts of intermediate oxidation products were detected during the reactions. For example, in the case of *anti-2*, the intermediate mono- and dioxidation products were not detected even in the initial stage of the reaction, but **11** was formed instead. This means that the mono- and dioxidation products are highly activated toward oxidation and are immediately oxidized to **11** after they are formed. This "domino oxidation" is explained by the fact that Si–Si bonds adjacent to oxygen are significantly activated toward oxidation by steric or electronic effects[47h] or by hydrogen bonding of the peracid to the siloxane oxygen which brings the peracid into close proximity to the adjacent Si–Si bonds.[47i] When all the cyclotetrasilane rings were oxidized to five-membered rings, further oxidation did not easily proceed, and the polyoxidation products containing an oxygen atom in each ring were obtained.

The structures of these oxidation products were confirmed by X-ray crystallography (Fig. 11). The oxygen atom of **8** connects two bridgehead silicon atoms with a Si–O bond length of 1.683(4) Å and a Si–O–Si bond angle of 120.2(4)°. The Si–O bond length is somewhat longer than those of

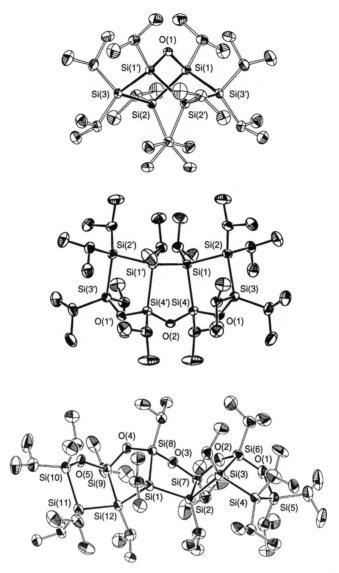

FIG. 11. Molecular structures of **8** (top), **11** (middle), and **13** (bottom). Thermal ellipsoids are drawn at the 30% probability level.

other disiloxanes ([($i$-Pr)$_2$Si]$_4$O: 1.638 and 1.654 Å,[50] Me$_3$SiOSiMe$_3$: 1.626(2) Å,[51] and Ph$_3$SiOSiPh$_3$: 1.616(1) Å[52]). The Si–O–Si bond angle is much smaller than for other disiloxanes ([($i$-Pr)$_2$Si]$_4$O: 133.6°,[50] Me$_3$SiOSiMe$_3$: 148.8(1)°,[51] and Ph$_3$SiOSiPh$_3$: 180°[52]). It is noted that the distance between two bridgehead silicon atoms (2.920 Å) is within the sum of the van der Waals radius[53] and the hard-sphere radius[54] of a silicon atom.

The domino oxidation products have a curved shape containing a polysiloxane chain as the outer arch and a polysilane chain as the inner arch (Fig. 11 (middle)). These molecules retain the anti structure of the starting ladder polysilanes, and each five-membered ring is catenated in a corrugated manner. Furthermore, the corrugated arch is twisted as shown in Fig. 11 (bottom). The twisted structures arise from the systematic catenation of the five-membered rings which have an intermediate structure between the envelope and twist forms. The Si–Si bonds of the inner polysilane chain are significantly long (**11**: 2.398(2)–2.409(2) Å, average 2.405 Å, **12**: 2.404(3)–2.420(3) Å, average 2.413 Å, **13**: 2.389(2)–2.435(3) Å, average 2.414 Å), while the other Si–Si bond lengths are normal (**11**: 2.388(2)–2.390(2) Å, average 2.389 Å, **12**: 2.375(3)–2.395(3) Å, average 2.386 Å, **13**: 2.390(3)–2.410(2) Å, average 2.398 Å). The enlarged steric repulsion among the isopropyl groups of the inner polysilane chain, which is caused by the introduction of oxygen atoms to the opposite chain, seems to be the origin of the elongation of the Si–Si bonds.

The oxidation products show interesting optical properties. In Fig. 12, the UV spectrum of **8** is shown together with those of **1** and **9** for comparison. Compound **8** shows new absorption bands at ca. 270–340 nm. The lowest energy absorption maximum (313 nm) lies in almost the same position as that of **1** (310 nm), and the molecular extinction coefficient ($\varepsilon$ 5100) is fairly large. These results are quite remarkable in light of prior reports that insertion of an oxygen atom to catenated silicon atoms interrupts $\sigma$ conjugation and results in the hypsochromic shift in the

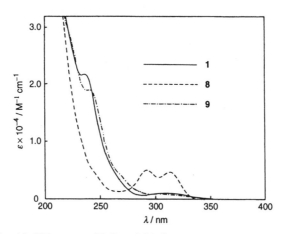

FIG. 12.  UV spectra of **1**, **8**, and **9** in hexane at room temperature.

absorption spectrum.[55] On the other hand, the UV spectrum of **9** does not show such unique absorption and resembles that of **1**, especially in the region of 290–350 nm. Therefore, the unique absorption bands of **8** seem to be due to the stereoelectronic effect of the oxygen atom at the 7-position. In order to explain such an effect, we carried out *ab initio* calculations (STO-3G). Although the lobes of the HOMO of **1** are preferentially localized in the central Si–Si bond, the lobes of the HOMO and the next HOMO of **8** are delocalized in both the *n* orbitals of the oxygen atom and the peripheral Si–Si σ orbitals. Especially in the HOMO, the *n* orbital of the oxygen atom is perpendicularly oriented to the Si–O–Si plane. This result shows that the *n* orbital of the oxygen atom interacts with the Si–Si σ orbitals, and the novel σ–*n* conjugation may be the origin of the new absorption bands of **8**.

Compound **8** shows relatively intense fluorescence in the region of 330–550 nm ($\lambda_{max} = 373$ nm) as shown in Fig. 13. The fluorescence quantum yield of **8** ($\phi_f = 0.014$) is far larger than that of **1**, while **9** shows relatively weak fluorescence ($\phi_f = 1.0 \times 10^{-3}$). The intense fluorescence of **8** corresponds to the relatively large extinction coefficient in the UV spectrum. The Stokes shift in the fluorescence of **8** ($5100\,\text{cm}^{-1}$) is not as large as that of **1** ($12,400\,\text{cm}^{-1}$). The difference in the Stokes shifts is explained by the difference between the structures of the ground state and the excited singlet state. The norbornane skeleton of **8** seems rigid, and the structural change by excitation may be relatively small. On the other hand, the structure of the excited singlet state of **1** is assumed to be significantly changed because the most strained bridgehead Si–Si bond becomes weakened by the electron transition from the HOMO.

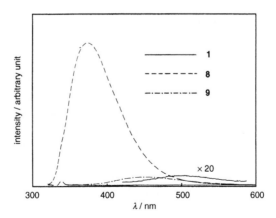

FIG. 13. Fluorescence spectra of **1**, **8**, and **9** in hexane at room temperature. The excitation wavelength is 310 nm.

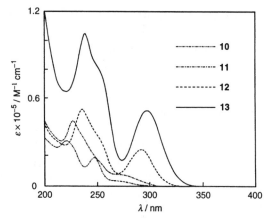

FIG. 14. UV spectra of **10–13** in hexane at room temperature.

In Fig. 14, the UV spectra of the domino oxidation products are shown. As the number of oxatetrasilacyclopentane rings increases, the lowest energy absorption maximum shifts bathochromically, and the molecular extinction coefficient becomes far larger (**10**: 270 nm ($\varepsilon$ 3200), **11**: 273 nm ($\varepsilon$ 7600), **12**: 292 nm ($\varepsilon$ 25000), **13**: 297 nm ($\varepsilon$ 51900). The intense absorption of the order $10^4$ is remarkable because these molecules contain no obvious chromophores which should give such intense absorption. Since the intense absorption is not observed in the ladder polysilanes, it is apparently due to the electronic effect of the oxygen atoms on the Si–Si $\sigma$ conjugation systems.

## B. Reduction to Radical Anions

The radical anions of polysilanes have been studied as unique $\sigma$-conjugated radical ion species.[2,4,56,57] Although many radical anions of polysilanes have been reported, most of them are unstable species that can only be observed at low temperatures. Quite recently, we found that the radical anions of longer ladder polysilanes are persistent at room temperature.[58]

The radical anions **1·⁻**, *anti*-**2·⁻**, *anti,anti*-**3·⁻**, and *anti,anti,anti*-**4·⁻** were generated by the reduction of the corresponding ladder polysilanes with potassium (Scheme 8). When the ladder polysilanes were treated with potassium in tetrahydrofuran (THF) at ca. −70 °C, the solutions were immediately colored: **1·⁻**, purple; *anti*-**2·⁻**, brown; *anti,anti*-**3·⁻**, blue; and *anti,anti,anti*-**4·⁻**, black. In the UV–visible–NIR spectra, several absorption bands appeared (Fig. 15). The intense absorption of *anti,anti*-**3·⁻** and *anti,anti,anti*-**4·⁻** in the near-infrared region is noted because it has been

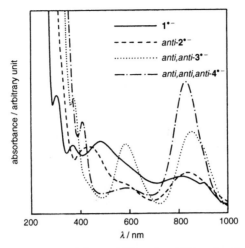

FIG. 15. UV–visible–NIR spectra of $\mathbf{1}^{\bullet-}$, *anti*-$\mathbf{2}^{\bullet-}$, *anti,anti*-$\mathbf{3}^{\bullet-}$, and *anti,anti,anti*-$\mathbf{4}^{\bullet-}$ in THF at $-70\ °C$.

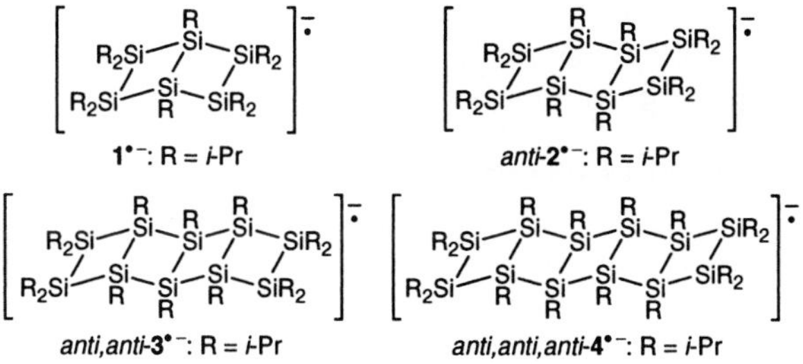

$\mathbf{1}^{\bullet-}$: R = *i*-Pr

*anti*-$\mathbf{2}^{\bullet-}$: R = *i*-Pr

*anti,anti*-$\mathbf{3}^{\bullet-}$: R = *i*-Pr

*anti,anti,anti*-$\mathbf{4}^{\bullet-}$: R = *i*-Pr

SCHEME 8

reported that the absorption of the radical anions of cyclopolysilanes usually falls in the visible region and does not extend into the near-infrared region (e.g., $[(t\text{-Bu})\text{-MeSi}]_4^{\bullet-}$, $\lambda_{max}$ 410 nm[57a]; $(\text{Me}_2\text{Si})_5^{\bullet-}$, $\lambda_{max}$ 645 nm[57b]; $(\text{Me}_2\text{Si})_6^{\bullet-}$, $\lambda_{max}$ 425 nm[57b]). The absorption in the near-infrared region is probably due to the closely stacked molecular orbitals of the highly conjugated polysilane systems[59].

The ESR spectra of the radical anions of the ladder polysilanes show a relatively broad signal with satellites (Fig. 16). From the intensity, the satellites are attributed to the $^{13}\text{C}$ nuclei at the $\alpha$-positions of the isopropyl groups rather than the $^{29}\text{Si}$ nuclei. The number of spin couplings is equal to the number of equivalent carbon atoms at the $\alpha$-positions, indicating that the spin is highly delocalized. These results resemble those of the radical anions of other cyclopolysilanes observed by ESR[57] and can be explained by

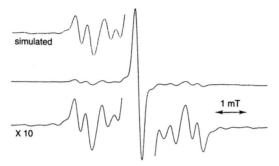

FIG. 16. ESR spectrum of *anti,anti,anti*-**4**·⁻ in THF at room temperature with the simulated satellite spectrum.

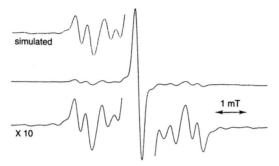

SCHEME 9

the model in which the spin is delocalized to the substituent carbon atoms by the interaction between the Si–Si σ* and Si–C σ* orbitals in the singly occupied molecular orbital (SOMO).[57j]

The structures of the radical anions were confirmed by the following experiment (Scheme 9). The reduction of the ladder polysilanes was monitored by UV–visible–NIR spectroscopy. When the absorption of the ladder polysilanes was completely replaced by the absorption of the radical anions, the sealed tube was opened. The radical anions were immediately oxidized, and the starting ladder polysilanes were recovered in high isolated yields. It is reasonable to conclude that the radical anions of the ladder polysilanes retain the ladder structure, and the Si–Si bond cleavage or skeletal rearrangement does not occur.

A remarkable feature of the radical anions is their stability. When the temperature of **1**·⁻ generated at −70 °C was raised above −10 °C, the ESR signals gradually decreased (Fig. 17 (left)). However, the radical anions become more stable as the number of cyclotetrasilane rings progressively increases. In the case of *anti,anti,anti*-**4**·⁻, the ESR signals do not decrease at

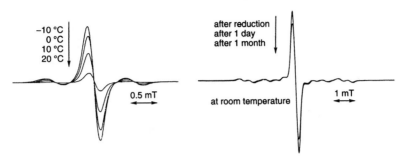

FIG. 17. Decay of the ESR signals of $1^{\bullet-}$ (left) and *anti,anti,anti*-$4^{\bullet-}$ (right) in THF.

all when the temperature is raised to room temperature. Furthermore, the ESR signals persist for several months at room temperature (Fig. 17 (right)). The intensity loss after one month is 19%, and the half-life is too long to be measured. This stability is striking because almost all the radical anions of cyclopolysilanes have been reported to be stable only below ca. −50 °C[2,4,57] except for $[(t\text{-Bu})\text{MeSi}]_4^{\bullet-}$, which persists for several days at room temperature.[57g] The unusual stability of *anti,anti,anti*-$4^{\bullet-}$ is explained by the highly delocalized spin mentioned above.

Another notable point of *anti,anti,anti*-$4^{\bullet-}$ is its generation with a weaker reducing agent. Compound *anti,anti,anti*-4 can be reduced even with lithium in THF at room temperature to produce *anti,anti,anti*-$4^{\bullet-}$, although the reduction is far slower (ca. one day) than that with potassium. The radical anion *anti,anti,anti*-$4^{\bullet-}$ shows UV–visible–NIR and ESR spectra identical to those in Fig. 15 and Fig. 16, and is persistent at room temperature. Although many radical anions of polysilanes have been generated by the reduction with potassium or sodium–potassium alloy,[2,4,57] no examples of the reduction with lithium have been reported to our knowledge. These results show that the lowest unoccupied molecular orbital (LUMO) of *anti,anti,anti*-4 is significantly stabilized by the interaction between the Si–Si $\sigma^*$ and Si–C $\sigma^*$ orbitals extended over the molecule.

## C. Ring-Opening Reactions

The Si–Si bonds of the ladder polysilanes are cleaved by electrophiles.[39,60,61] When **1** was allowed to react with an excess amount of $\text{PdCl}_2(\text{PhCN})_2$, *trans*-**7** and *cis*-**7** were obtained in 36 and 15% yields, respectively (Scheme 10).[39] Other chlorinated products were not formed, indicating the bridgehead Si–Si bond was selectively cleaved. The selectivity is rationalized by the coordination of the bridgehead Si–Si bond, in which

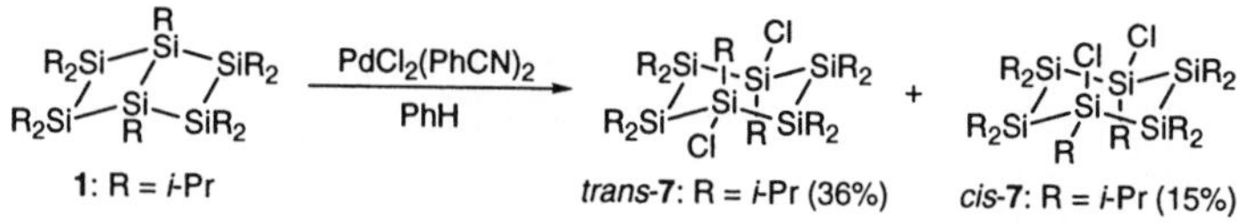

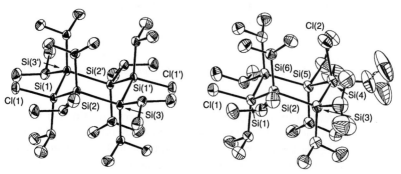

**SCHEME 10**

FIG. 18. Molecular structures of *trans*-7 (left) and *cis*-7 (right). Thermal ellipsoids are drawn at the 30% probability level.

the electrons of the HOMO are localized, to the palladium complex followed by oxidative cleavage.

The stereochemistry of *trans*-7 and *cis*-7 was determined by X-ray crystallography (Fig. 18). The cyclohexasilane rings of both isomers exist in distorted chair conformations. The chlorine atoms of *trans*-7 occupy equatorial positions, and the geminal isopropyl groups occupy axial positions. The 1,3-diaxial interaction among the isopropyl groups seems less important probably due to the long Si–Si bonds. If the isopropyl groups occupied equatorial positions, steric repulsion among the isopropyl group and four vicinal isopropyl groups would be large because of the quadruple gauche interaction. This explanation is supported by MM2 calculations:[62] the conformation of *trans*-7 shown in Fig. 18 (left) is more stable than that with the chlorine atoms at axial positions and the isopropyl groups at equatorial positions. Similar results have also been reported in *all-trans*-1,2,3,4,5,6-hexaisopropylcyclohexane, in which all isopropyl groups occupy axial positions.[63]

The ladder polysilanes can be used as useful probes for obtaining information on the stereochemistry of Si–Si bond cleavage reactions.[60] Watanabe and co-workers[64] and West and co-workers[65] reported that the strained Si–Si bonds of cyclotetrasilanes such as $[(i\text{-Pr})_2Si]_4$ and $(Et_2Si)_4$ are cleaved by hydrogen chloride, hydrochloric acid, and hydrobromic acid. However, to the best of our knowledge, no report has dealt with the stereochemistry of the cleavage reactions. Ladder polysilanes seem to be a suitable system for studying the stereochemistry, while stereochemical information is not easily accessible for linear and monocyclic polysilanes.

The reaction of **1** with hydrobromic acid gave quantitatively the cis adduct **14** (Scheme 11). The reaction of **1** with hydrochloric acid gave the cis adduct **15** in 79% yield and the trans adduct **16** in 18% yield. In these reactions, no other Si–Si bond cleavage products were obtained. The ladder polysilane **1** did not react with hydrofluoric acid. The reactions of 1,4-di-*tert*-butyl-2,2,3,3,5,5,6,6-octaisopropylbicyclo[2.2.0]hexasilane with hydro-bromic acid and hydrochloric acid were attempted, but no reactions took place. This result is ascribed to steric hindrance by the *tert*-butyl groups on the bridgehead silicon atoms.

The stereoselectivity observed in these reactions is explained by the following reaction mechanism (Scheme 12). During the first step of the reaction, the bridgehead Si–Si bond is attacked by hydrogen halide or a hydrogen ion to give the silyl cation intermediate in which the geminal isopropyl group of the hydrogen atom occupies an axial position. The silyl cation is attacked by a halide ion from two directions (paths a and b). In the path a, the halide ion can approach the silyl cation center without significant steric hindrance, while the halide ion suffers significant steric repulsion by two neighboring axial isopropyl groups in the path b. Therefore, the cis adduct **15** is formed preferentially in the reaction with hydrochloric acid. For hydrobromic acid, the steric repulsion in path b is much greater than that of a chloride ion, and only the cis adduct **14** is formed.

The ring-opening reactions of a tricyclic ladder polysilane lead to novel bicyclo[3.3.0]octasilane and bicyclo[4.2.0]octasilane systems.[61] The reaction of *anti*-**2** with a small excess amount of PdCl$_2$(PhCN)$_2$ gave *trans*-**17** and *trans*-**18** in 36 and 25% yields, respectively (Scheme 13). The formation of *trans*-**17** shows that both bridgehead Si–Si bonds of *anti*-**2** were cleaved and rearranged to the bicyclo[3.3.0]octasilane system, while in *trans*-**18**, one of the bridgehead Si–Si bonds of *anti*-**2** was simply cleaved

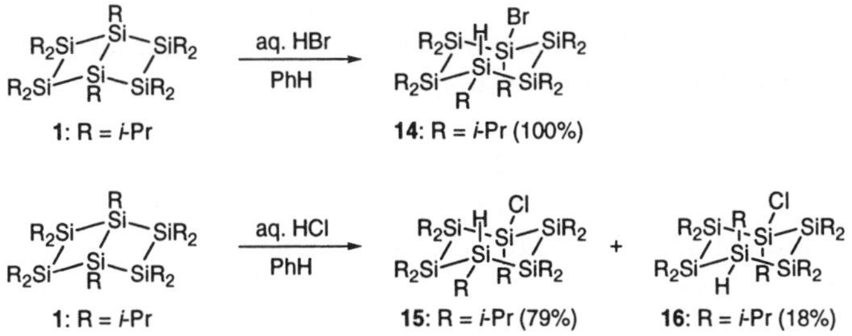

**SCHEME 11**

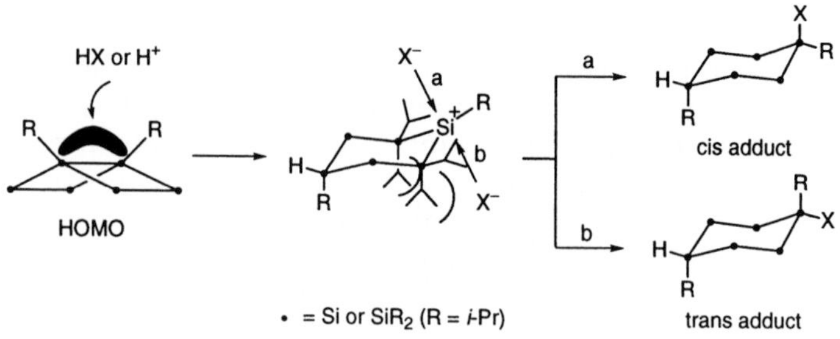

• = Si or SiR₂ (R = *i*-Pr)

Sᴄʜᴇᴍᴇ 12

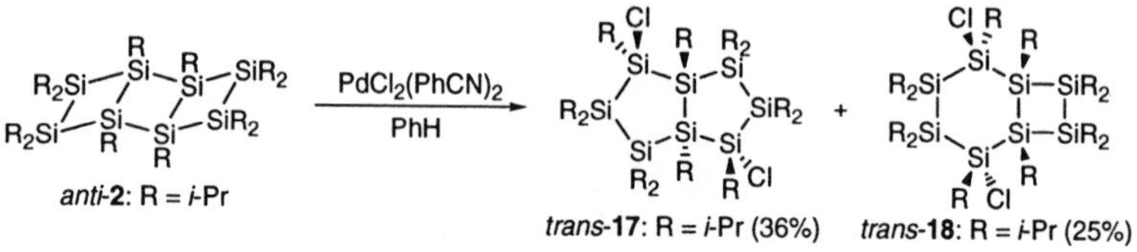

*anti*-2: R = *i*-Pr

*trans*-17: R = *i*-Pr (36%)    *trans*-18: R = *i*-Pr (25%)

Sᴄʜᴇᴍᴇ 13

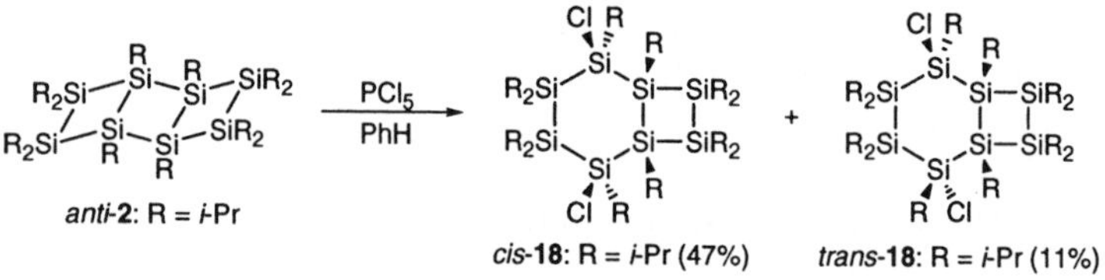

*anti*-2: R = *i*-Pr

*cis*-18: R = *i*-Pr (47%)    *trans*-18: R = *i*-Pr (11%)

Sᴄʜᴇᴍᴇ 14

to form the bicyclo[4.2.0]octasilane system. Similar rearrangement has been reported in the ring-opening reactions of octakis(1,1,2-trimethylpropyl)-octasilacubane.[16,17] The formation of other ring-opening products was negligible, indicating the high reactivity of the bridgehead Si–Si bonds of *anti*-2.

The ring-opening reaction of *anti*-2 also proceeded using PCl₅ to give *cis*-18 and *trans*-18 in 47 and 11% yields, respectively (Scheme 14). In this reaction, rearrangement products such as *trans*-17 were not formed, and *cis*-18 was produced in preference to *trans*-18 in contrast with the reaction using PdCl₂(PhCN)₂. These results seem to reflect different mechanisms in the ring-opening reactions of *anti*-2 with PdCl₂(PhCN)₂ and PCl₅, although detailed mechanisms cannot be clearly explained at present.

Conformational analysis of the novel bicyclo[3.3.0]octasilane and bicyclo[4.2.0]octasilane systems was carried out by X-ray crystallography (Figs. 19–21). The bicyclo[3.3.0]octasilane framework of *trans*-17 has a trans-fused structure of cyclopentasilane rings. The trans-fused structure is

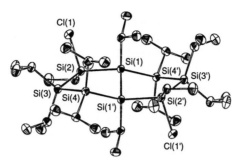

FIG. 19. Molecular structure of *trans*-**17**. Thermal ellipsoids are drawn at the 30% probability level.

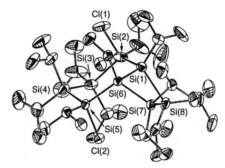

FIG. 20. Molecular structure of *trans*-**18**. Thermal ellipsoids are drawn at the 30% probability level.

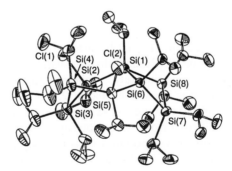

FIG. 21. Molecular structure of *cis*-**18**. Thermal ellipsoids are drawn at the 30% probability level.

notable because in bicyclo[3.3.0]octane, the cis-fused structure has been reported to be more stable than the trans-fused structure.[24,66] In fact, MM2 calculations show that the cis-fused isomer of *trans*-**17** is more stable than *trans*-**17**. Therefore, the ring-opening reaction is considered to be kinetically controlled. The cyclopentasilane rings have an intermediate conformation between the envelope and twist forms. The peripheral Si–Si bonds (2.422(2)–2.433(2) Å, average 2.428 Å) are longer than the bridgehead Si–Si bond (2.373(2) Å), probably due to the steric hindrance of the isopropyl groups.

The cyclohexasilane ring of *trans*-**18** has a chair form and both chlorine atoms occupy axial positions. The cyclotetrasilane ring has a folded structure with the fold angles of 33.0 and 33.6°. The structure of the silicon framework of *trans*-**18** resembles that of bicyclo[4.2.0]octane, in which the cyclohexane ring has a chair form and the cyclobutane ring has a folded structure.[67]

The X-ray analysis of *cis*-**18** showed a unique structure which is quite different from that of *trans*-**18**. The cyclohexasilane ring does not adopt a well-known conformation such as chair, boat, twist-boat, and half-chair forms. In this cyclohexasilane ring, the Si(1)–Si(6)–Si(5)–Si(4) atoms construct a partial boat form, and the Si(1)–Si(2)–Si(3)–Si(4) atoms construct a partial twist-boat form. Therefore, the cyclohexasilane ring can be regarded as the half-twist-boat form. The cyclotetrasilane ring has a folded structure with the fold angles of 32.5 and 33.8°. The Si–Si bond lengths (2.390(6)–2.430(4) Å, average 2.414 Å) are longer than those of *trans*-**18** (2.381(3)–2.427(3) Å, average 2.402 Å). Especially, the bridgehead Si–Si bond (2.427(5) Å) is far longer than that of *trans*-**18** (2.388(2) Å). These structural features reveal that *cis*-**18** is a highly strained molecule compared with *trans*-**18**.

## D. *Thermal Isomerization*

The thermal isomerization of *syn*-**2** has been studied in detail.[68] When a solution of *syn*-**2** in decahydronaphthalene was heated at 220 °C, it isomerized to *anti*-**2** quantitatively (Scheme 15). When *anti*-**2** in decahydronaphthalene was likewise heated at 220 °C, no isomerization of *anti*-**2** to *syn*-**2** was observed, and *anti*-**2** was recovered. The thermolysis pathway is partially different from that of the corresponding ladderane: *syn*-tricyclo[4.2.0.0²,⁵]octane was reported to give 1,5-cyclooctadienes, *anti*-tricyclo[4.2.0.0²,⁵]octane, and 1,2-divinylcyclobutanes in 51, 41, and 8% yields, respectively.[29c,d]

In order to obtain thermodynamic parameters for the isomerization process, a kinetic study was carried out. The thermal isomerization of *syn*-**2** to *anti*-**2** shows a first-order dependence on the concentration of *syn*-**2** in accord with Eq. (1).

$$\ln (1 + [anti\text{-}\mathbf{2}]/[syn\text{-}\mathbf{2}]) = kt \qquad (1)$$

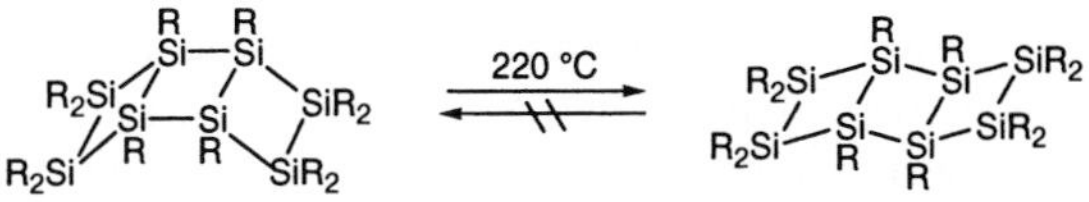

**syn-2**: R = *i*-Pr                                    **anti-2**: R = *i*-Pr

SCHEME 15

TABLE II

THERMODYNAMIC PARAMETERS FOR THERMAL
ISOMERIZATION OF *SYN*-**2** TO *ANTI*-**2**

| | |
|---|---|
| $k/s^{-1}$ | $(3.53 \pm 0.09) \times 10^{-5}$ (200 °C) |
| | $(5.68 \pm 0.18) \times 10^{-5}$ (205 °C) |
| | $(9.32 \pm 0.38) \times 10^{-5}$ (210 °C) |
| | $(1.44 \pm 0.06) \times 10^{-4}$ (215 °C) |
| | $(2.17 \pm 0.09) \times 10^{-4}$ (220 °C) |
| $\Delta G^{\ddagger}/\text{kcal mol}^{-1}$ | $37.77 \pm 0.02$ (200 °C) |
| | $37.73 \pm 0.03$ (205 °C) |
| | $37.66 \pm 0.04$ (210 °C) |
| | $37.63 \pm 0.04$ (215 °C) |
| | $37.63 \pm 0.04$ (220 °C) |
| $E_{\text{a}}/\text{kcal mol}^{-1}$ | $42.3 \pm 0.4$ |
| $\log A/s^{-1}$ | $15.1 \pm 0.2$ |
| $\Delta H^{\ddagger}/\text{kcal mol}^{-1}$ | $41.4 \pm 0.4$ |
| $\Delta S^{\ddagger}/\text{kcal mol}^{-1}\text{K}^{-1}$ | $(7.6 \pm 0.8) \times 10^{-3}$ |

The plot of ln $(1 + [anti$-$2]/[syn$-$2])$ vs. $t$ shows good linearity in the
temperature range of 200–220 °C, and the rate constants are listed in
Table II. The thermodynamic parameters determined by use of the following equations are also listed in Table II.

$$\ln k = -E_{\text{a}}/RT + \ln A \tag{2}$$

$$\ln k/T = -\Delta H^{\ddagger}/RT + \Delta S^{\ddagger}/R + \ln k/h \tag{3}$$

$$\Delta G^{\ddagger} = RT(\ln kT/h - \ln k) \tag{4}$$

In the thermal isomerization of *syn*-**2** to *anti*-**2**, a relatively high
temperature (200–220 °C) is necessary because of the large activation energy
(42.3 kcal mol$^{-1}$). Since the heat of formation of *syn*-**2** calculated by the
MM2 method is larger than that of *anti*-**2** by 13.4 kcal mol$^{-1}$, the activation
energy for the isomerization of *anti*-**2** to *syn*-**2** is estimated to be greater than
55 kcal mol$^{-1}$, so that the isomerization of *anti*-**2** to *syn*-**2** cannot take place
at 220 °C. In contrast, it has been reported that the activation energy for the

thermal isomerization of *syn*-tricyclo[4.2.0.0$^{2,5}$]octane to *anti*-tricy-clo[4.2.0.0$^{2,5}$]octane is relatively low (31.4 kcal mol$^{-1}$), and that isomerization occurs at lower temperatures (117–146 °C).[29d] At this time, the mechanism of the thermal isomerization of *syn*-**2** to *anti*-**2** is not clear. However, if it is assumed to proceed *via* a biradical intermediate, as reported in the isomerization of *syn*-tricyclo[4.2.0.0$^{2,5}$]octane,[29c,d] the activation energy approximately corresponds to the bond dissociation energy. Although the bond dissociation energy of the bridgehead Si–Si bond of *syn*-**2** is much smaller than for usual Si–Si bonds (ca. 70–80 kcal mol$^{-1}$)[69–71] because of ring strain, it is significantly larger than that of the bridgehead C–C bond of *syn*-tricyclo[4.2.0.0$^{2,5}$]octane. The large bond dissociation energy of the bridgehead Si–Si bond of *syn*-**2** may be attributed in part to the reduced ring strain in ladder polysilanes compared with that in ladderanes.[72] The log *A* value for the thermal isomerization of *syn*-**2** to *anti*-**2** (15.1 s$^{-1}$) is slightly larger than that for the thermal isomerization of *syn*-tricyclo[4.2.0.0$^{2,5}$]octane to *anti*-tricyclo[4.2.0.0$^{2,5}$]octane (13.4 s$^{-1}$). The activation entropy (7.6 × 10$^{-3}$ kcal mol$^{-1}$ K$^{-1}$) is rather small in accord with the fact that the thermal isomerization is a unimolecular reaction.

## E. *Photolysis*

The photochemical cleavage of Si–Si bonds of cyclotetrasilanes has been reported to generate several reactive intermediates. For example, Nagai and co-workers reported that silylene and cyclotrisilane are generated during the photolysis of a cyclotetrasilane with a folded structure.[73] Shizuka, Nagai, West, and co-workers reported that the photolysis of planar cyclotetrasilanes gives two molecules of disilene.[74]

The tricyclic ladder polysilanes were found to be good precursors of cyclotetrasilenes.[75,76] Upon irradiation of hexane solutions of *anti*-**2** and **6** in the presence of methanol, 2,3-dimethyl-1,3-butadiene, and anthracene with a high-pressure mercury lamp through a filter, the cyclotetrasilane derivatives **21**–**26** were formed (Scheme 16). In these reactions, the trapping products of the dialkylsilylene and tetraalkyldisilene inter-mediates were not detected. These results indicate that two peripheral Si–Si bonds in the central cyclotetrasilane ring are selectively cleaved to afford two molecules of the cyclotetrasilene intermediates **19** and **20**. It is interesting that silylene is not formed, but cyclotetrasilene is produced from *anti*-**2** and **6**, in which each cyclotetrasilane ring has a folded structure[38].

The site selectivity in the Diels-Alder reactions of **19** and **20** with anthracene is especially noteworthy. The cycloaddition of **19** takes place at

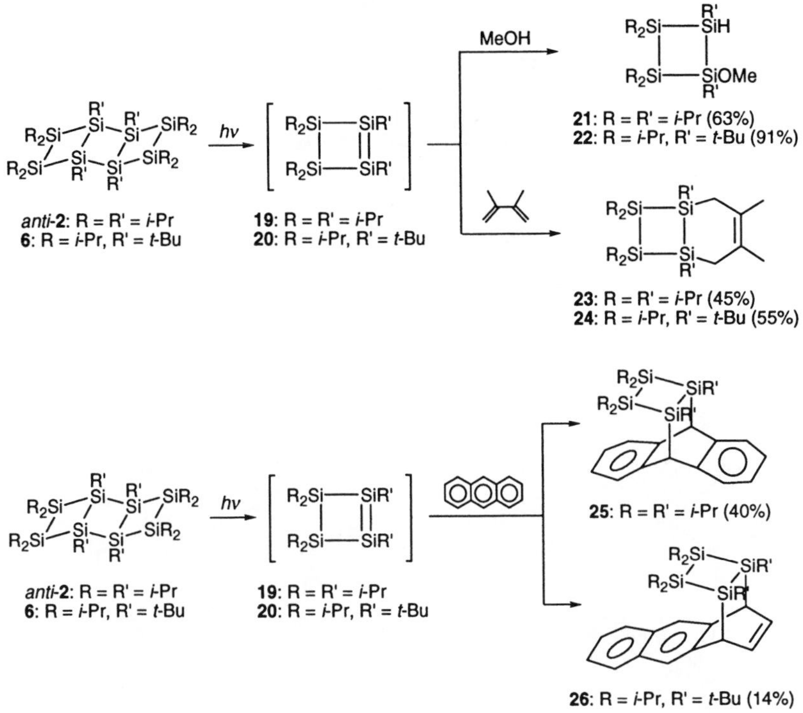

SCHEME 16

the 9,10-positions of anthracene according to the frontier orbital theory.[77] However, in the case of **20**, the cycloaddition at the 9,10-positions seems unfavorable because of the steric hindrance between the *tert*-butyl groups and a benzene ring. Avoiding such steric hindrance, the cycloaddition of **20** took place at the 1,4-positions to give **26**. To our knowledge, this is the first example of a Diels-Alder reaction of anthracene at the 1,4-positions.

The structures of **24–26** were determined by X-ray crystallography (Figs. 22 and 23). Compound **24** has a cis-fused bicyclic structure. The cyclotetrasilane ring has a moderately folded structure with fold angles of 14.0 and 14.3°. The disilacyclohexene ring has a half-boat structure which is a transition state between the two stable half-chair structures of cyclohexene.[78] The bridgehead Si–Si bond is significantly short (2.349(1) Å) compared with other Si–Si bonds (2.382(1)–2.398(1) Å, average 2.393 Å). The 2-butene-1,4-diyl group does not seem long enough to connect the bridgehead silicon atoms without distortion.

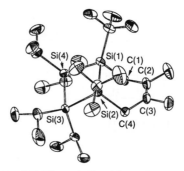

FIG. 22. Molecular structure of **24**. Thermal ellipsoids are drawn at the 30% probability level.

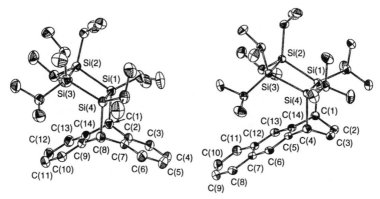

FIG. 23. Molecular structures of **25** (left) and **26** (right). Thermal ellipsoids are drawn at the 30% probability level.

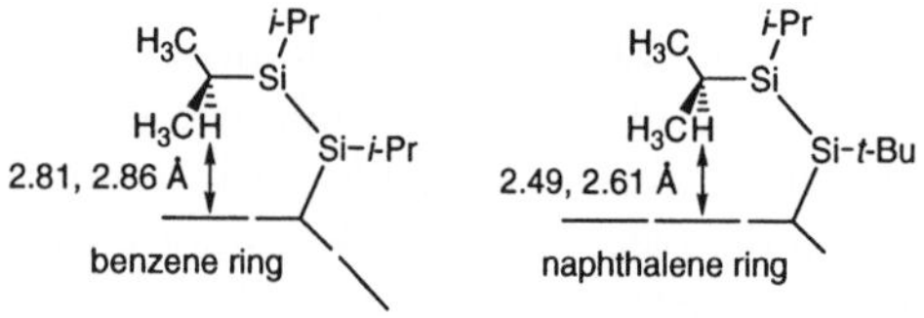

SCHEME 17

The cycloadducts **25** and **26** have unique structures. Two isopropyl groups on the cyclotetrasilane ring are located over the benzene ring or the naphthalene ring. The short distances between the methine protons of the two isopropyl groups and the aromatic rings (**25**: 2.81 and 2.86 Å, **26**: 2.49 and 2.61 Å) cause an upfield shift of the $^1$H NMR signals of the methine protons (**25**: $\delta$ 0.18 ppm, **26**: $\delta$ −0.17 ppm) (Scheme 17).

# VI

# CONCLUSION

Fundamental knowledge on the structures and properties of the ladder polysilanes has accumulated in our research for the past 15 years. Some results were unpredictable, including the silicon double helix structure, the domino oxidation, the formation of persistent radical anions, the Diels-Alder reactions at the 1,4-positions of anthracene, etc. These results let us recognize that the construction of novel structures will open the new chemistry.

In spite of our research explained in this review, many unsolved problems remain. For example, the double helix structure presents the problem of separation of chiral helical ladder polysilanes. The interaction between Si–Si $\sigma$ bonds and oxygen atoms in the oxidation products of the ladder polysilanes shows that it might be necessary to reinvestigate other polysilanes containing oxygen atoms. Furthermore, the properties of the ladder polysilane high polymers have not been fully investigated. Further studies are necessary, and some projects are now in progress.

### ACKNOWLEDGEMENTS

The authors express sincere thanks to the co-workers and colleagues, those names are cited in the references. In addition, financial support to our work from the Ministry of Education, Culture, Sports, Science, and Technology, Japan, Core Research for Evolutional Science and Technology (CREST), and the Toray Science Foundation is gratefully acknowledged.

### REFERENCES

(1) For reviews, see: (a) West, R.; Carberry, E. *Science* **1975**, *189*, 179; (b) West, R. *Pure Appl. Chem.* **1982**, *54*, 1041; (c) Bock, H. *Angew. Chem., Int. Ed. Engl.* **1989**, *28*, 1627; (d) West, R. In *The Chemistry of Organic Silicon Compounds*; Patai, S.; Rappoport, Z.; Eds.; Wiley: Chichester, 1989; Ch. 19; (e) Tsumuraya, T.; Batcheller, S.A.; Masamune, S. *Angew. Chem., Int. Ed. Engl.* **1991**, *30*, 902; (f) Sekiguchi, A.; Sakurai, H. *Adv. Organomet. Chem.* **1995**, *37*, 1; (g) Hengge, E.; Janoschek, R. *Chem. Rev.* **1995**, *95*, 1495.

(2) West, R.; Indriksons, A. *J. Am. Chem. Soc.* **1972**, *94*, 6110.

(3) Stallings, W.; Donohue, J. *Inorg. Chem.* **1976**, *15*, 524.

(4) Jenkner, P. K.; Hengge, E.; Czaputa, R.; Kratky, C. *J. Organomet. Chem.* **1993**, *446*, 83.

(5) Boudjouk, P.; Sooriyakumaran, R. *J. Chem. Soc., Chem. Commun.* **1984**, 777.

(6) (a) Masamune, S.; Kabe, Y.; Collins, S.; Williams, D. J.; Jones, R. *J. Am. Chem. Soc.* **1985**, *107*, 5552; (b) Jones, R.; Williams, D. J.; Kabe, Y.; Masamune, S. *Angew. Chem., Int. Ed. Engl.* **1986**, *25*, 173.

(7) Kabe, Y.; Kawase, T.; Okada, J.; Yamashita, O.; Goto, M.; Masamune, S. *Angew. Chem., Int. Ed. Engl.* **1990**, *29*, 794.

(8)  Kabe, Y.; Kuroda, M.; Honda, Y.; Yamashita, O.; Kawase, T.; Masamune, S. *Angew. Chem., Int. Ed. Engl.* **1988**, *27*, 1725.

(9)  Matsumoto, H.; Higuchi, K.; Hoshino, Y.; Koike, H.; Naoi, Y.; Nagai, Y. *J. Chem. Soc., Chem. Commun.* **1988**, 1083.

(10)  Furukawa, K.; Fujino, M.; Matsumoto, N. *Appl. Phys. Lett.* **1992**, *60*, 2744.

(11)  Sekiguchi, A.; Yatabe, T.; Kamatani, H.; Kabuto, C.; Sakurai, H. *J. Am. Chem. Soc.* **1992**, *114*, 6260.

(12)  Matsumoto, H.; Higuchi, K.; Kyushin, S.; Goto, M. *Angew. Chem., Int. Ed. Engl.* **1992**, *31*, 1354.

(13)  Sekiguchi, A.; Yatabe, T.; Kabuto, C.; Sakurai, H. *J. Am. Chem. Soc.* **1993**, *115*, 5853.

(14)  Wiberg, N.; Finger, C. M. M.; Polborn, K. *Angew. Chem., Int. Ed. Engl.* **1993**, *32*, 1054.

(15)  Ishikawa, M.; Watanabe, M.; Iyoda, J.; Ikeda, H.; Kumada, M. *Organometallics* **1982**, *1*, 317.

(16)  Unno, M.; Higuchi, K.; Ida, M.; Shioyama, H.; Kyushin, S.; Matsumoto, H.; Goto, M. *Organometallics* **1994**, *13*, 4633.

(17)  Unno, M.; Shioyama, H.; Ida, M.; Matsumoto, H. *Organometallics* **1995**, *14*, 4004.

(18)  For previous reviews of ladder polysilanes, see: (a) Kyushin, S.; Matsumoto, H.; Kanemitsu, Y.; Goto, M. *J. Phys. Soc. Jpn.* **1994**, *63*, Suppl. B, 46; (b) Matsumoto, H.; Kyushin, S.; Unno, M.; Tanaka, R. *J. Organomet. Chem.* **2000**, *611*, 52.

(19)  Zelinsky, N. D.; Kozeschkow, K. A. *Ber.* **1927**, *60*, 1102.

(20)  Avram, M.; Dinulescu, I. G.; Marica, E.; Mateescu, G.; Sliam, E.; Nenitzescu, C. D. *Chem. Ber.* **1964**, *97*, 382.

(21)  Srinivasan, R.; Hill, K. A. *J. Am. Chem. Soc.* **1965**, *87*, 4653.

(22)  Andersen, B.; Srinivasan, R. *Acta Chem. Scand.* **1972**, *26*, 3468.

(23)  Andersen, B.; Fernholt, L. *Acta Chem. Scand.* **1970**, *24*, 445.

(24)  Burkert, U.; Allinger, N. L. *Molecular Mechanics*; the American Chemical Society: Washington, D. C., 1982; Ch. 4.

(25)  Miller, M. A.; Schulman, J. M. *J. Mol. Struct. (Theochem)* **1988**, *163*, 133.

(26)  (a) Cremer, S.; Srinivasan, R. *Tetrahedron Lett.* **1960**, 24; (b) Steel, C.; Zand, R.; Hurwitz, P.; Cohen, S. G. *J. Am. Chem. Soc.* **1964**, *86*, 679; (c) Scherer, K. V., Jr. *Tetrahedron Lett.* **1966**, 5685; (d) Srinivasan, R. *Int. J. Chem. Kinet.* **1969**, *1*, 133; (e) van Bekkum, H.; van Rantwijk, F.; van Minnen-Pathuis, G.; Remijnse, J. D.; van Veen, A. *Rec. Trav. Chim. Pays-Bas* **1969**, *88*, 911; (f) Paquette, L. A.; Schwartz, J. A. *J. Am. Chem. Soc.* **1970**, *92*, 3215; (g) Owsley, D. C.; Bloomfield, J. J. *J. Am. Chem. Soc.* **1971**, *93*, 782; (h) Cain, E. N. *Tetrahedron Lett.* **1971**, 1865; (i) Cain, E. N.; Solly, R. K. *Int. J. Chem. Kinet.* **1972**, *4*, 159; (j) Cain, E. N.; Solly, R. K. *J. Am. Chem. Soc.* **1972**, *94*, 3830; (k) Cain, E. N.; Solly, R. K. *Aust. J. Chem.* **1972**, *25*, 1443; (l) Goldstein, M. J.; Benzon, M. S. *J. Am. Chem. Soc.* **1972**, *94*, 5119; (m) Cain, E. N.; Solly, R. K. *J. Am. Chem. Soc.* **1973**, *95*, 4791; (n) Sinnema, A.; van Rantwijk, F.; de Koning, A. J.; van Wijk, A. M.; van Bekkum, H. *J. Chem. Soc., Chem. Commun.* **1973**, 364.

(27)  Vogel, E.; Roos, O.; Disch, K.-H. *Justus Liebigs Ann. Chem.* **1962**, *653*, 55.

(28)  Belluš, D.; Mez, H.-C.; Rihs, G.; Sauter, H. *J. Am. Chem. Soc.* **1974**, *96*, 5007.

(29)  (a) Martin, H.-D.; Eisenmann, E. *Tetrahedron Lett.* **1975**, 661; (b) Martin, H.-D.; Heiser, B.; Kunze, M. *Angew. Chem., Int. Ed. Engl.* **1978**, *17*, 696; (c) Martin, H.-D.; Eisenmann, E.; Kunze, M.; Bonačic-Koutecký, V. *Chem. Ber.* **1980**, *113*, 1153; (d) Walsh, R.; Martin, H.-D.; Kunze, M.; Oftring, A.; Beckhaus, H.-D. *J. Chem. Soc., Perkin Trans. 2* **1981**, 1076.

(30)  Mehta, G.; Viswanath, M. B.; Sastry, G. N.; Jemmis, E. D.; Reddy, D. S. K.; Kunwar, A. C. *Angew. Chem., Int. Ed. Engl.* **1992**, *31*, 1488.

(31)  Warrener, R. N.; Abbenante, G.; Kennard, C. H. L. *J. Am. Chem. Soc.* **1994**, *116*, 3645.

(32) Oevering, H.; Paddon-Row, M. N.; Heppener, M.; Oliver, A. M.; Cotsaris, E.; Verhoeven, J. W.; Hush, N. S. *J. Am. Chem. Soc.* **1987**, *109*, 3258.

(33) Craig, D. C.; Lawson, J. M.; Oliver, A. M.; Paddon-Row, M. N. *J. Chem. Soc., Perkin Trans. 1* **1990**, 3305.

(34) Feldman, K. S.; Bobo, J. S.; Tewalt, G. L. *J. Org. Chem.* **1992**, *57*, 4573.

(35) Sekiguchi, A.; Naito, H.; Kabuto, C.; Sakurai, H. *Nippon Kagaku Kaishi* **1994**, 248.

(36) Sita, L. R.; Bickerstaff, R. D. *J. Am. Chem. Soc.* **1989**, *111*, 3769.

(37) Matsumoto, H.; Miyamoto, H.; Kojima, N.; Nagai, Y. *J. Chem. Soc., Chem. Commun.* **1987**, 1316.

(38) Kyushin, S.; Kawabata, M.; Yagihashi, Y.; Matsumoto, H.; Goto, M. *Chem. Lett.* **1994**, 997.

(39) Kyushin, S.; Yamaguchi, H.; Okayasu, T.; Yagihashi, Y.; Matsumoto, H.; Goto, M. *Chem. Lett.* **1994**, 221.

(40) Kumar, K.; Litt, M. H. *J. Polym. Sci., Polym. Lett. Ed.* **1988**, *26*, 25.

(41) Matsumoto, H.; Miyamoto, H.; Kojima, N.; Nagai, Y.; Goto, M. *Chem. Lett.* **1988**, 629.

(42) Matsumoto, H. et al., to be submitted.

(43) Watanabe, H.; Kato, M.; Okawa, T.; Kougo, Y.; Nagai, Y.; Goto, M. *Appl. Organomet. Chem.* **1987**, *1*, 157.

(44) Gilman, H.; Atwell, W. H.; Schwebke, G. L. *J. Organomet. Chem.* **1964**, *2*, 369.

(45) Bock, H.; Enßlin, W. *Angew. Chem., Int. Ed. Engl.* **1971**, *10*, 404.

(46) Watanabe, H.; Yoshizumi, K.; Muraoka, T.; Kato, M.; Nagai, Y.; Sato, T. *Chem. Lett.* **1985**, 1683.

(47) (a) Sakurai, H.; Imoto, T.; Hayashi, N.; Kumada, M. *J. Am. Chem. Soc.* **1965**, *87*, 4001; (b) Tamao, K.; Kumada, M.; Sugimoto, T. *J. Chem. Soc., Chem. Commun.* **1970**, 285; (c) Tamao, K.; Kumada, M.; Ishikawa, M. *J. Organomet. Chem.* **1971**, *31*, 17; (d) Tamao, K.; Kumada, M. *J. Organomet. Chem.* **1971**, *31*, 35; (e) Nakadaira, Y.; Sakurai, H. *J. Organomet. Chem.* **1973**, *47*, 61; (f) Sakurai, H.; Kamiyama, Y. *J. Am. Chem. Soc.* **1974**, *96*, 6192; (g) Dixon, T. A.; Steele, K. P.; Weber, W. P. *J. Organomet. Chem.* **1982**, *231*, 299; (h) Helmer, B. J.; West, R. *Organometallics* **1982**, *1*, 1463; (i) Alnaimi, I. S.; Weber, W. P. *Organometallics* **1983**, *2*, 903; (j) Weidenbruch, M.; Schäfer, A. *J. Organomet. Chem.* **1984**, *269*, 231; (k) Razuvaev, G. A.; Brevnova, T. N.; Semenov, V. V. *J. Organomet. Chem.* **1984**, *271*, 261.

(48) Kyushin, S.; Sakurai, H.; Yamaguchi, H.; Goto, M.; Matsumoto, H. *Chem. Lett.* **1995**, 815.

(49) Kyushin, S.; Tanaka, R.; Arai, K.; Sakamoto, A.; Matsumoto, H. *Chem. Lett.* **1999**, 1297.

(50) (a) Watanabe, H. et al., unpublished result; (b) Adachi, T., Master Thesis of Gunma University (1993).

(51) Barrow, M. J.; Ebsworth, E. A. V.; Harding, M. M. *Acta Crystallogr.* **1979**, *B35*, 2093.

(52) Glidewell, C.; Liles, D. C. *Acta Crystallogr.* **1978**, *B34*, 124.

(53) Bondi, A. *J. Phys. Chem.* **1964**, *68*, 441.

(54) Glidewell, C. *Inorg. Chim. Acta* **1979**, *36*, 135.

(55) Trefonas, P., III; West, R. *J. Polym. Sci., Polym. Lett. Ed.* **1985**, *23*, 469.

(56) For reviews, see References 1a, 1b, 1d, and 1g.

(57) (a) Husk, G. R.; West, R. *J. Am. Chem. Soc.* **1965**, *87*, 3993; (b) Carberry, E.; West, R.; Glass, G. E. *J. Am. Chem. Soc.* **1969**, *91*, 5446; (c) West, R.; Kean, E. S. *J. Organomet. Chem.* **1975**, *96*, 323; (d) Biernbaum, M.; West, R. *J. Organomet. Chem.* **1977**, *131*, 179; (e) Kira, M.; Bock, H.; Hengge, E. *J. Organomet. Chem.* **1979**, *164*, 277; (f) Buchanan, A. C., III; West, R. *J. Organomet. Chem.* **1979**, *172*, 273; (g) Helmer, B. J.; West, R. *Organometallics* **1982**, *1*, 1458; (h) Katti, A.; Carlson, C. W.; West, R. *J. Organomet. Chem.* **1984**, *271*, 353; (i) Wadsworth, C. L.; West, R.; Nagai, Y.; Watanabe, H.; Muraoka, T. *Organometallics* **1985**,

*4*, 1659; (j) Wadsworth, C. L.; West, R. *Organometallics* **1985**, *4*, 1664; (k) Kirste, B.; West, R.; Kurreck, H. *J. Am. Chem. Soc.* **1985**, *107*, 3013; (l) Wadsworth, C. L.; West, R.; Nagai, Y.; Watanabe, H.; Matsumoto, H.; Muraoka, T. *Chem. Lett.* **1985**, 1525.

(58) Kyushin, S.; Miyajima, Y.; Matsumoto, H. *Chem. Lett.* **2000**, 1420.

(59) (a) Bock, H.; Solouki, B. In *The Chemistry of Organic Silicon Compounds*; Patai, S.; Rappoport, Z.; Eds.; Wiley: Chichester, 1989; Ch. 9; (b) Bock, H.; Solouki, B. *Chem. Rev.* **1995**, *95*, 1161.

(60) Meguro, A.; Sakurai, H.; Kato, K.; Kyushin, S.; Matsumoto, H. *Chem. Lett.* **2001**, 1212.

(61) Kyushin, S.; Sakurai, H.; Yamaguchi, H.; Matsumoto, H. *Chem. Lett.* **1996**, 331.

(62) Frierson, M. R.; Imam, M. R.; Zalkow, V. B.; Allinger, N. L. *J. Org. Chem.* **1988**, *53*, 5248.

(63) Goren, Z.; Biali, S. E. *J. Am. Chem. Soc.* **1990**, *112*, 893.

(64) Watanabe, H.; Muraoka, T.; Kageyama, M.; Nagai, Y. *J. Organomet. Chem.* **1981**, *216*, C45.

(65) Carlson, C. W.; West, R. *Organometallics* **1983**, *2*, 1801.

(66) Chang, S.; McNally, D.; Shary-Tehrany, S.; Hickey, M. J.; Boyd, R. H. *J. Am. Chem. Soc.* **1970**, *92*, 3109.

(67) Spelbos, A.; Mijlhoff, F. C.; Bakker, W. H.; Baden, R.; van den Enden, L. *J. Mol. Struct.* **1977**, *38*, 155.

(68) Kyushin, S.; Yagihashi, Y.; Matsumoto, H. *J. Organomet. Chem.* **1996**, *521*, 413.

(69) Davidson, I. M. T.; Howard, A. V. *J. Chem. Soc., Faraday Trans. 1* **1975**, *71*, 69.

(70) Walsh, R. *Acc. Chem. Res.* **1981**, *14*, 246.

(71) Walsh, R. In *The Chemistry of Organic Silicon Compounds*; Patai, S.; Rappoport, Z.; Eds.; Wiley: Chichester, 1989; Ch. 5.

(72) (a) Nagase, S.; Kudo, T. *J. Chem. Soc., Chem. Commun.* **1988**, 54; (b) Nagase, S.; Kudo, T. *J. Chem. Soc., Chem. Commun.* **1990**, 630.

(73) Watanabe, H.; Kougo, Y.; Nagai, Y. *J. Chem. Soc., Chem. Commun.* **1984**, 66.

(74) Shizuka, H.; Murata, K.; Arai, Y.; Tonokura, K.; Tanaka, H.; Matsumoto, H.; Nagai, Y.; Gillette, G.; West, R. *J. Chem. Soc., Faraday Trans. 1* **1989**, *85*, 2369.

(75) Kyushin, S.; Meguro, A.; Unno, M.; Matsumoto, H. *Chem. Lett.* **2000**, 494.

(76) For stable cyclotetrasilenes, see: (a) Kira, M.; Iwamoto, T.; Kabuto, C. *J. Am. Chem. Soc.* **1996**, *118*, 10303; (b) Wiberg, N.; Auer, H.; Nöth, H.; Knizek, J.; Polborn, K. *Angew. Chem., Int. Ed. Engl.* **1998**, *37*, 2869.

(77) Fukui, K. In *Molecular Orbitals in Chemistry, Physics, and Biology*; Löwdin, P.-O.; Pullman, B.; Eds.; Academic Press: New York, 1964; p. 513.

(78) Kagan, H. B. *La stéréochimie organique*; Presses Universitaires de France: Paris, 1975; Ch. 3.

# Structure–Reactivity Relationships in the Cyclo-Oligomerization of 1,3-Butadiene Catalyzed by Zerovalent Nickel Complexes

## SVEN TOBISCH

*Institut für Anorganische Chemie der Martin-Luther-Universität Halle-Wittenberg, Fachbereich Chemie, Kurt-Mothes-Straße 2, D-06120 Halle, Germany*

ADVANCES IN ORGANOMETALLIC CHEMISTRY
VOLUME 49 ISSN 0065-3055/DOI 10.1016/S0065-3055(03)49005-9

©2003 Elsevier Science (USA)
All rights reserved.

# I

# INTRODUCTION

The catalytic cyclo-oligomerization of 1,3-butadiene mediated by transition-metal complexes is one of the key reactions in homogeneous catalysis.[1] Several transition metal complexes and Ziegler–Natta catalyst systems have been established that actively catalyze the stereoselective cyclo-oligomerization of 1,3-dienes.[2] Nickel complexes, in particular, have been demonstrated to be the most versatile catalysts.[3]

The catalytic cyclo-oligomerization of 1,3-butadiene was first reported by Reed in 1954 using modified Reppe catalysts.[4] Wilke *et al.*, however, demonstrated in pioneering, comprehensive and systematic mechanistic investigations, the implications, versatility and the scope of the nickel-catalyzed 1,3-diene cyclo-oligomerization reactions.[3,5]

Depending on the actual structure of the active catalyst species, nickel complexes catalyze the formation of $C_8$- and $C_{12}$-cyclo-olefins along different reaction channels, that involve the linkage of two and three butadiene moieties, respectively. The zerovalent "ligand-stabilized" [$Ni^0(butadiene)_2L$] complex is known as the active catalyst that predominantly leads to the formation of $C_8$-cyclo-bisolefin products, where the ligand L is typically an alkyl-/arylphosphine $PR_3$ or phosphite $P(OR)_3$, respectively.[6] This type of catalyst cyclodimerizes 1,3-butadiene under moderate reaction cond-itions to a mixture of three principal $C_8$-cyclo-oligomers:[7] namely *cis,cis*-cycloocta-1,5-diene (*cis,cis*-COD), 4-vinylcyclohexene (VCH), and *cis*-1, 2-divinylcyclobutane (*cis*-1,2-DVCB) (cf. Chart 1).[6] The composition of the cyclodimers was shown to be influenced by the properties of the ancillary ligand L, with *cis,cis*-COD being the only $C_8$-cyclo-oligomer that can be formed in a quantitative manner ($C_8$-product selectivity > 95%).[6,8] In the absence of a strongly coordinating ligand L, which can easily be displaced by incoming butadiene (so-called 'bare' nickel complexes), a third butadiene can participate in the cyclo-oligomerization process. For this type of catalyst the zerovalent [$Ni^0(butadiene)_x$] complex represents the catalytically active species, giving rise to $C_{12}$-cyclo-trisolefins as the favored products.[9] Among the several stereoisomers of the 1,5,9-cyclododecatriene (CDT) product of 1,3-butadiene cyclotrimerization (Chart 2), only three of the four possible isomers, namely *all-t*-CDT, *c,c,t*-CDT and *c,t,t*-CDT are formed. *All-t*-CDT

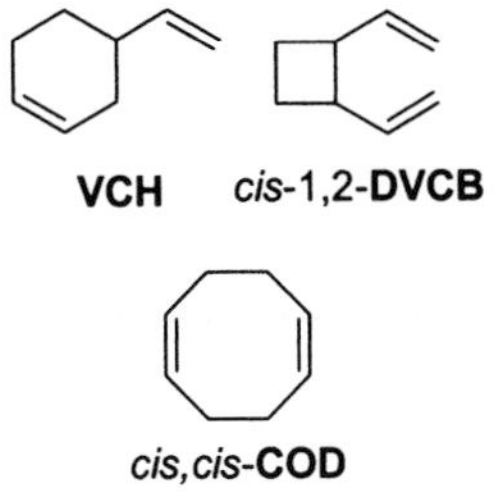

**VCH**    *cis*-1,2-**DVCB**

*cis,cis*-**COD**

**Chart 1**

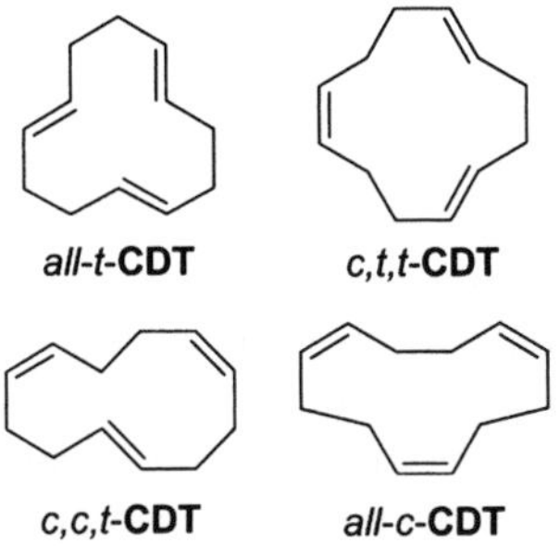

*all-t*-**CDT**    *c,t,t*-**CDT**

*c,c,t*-**CDT**    *all-c*-**CDT**

**Chart 2**

is the major $C_{12}$-cyclo-oligomer ($C_{12}$-product selectivity  >90%), while *all-c*-CDT is not formed in the catalytic cyclo-oligomerization reaction.[9]

From a mechanistic point of view, the nickel-catalyzed cyclo-oligomerization of 1,3-butadiene is probably one of the most thoroughly investigated reactions in the whole of homogeneous transition-metal catalysis. Two fundamental catalytic principles were observed here for the first time: (i) that the electronic and steric properties of the ancillary ligand L decisively regulate the $C_8$-cyclodimer product selectivity (nowadays called "ligand tailoring"), and (ii) that the reactivity of the cyclo-oligomerization process is closely connected to its selectivity. Furthermore, the isolation and characterization of a bis($\eta^3$)-dodecatrienediyl–Ni$^{II}$ species as a reactive intermediate of the $C_{12}$-cyclo-oligomer generating reaction channel,[10] and the detailed examination of individual elementary steps by stoichiometric reactions demonstrated important cornerstones in the development of the mechanistic understanding of homogeneous catalysis.

Although the pioneering, systematic, and comprehensive experimental work of Wilke *et al.*[3,5] has led to a thorough understanding of the nickel-catalyzed cyclo-oligomerization reaction of 1,3-butadiene, there are still some essential mechanistic details that are not yet firmly established (*vide infra*). In the following account, we summarize recent progress in the computational modeling of the nickel-catalyzed cyclo-oligomerization of 1,3-butadiene.[11] On the basis of the original proposal of Wilke *et al.*, we shall present a theoretically well-founded, complete and refined mechanistic view of the cyclo-oligomerization processes as the result of a detailed computational exploration of all the critical elementary steps involved in the whole catalytic cycle for the $C_8$- and $C_{12}$-cyclo-oligomer generating reaction channels. The computational modeling at the atomic level provides novel insights into the structure–reactivity relationships of the nickel-catalyzed cyclo-oligomerization reaction and makes a substantial contribution in rationalizing the decisive factors that regulate the cyclo-oligomer product selectivity.

This article is organized as follows. In Section 2 we outline the known mechanistic details for the $C_8$- and $C_{12}$-cyclo-oligomer generating channels together with the catalytic cycles proposed by Wilke *et al.*, followed by a brief description of computational approach employed and the chosen catalyst models in Section 3. The results of the theoretical exploration of critical elementary steps of the complete catalytic cycles are presented in Section 4. The influence of electronic and steric properties of the ancillary ligand L on the thermodynamic and kinetic aspects of individual steps of the $C_8$-channel is analyzed in Section 5. In Section 6 we propose theoretically well-founded catalytic schemes for the $C_8$- and $C_{12}$-cyclo-oligomer generation channels, followed by elucidation of the regulation of the product selectivity for the two channels, respectively, Section 7, and rationalization of the $C_8$:$C_{12}$-cyclo-oligomer selectivity for the two major types of catalysts, that arise form the subtle interplay between the two channels, Section 8.

## II

## CATALYTIC REACTION CYCLES FOR THE NICKEL-CATALYZED
## CYCLO-OLIGOMERIZATION OF 1,3-BUTADIENE

The nickel-catalyzed cyclo-oligomerization of 1,3-dienes has been unequivocally established to occur in a multistep fashion,[3,12] and not via a one-step concerted suprafacial fusion of two or three 1,3-diene moieties.[13] The nickel atom template undergoes a repeated change in its formal oxidation state; namely $[Ni^0] \rightleftharpoons [Ni^{II}]$, during the multistep addition elimination mechanism. The isolation and characterization of octadienediyl–$Ni^{II}$ and dodecatrienediyl–$Ni^{II}$ species as reactive intermediates of the nickel mediated linkage of two or three butadienes, respectively,[10,14,15] convincingly support the hypothesis that cyclo-oligomerization processes proceed in a multistep

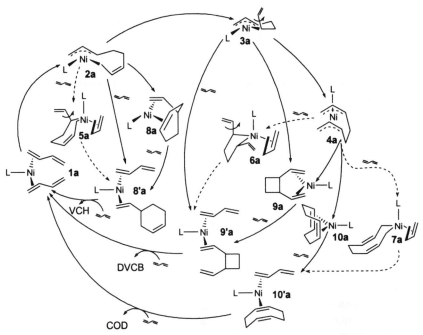

**Scheme 1.** Catalytic cycle of the [Ni⁰L]-catalyzed cyclo-oligomerization of 1,3-butadiene affording $C_8$-cyclo-oligomer products (according to Wilke et al.).[3,14]

fashion, thus clearly eliminating a single-step concerted process as a possibility. Further evidence in favor of a multistep mechanism arises from the observed stereochemistry of the cyclodimers that are formed from substituted 1,3-dienes[16] as well as from deuterium-labeled butadienes.[17]

Scheme 1 displays the general catalytic cycle for the $C_8$-cyclo-oligomer generating reaction channel proposed by Wilke et al.[3,14] Starting from the [Ni⁰(butadiene)$_2$L] active catalyst **1a**, the two-coordinated butadiene moieties undergo oxidative coupling giving rise to the [Ni$^{II}$(octadienediyl)L] complex. The [Ni$^{II}$(octadienediyl)L] complex is the crucial species of the catalytic cycle and exists in several configurations, that are distinguished by the different coordination modes of the octadienediyl framework to nickel, namely, the $\eta^3,\eta^1$ species **2a** and **3a**, the bis($\eta^3$) species **4a**, and the bis($\eta^1$) species **5a**, **6a** and **7a**. The isomerization of the terminal allylic groups of the [Ni$^{II}$(octadienediyl)L] complex represents a further critical elementary process. The cyclodimer products are formed along three reaction routes via reductive elimination under ring closure starting from different [Ni$^{II}$(octadienediyl)L] species. This results in the formation of the [Ni⁰(cyclodimer)L] complexes **8a**, **9a**, and **10a**, respectively, which may be stabilized by coordination of an additional butadiene. Expulsion of the cyclodimers in a subsequent substitution with incoming butadiene, which is supposed to proceed without a significant barrier, regenerates the active catalyst **1a**, thus completing the catalytic cycle. VCH is formed along the reaction route starting from the $\eta^3,\eta^1(C^1)$ species **2a**, while the bis($\eta^3$) species **4a** acts as precursor for the formation of *cis,cis*-COD. Two plausible reaction routes are conceivable for the formation of *cis*-1,2-DVCB, with the $\eta^3,\eta^1(C^3)$ and bis($\eta^3$) species **3a** and **4a**, respectively, representing the precursor.

Two different mechanistic proposals have emerged for the reductive elimination under ring closure, that are distinguished in the suggested coordination mode of the allylic groups in the involved key intermediates. According to the first mechanism,[18] the reductive elimination is believed to proceed via direct paths commencing from the $\eta^3$-$\pi$-octadienediyl–Ni[II] species (indicated by solid lines for the reductive elimination processes in Scheme 1), i.e., along the direct **2a** → **8a** path for the VCH generating route. In the second mechanism,[5,14] bis($\eta^1$-$\sigma$)-octadienediyl–Ni[II] species are suggested as key intermediates (indicated by dashed lines for the reductive elimination processes in Scheme 1), hence VCH should be formed following the **2a** → **5a** → **8′a** path.

The proposed catalytic cycle in Scheme 1 has been decisively supported by the stoichiometric cyclodimerization reaction.[14] All the several octa-dienediyl–Ni[II] forms, **2a–7a**, are supposed to be in a dynamic equilibrium, since there is no experimental evidence for a significant kinetic barrier associated with their mutual interconversion. For butadiene, as well as substituted 1,3-dienes, the $\eta^3$-*syn*,$\eta^1(C^1)$,$\Delta$-*cis* isomer of **2a** is confirmed to be the favorable initial oxidative coupling species that readily rearranges into **4a**, with the bis($\eta^3$-*syn*) isomer being thermodynamically preferred.[14,15] The position of the equilibrium between **2a** and **4a** has been shown to be strongly dependent on the properties of the ancillary ligand L. For strong $\sigma$-donors ($L = PCy_3$, $PPr^i_3$), **2a** was exclusively detected by NMR, whereas **4a** becomes thermodynamically favored for weak $\sigma$-donors ($L = PPh_3$) as well as for strong $\pi$-acceptor ligands ($L = P(OC_6H_4$-$o$-$Ph)_3$).[14] The favorable stereoisomers of **2a** and **4a** (*vide supra*), established by following the stoichiometric process through NMR, makes an interconversion of the configuration of the terminal allylic group in the [Ni[II](octadienediyl)L] complex indispensable, which is indicated to be a facile process. In later stages of the stoichiometric reaction, *cis,cis*-COD is formed via reductive elimination[14] instead of the directly related stereoisomer of COD, namely *trans,trans*-COD. This underlines the fact that allylic isomerization represents an important elementary step in the reaction course. The $\eta^3,\eta^1(C^1)$ species **2a** has been identified as the precursor for the reductive

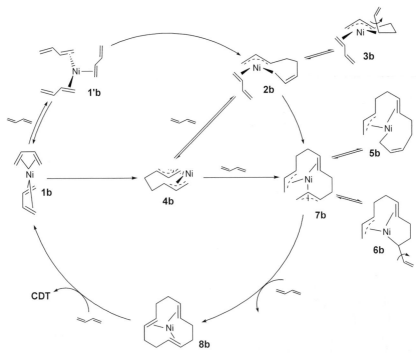

**Scheme 2.** Catalytic cycle of the [Ni⁰]-catalyzed cyclo-oligomerization of 1,3-butadiene affording C$_{12}$-cyclo-oligomer products (according to Wilke et al.).[3,9]

elimination route affording VCH, and indirect evidence has been provided showing that the bis(η³) species **4a** acts as precursor for the formation of *cis*-1,2-DVCB.[14] The reversibility of both the reductive elimination and the oxidative coupling has been demonstrated.[14,15a,19]

The general catalytic cycle proposed by Wilke *et al.* for the C$_{12}$-cyclo-oligomer generating reaction channel is shown in Scheme 2.[3,9] The [Ni⁰(butadiene)$_x$] active catalyst complex can exist in several forms of either the [Ni⁰(butadiene)$_2$] (**1b**) or the [Ni⁰(butadiene)$_3$] (**1′b**) species. These undergoes oxidative coupling of two coordinated butadienes, affording the octadienediyl–Ni$^{II}$ complex, which may be coordinatively saturated by complexation of additional butadiene. Two different forms of the octadi-enediyl–Ni$^{II}$ complex can be conceived to have been formed as the initial coupling product: (i) the η³,η¹(C¹) species **2b**, along a path commencing from the [Ni⁰(butadiene)$_3$] species **1′b**, and (ii) the bis(η³) species **4b** along the **1b** → **4b** path that involves bis(butadiene)–Ni species. The octadiene-diyl–Ni$^{II}$ species **2b** and **4b** are supposedly in an equilibrium with the η³,η¹(C³) species **3b**. Butadiene insertion into the allyl–Ni$^{II}$ bond in either of the octadienediyl–$Ni^{II}$ species leads to the [$Ni^{II}$(dodecatrienediyl)] complex. Similar to the octadienediyl–$Ni^{II}$ complex, the dodecatrienediyl–$Ni^{II}$ complex is present in the $\eta^3,\eta^1$ configuration **5b** and **6b**, and the bis($\eta^3$) configuration **7b** as well. For the sake of clarity, bis($\eta^1$) species of the octadienediyl–$Ni^{II}$ and the dodecatrienediyl–$Ni^{II}$ complex are not included in Scheme 2, although they can serve as possible intermediates for allylic isomerization and reductive elimination.

The [$Ni^0$(CDT)] product complex **8b** is formed via reductive elimination under ring closure starting from the dodecatrienediyl–$Ni^{II}$ complex. The formation of the several isomers of CDT occurs via competing paths for reductive elimination that involves different stereoisomers. Displacement of the cyclotrimer product in subsequent consecutive substitution steps with butadiene, which is supposed to take place without a significant barrier, regenerates the [$Ni^0$(butadiene)$_x$] active catalyst; thus completing the catalytic cycle.

Although the [$Ni^0$(butadiene)$_x$] active catalyst has never been experimentally characterized, formal $16e^-$ or $18e^-$ species of **1b** and **1'b**, respectively, are likely candidates. A [$Ni^0(\eta^4$-$cis$-2,3-dimethylbutadiene)$_2$] complex, with the two $cis$-butadienes coordinated in a tetrahedral manner, is well known[15e,20] and has been shown to yield an $\eta^3,\eta^1(C^1)$-octadienediyl–$Ni^{II}$ complex in the reaction with donor phosphines (e.g., $PCy_3$).[15e] The [$Ni^{II}$(bis($\eta^3$),$\Delta$-dodecatrienediyl)] intermediate has been isolated in the stoichiometric reaction of zerovalent 'bare' nickel complexes with butadiene at $-40\,°C$,[9a,10] convincingly supporting the suggested catalytic cycle in Scheme 2. NMR spectroscopic investigation has confirmed two energetically close lying stereoisomers of the isolated [$Ni^{II}$(bis($\eta^3$-$anti$),$\Delta$-$trans$-dodecatrienediyl)] intermediate, that are distinguished by opposite enantiofaces of the coordinated olefinic double bond.[20,21] A facile allylic isomerization prior to reductive elimination has been observed in the reaction of the dodecatrienediyl–$Ni^{II}$ intermediate with $PMe_3$ at low temperature by NMR. Reductive elimination occurs only at elevated temperatures affording a mixture of all-$t$-CDT and $c,t,t$-CDT.[22] The formation of respective $c,c,t$-CDT (direct product without prior isomerization) is indicated to be kinetically impeded, due to the $trans$ orientation of the terminal carbon of the two $anti$-allylic groups. For reductive elimination along feasible pathways, one or both allylic groups have to undergo prior facile isomerization, thus leading to $c,t,t$-CDT and all-$t$-CDT, respectively. Starting from the dodecatrienediyl–$Ni^{II}$ complex, the formation of the twelve-membered cycle has been demonstrated in stoichiometric reactions to be facilitated by the presence of donor phosphines (i.e., $PMe_3$, $PEt_3$, $PPh_3$) and also by excess butadiene.[9a,23] The formal $16e^-$ [$Ni^0$(CDT)] product, a well established zerovalent nickel complex (in particular, the *all-t*-CDT isomer),[24] is known to form stable adducts with donor ligands.[1a,2b,15e,25]

As one of the greatest accomplishments of Wilke and co-workers, their systematic and comprehensive investigations led to a fundamental under-standing of how the nickel-catalyzed cyclo-oligomerization of 1,3-butadiene operates. However, there are some important and intriguing, but still not firmly resolved, mechanistic aspects that have been the objectives of recent theoretical mechanistic investigations.[11] (i) What are the thermodynamically favorable and the catalytically active forms of the catalyst complexes for the two reaction channels? (ii) In what preferred fashion does the oxidative coupling process take place and which of the several octadienediyl–Ni[II] species is formed as the initial coupling product? (iii) What role does allylic isomerization play in the catalytic reaction course? (iv) Which of the two proposed mechanisms for reductive elimination operates? (v) Which of the elementary steps is rate-determining? (vi) What are the critical factors that regulate the cyclo-oligomer selectivity for the $C_8$- and $C_{12}$-product channels, respectively? (vii) How do the steric and electronic properties of the ancillary ligand L influence the thermodynamic and kinetic aspects of individual steps of the $C_8$-channel and thereby influence the cyclodimer product selectivity, as well as the $C_8:C_{12}$-cyclo-oligomer product ratio in the [Ni$^0$L]-catalyzed cyclo-oligomerization? The following sections review theoretical results aimed at addressing these questions.

# III

# COMPUTATIONAL MODELS AND METHODS

## A. *Models*

The complete catalytic cycles for the $C_8$- and $C_{12}$-cyclo-oligomer generation channels consisting of the crucial elementary steps displayed in Schemes 1 and 2, respectively, have been computationally investigated for the respective [Ni$^0$(butadiene)$_2$L] and [Ni$^0$(butadiene)$_x$] active catalysts. Several conceivable routes for the critical steps of the $C_8$-product channel have first been explored for the generic catalyst (L = PH$_3$) to clarify the most feasible route for each of the elementary processes. The role of electronic and steric effects has further been elucidated for six real [Ni$^0$(butadiene)$_2$L] catalysts. The catalysts chosen contain ligands L with a broad range of electronic and steric properties through L = PMe$_3$, **I**; L = PPh$_3$, **II**; L = PPr$_3^i$, **III**; L = P(OPh)$_3$, **IV**, L = P(OMe)$_3$, **V**; and L = PBu$_3^t$, **VI**.

To rationalize the electronic and steric properties of $PR_3/P(OR)_3$ ligands, several models have been proposed. The cone angle $\theta$ proposed by Tolman[26] is still one of the most popular concepts in coordination chemistry used to quantify the steric demand of the phosphine ligands. To describe the $\sigma$-donor/$\pi$-acceptor ability of phosphine ligands, Tolman introduced the electronic parameter $\chi$, which was based on the values of the $a_1CO$ stretching frequency in $Ni(CO)_3PR_3$ complexes.[26] In recent theoretical investigations[27] it was demonstrated that the energy of the lone-pair at phosphorus, which is $E_{HOMO}$ in free phosphines, correlates with experimental proton affinities; a parameter that has been employed widely to measure $\sigma$-basicity.[28] Furthermore, $E_{LUMO}$ of the free phosphine was found to compare well with the back-donation component in $Fe(CO)_4PR_3$ complexes.[27a] Therefore, the lone-pair energies, $E_{HOMO}$, and $E_{LUMO}$ of free phosphines can serve as a measure of the $\sigma$-donor and $\pi$-acceptor strengths, respectively.[27] This should provide a reasonable ordering scheme of the electronic properties of phosphine ligands, although their actual $\sigma$-donor/$\pi$-acceptor ability decisively depends on the respective metal complex fragment. We shall note here that the correlation found for $E_{HOMO/LUMO}$ holds for a broad range of different ligands, with the only exception observed being $PH_3$. Accordingly, the $\sigma$-donor strength of the actual ligands L investigated here should decrease in the following order, starting with the strongest $\sigma$-donor: $PBu_3^t \sim PPr_3^i > PMe_3 > PPh_3 > P(OMe)_3 \sim P(OPh)_3$. Moreover, the $\pi$-acceptor strength is likely to decrease in the following order: $P(OPh)_3 > P(OMe)_3 > PPh_3 > PMe_3 \sim PPr_3^i \sim PBu_3^t$. According to the proposed cone angles $\theta$ the steric demand decreases in the following order, starting from the most bulky ligand $PBu_3^t > PPr_3^i > PPh_3 > P(OPh)_3 > PMe_3 > P(OMe)_3$.

## B. *Stereoisomers of the Key Species*

The enantioface and also the configuration (*s-trans*, *s-cis*) of the prochiral butadienes involved in the several elementary steps are of crucial importance for the stereocontrol of the cyclo-oligomer formation. Oxidative coupling, for example, can occur between two *cis*-butadienes, two *trans*-butadienes or between *cis*- and *trans*-butadiene with either the same or the opposite enantioface of the two butadienes involved. The several stereoisomers are exemplified for the $[Ni^0(butadiene)_2L]$ active catalysts for cyclodimer formation, that are schematically depicted in Fig. 1, together with the related stereoisomers of the $\eta^3,\eta^1(C^1)$ and bis($\eta^3$) octadienediyl–$Ni^{II}$ species **2a** and **4a**, respectively. For each of the individual elementary steps there are several stereochemical pathways, which are exemplified in Fig. 1 for the

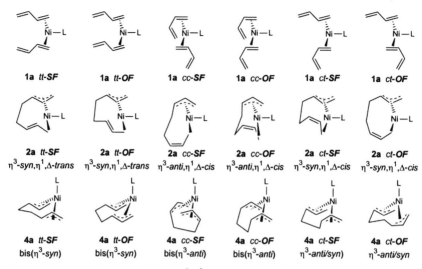

FIG. 1. Stereoisomeric forms of the [Ni$^0$(η$^2$-butadiene)$_2$L] active catalyst complex **1a** of the C$_8$-cyclodimer reaction channel and the related stereoisomers of the η$^3$,η$^1$(C$^1$), **2a**, and bis(η$^3$), **4a**, octadienediyl–Ni$^{II}$ species. SF and OF denotes the coordination of two *cis*-butadienes (*cc*), of two *trans*-butadienes (*tt*), or of *cis/trans*-butadienes (*ct*) in **1a** with the same or the opposite enantioface, respectively.

C$_8$-product channel. In the account given here, the focus is on the most feasible of the several stereochemical pathways for individual elementary processes.[29a]

## C. Methods

Currently, the density functional theory (DFT) method has become the method of choice for the study of reaction mechanism with transition-metals involved. Gradient corrected DFT methods are of particular value for the computational modeling of catalytic cycles. They have been demonstrated in numerous applications for several elementary processes, to be able to provide quantitative information of high accuracy concerning structural and energetic properties of the involved key species and also to be capable of treating large model systems.[30]

For all atoms a standard all electron basis set of triple-ζ quality for the valence electrons augmented with polarization functions was used in the calculations.[29b] The local exchange–correlation potential by Slater[31a,31b] and Vosko *et al*.[31c] was augmented with gradient-corrected functionals for electron exchange according to Becke[31d] and correlation according to Perdew[31e,31f] in a self-consistent fashion. This gradient-corrected density functional is usually termed BP86 in the literature and has proven to be reliable for both geometries and energetics of transition metal complexes. In recent benchmark computational studies, it was demonstrated that the BP86 functional gives results in excellent agreement with the best wave function-based methods available today, for the class of reactions investigated here.[32] Activation barriers for monomer insertion and monomer uptake energies are reproduced with an accuracy of $\sim 2\,kcal\,mol^{-1}$ when compared with sophisticated wave function-based methods. Due to the similar structure of key species for competing oxidative coupling, butadiene insertion, reductive elimination, and isomerization steps of the cyclo-oligomerization reactions investigated here, a higher accuracy could be expected for the relative barriers calculated. This particularly holds true for relative elimination barriers connected with competing routes and/or stereochemical pathways.

A combined DFT and molecular mechanics (QM/MM) approach[33] was employed in addition to the pure DFT treatment, QM(DFT), for the $[Ni^0(butadiene)_2L]$ catalysts I–VI, aimed at elucidating the role played by steric and electronic properties of the ancillary ligand L on thermodynamic and kinetic aspects of critical elementary steps of the $C_8$-channel.[29c] The combined QM/MM methodology has successfully been applied for studying transition metal catalyzed reactions and represents a reliable tool for quantifying steric effects.[34] In this approach the chemically important region of the catalyst, that for instance involves bond breaking and forming, is treated by a high level method, e.g., QM(DFT), while the effects of the remaining parts of the catalyst, which may constitute of a large ancillary ligand sphere, are considered by the computationally less demanding molecular mechanics (MM) method. The QM(DFT) part consisted of the generic $[(C_8H_{12})NiPH_3]$ catalyst species in which the substituents on the phosphorous atom were replaced by hydrogen atoms. The alkyl and aryl groups attached to phosphorous for the actual $PR_3/P(OR)_3$ ligands of catalysts I–VI were described by a MM3 molecular mechanics force field[35] without the electrostatic contributions. The QM(DFT) and MM parts were coupled self-consistently according to the scheme proposed by Maseras and Morokuma,[33d] where both regions interact with one another exclusively via steric potentials, while the direct electronic interactions between the atoms of the two regions are not taken into account. Therefore, this scheme introduces only the steric effects of the atoms in the MM region to QM(DFT) part. On the one hand, this QM/MM method is certainly not appropriate to provide a balanced description of competing elementary steps for catalysts I–VI, since the important electronic effects of the actual $PR_3/P(OR)_3$ ligands do not come into play. On the other hand, however, this approach allows a straightforward, plausible quantitative separation between electronic and steric effects.

# IV

# THEORETICAL EXPLORATION OF CRITICAL ELEMENTARY
# REACTION STEPS

The theoretical mechanistic investigation of the catalytic cyclo-oligomerization reaction starts with a careful examination of critical elementary processes of the $C_8$- and $C_{12}$-cyclo-oligomer production channels with the generic $[Ni^0(butadiene)_2PH_3]$ complex[11a] and the $[Ni^0(butadiene)_x]$ complex[11c] representing the respective active catalyst complexes. This examination is aimed at enlightening the crucial aspects of each of the individual steps for the two channels and proposing the most feasible of the several conceivable paths.

## A. Oxidative Coupling

Several forms are imaginable for the $[Ni^0(butadiene)_2L]$ and $[Ni^0(butadiene)_x]$ active catalysts, depending on the monodentate ($\eta^2$) or the bidentate ($\eta^4$) coordination mode of butadiene from either its s-*cis* or its s-*trans* configuration. The two butadienes can be coordinated in bis($\eta^2$), $\eta^4$, $\eta^2$, and bis($\eta^4$) modes for the $PR_3/P(OR)_3$-stabilized catalyst complex, giving rise to formal 16e⁻, 18e⁻, and 20e⁻ species. On the other hand, bis($\eta^4$)- and $\eta^4,\eta^2$-butadiene species and also tris($\eta^2$)- and $\eta^4$,bis($\eta^2$)-butadiene compounds are possible species for the $[Ni^0(butadiene)_2]$ and $[Ni^0(butadiene)_3]$ forms for the $[Ni^0(butadiene)_x]$ active catalyst. In general, for butadiene to coordinate in a bidentate fashion, the $\eta^4$-*cis* mode is thermodynamically favorable relative to the $\eta^4$-*trans* mode, while the $\eta^2$-*trans* mode prevails for monodentate coordination.

Among the several possible forms of the respective catalyst complexes, all of which are in equilibrium, the formal 16e⁻ trigonal planar $[Ni^0(\eta^2$-butadiene)_2L]$ (**1a**, Scheme 1) and $[Ni^0(\eta^2$-butadiene)_3]$ (**1'b**, Scheme 2) compounds, with butadiene preferably coordinated in the $\eta^2$-*trans* mode, represent the thermodynamically favorable form of the active catalyst for the $C_8$- and $C_{12}$-product channels, respectively. Although the species **1a** and **1'b** are predicted to be the prevalent forms of the respective catalyst complexes, the $[Ni^0(\eta^4$-*cis*-butadiene)_2]$ compound **1b**, in particular, should also be present in appreciable concentrations together with **1'b** since both forms differ only by $\sim 2\,kcal\,mol^{-1}$ in free energy. On the other hand, $PR_3/P(OR)_3$-stabilized $\eta^4,\eta^2$-butadiene forms are expected to be sparsely populated, since they are $\sim 6\,kcal\,mol^{-1}$ higher in free energy relative to **1a**.

The thermodynamically favored $[Ni^0(\eta^2\text{-butadiene})_2L]$ species **1a** and $[Ni^0(\eta^2\text{-butadiene})_3]$ species **1'b** also represent the active catalyst forms for the $C_8$- and $C_{12}$-product channels, since they act as the precursors for the most feasible path for oxidative coupling. The several other forms of the catalyst complex, however, are found either to participate along paths that are kinetically disfavored or these forms approach the bis($\eta^2$-butadiene) or tris($\eta^3$-butadiene) species, respectively, in the vicinity of the transition state.[11] In particular, the reaction paths connecting the active catalysts in a direct way together with the bis($\eta^3$)-octadienediyl species **4a**, **4b**, respectively, have been shown to be unfeasible. For the $PR_3/P(OR)_3$-stabilized forms, severe repulsive interactions between the reacting butadiene moieties are involved in the initial stages of the process, thereby giving rise to an expected high barrier. For the $C_{12}$-channel as well, the **1b** $\rightarrow$ **4b** path is seen to be kinetically disabled.[11]

For both the reaction channels, oxidative coupling is found to preferably proceed via establishment of a new C–C $\sigma$-bond between the terminal noncoordinated carbons $C^4$ and $C^5$ of two $\eta^2$-butadienes, (Fig. 2), that occurs at a distance of $\sim 2.2$–$2.3$ Å in the educt-like transition states TS[**1a**–**2a**] and TS[**1'b**–**2b**], respectively. The $\eta^3,\eta^1(C^1)$-octadienediyl–$Ni^{II}$ species **2a**, **2b** are formed as the initial coupling products. The activation energy as well as the thermodynamic driving force for oxidative coupling is primarily determined by the configuration and the enantioface of the two reacting butadiene moieties involved in the process, while the ancillary butadiene has a minor influence on the energetics for **1'b** $\rightarrow$ **2b**. The coupling of $\eta^2\text{-}trans/\eta^2\text{-}cis$ butadiene of opposite enantiofaces is favorable, both kinetically by the overall lowest barrier among the several stereochemical pathways, and also thermodynamically by the formation of the most stable $\eta^3\text{-}syn,\eta^1(C^1),\Delta\text{-}cis$ isomer of the initial coupling species. Similar activation barriers have to be overcome for the oxidative coupling step of the $C_8$- and $C_{12}$-cyclo-oligomer channel, that amounts to 13.6 and 12.6 kcal mol$^{-1}$ ($\Delta G^{\ddagger}$) for **1a** $\rightarrow$ **2a**, and **1'b** $\rightarrow$ **2b**, respectively, relative to the related favorable educt species $[Ni^0(\eta^2\text{-}trans\text{-butadiene})_2PH_3]$ **1a** and $[Ni^0(\eta^2\text{-}trans\text{-butadiene})_3]$ **1'b**. Overall, oxidative coupling along the most feasible pathway is a thermoneutral process. Thus, oxidative coupling is indicated to be a reversible process with a moderate kinetic barrier affording the $\eta^3\text{-}syn,\eta^1(C^1),\Delta\text{-}cis$-octadienediyl–$Ni^{II}$ isomer of **2a** and **2b** as the initial coupling product, that also represents the favorable isomer of $\eta^3,\eta^1$ species of the octadienediyl–$Ni^{II}$ complex. All these aspects are consistent with the experimental observation of the stoichiometric cyclodimerization reaction.[14]

For oxidative coupling to occur along the preferred **1a** $\rightarrow$ **2a**, and **1'b** $\rightarrow$ **2b** paths, the coupling of either two $\eta^2\text{-}cis$- or $\eta^2\text{-}trans$-butadienes

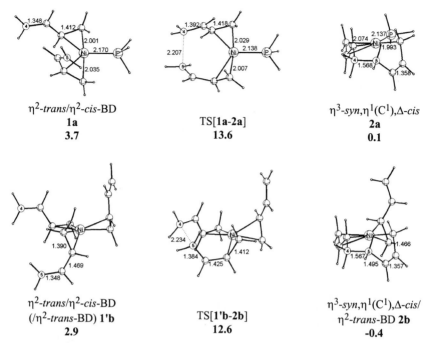

$\eta^2$-*trans*/$\eta^2$-*cis*-BD
**1a**
**3.7**

TS[**1a-2a**]
**13.6**

$\eta^3$-*syn*,$\eta^1$($C^1$),$\Delta$-*cis*
**2a**
**0.1**

$\eta^2$-*trans*/$\eta^2$-*cis*-BD
(/$\eta^2$-*trans*-BD) **1'b**
**2.9**

TS[**1'b-2b**]
**12.6**

$\eta^3$-*syn*,$\eta^1$($C^1$),$\Delta$-*cis*/
$\eta^2$-*trans*-BD **2b**
**-0.4**

FIG. 2. Selected geometric parameters (Å) of the optimized structures of the key species for oxidative coupling for the catalytically active generic [$Ni^0$($\eta^2$-butadiene)$_2$PH$_3$] species **1a** and the [$Ni^0$($\eta^2$-butadiene)$_3$] species **1'b** of the C$_8$- and C$_{12}$-product channel, respectively, via the most feasible pathway for $\eta^2$-*trans*/$\eta^2$-*cis*-butadiene coupling (of opposite enantiofaces) along **1a → 2a** and **1'b → 2b**. Free energies ($\Delta G$, $\Delta G^\ddagger$ in kcal mol$^{-1}$) are given relative to the favorable stereoisomer of the respective bis($\eta^2$-*trans*-butadiene) and tris($\eta^2$-*trans*-butadiene) precursors **1a** and **1'b**.

is kinetically impeded by higher barriers, relative to the favorable $\eta^2$-*cis*/$\eta^2$-*trans*-butadiene coupling. Highly strained, energetically disfavored $\eta^3$-*syn*,$\eta^1$($C^1$),$\Delta$-*trans* species are generated as initial coupling products of two $\eta^2$-*trans*-butadienes. Thus, these stereochemical pathways are indicated to be unlikely as well due to a reverse reductive decoupling process that becomes significantly more facile than the oxidative coupling. The enantiofaces of the two $\eta^2$-butadienes involved in the coupling have a pronounced influence on the kinetic barriers. Two $\eta^2$-*cis* and two $\eta^2$-*trans*-butadienes preferably couple with identical enantiofaces, while the coupling of opposite enantiofaces is favorable for mixed $\eta^2$-*trans*/$\eta^2$-*cis* butadiene. The preference for these stereochemical pathways is understandable from simple molecular orbital arguments, which indicate for these cases a reduced repulsive interaction between the $2\pi$-butadiene MO's in the C–C σ-bond formation process, when compared to the alternative stereoisomers (cf. Fig. 1).


### B. Thermodynamic Stability of the Several Species of the Octadienediyl–Ni$^{II}$ Complex

The [Ni$^{II}$(octadienediyl)L] complex is the crucial compound of the C$_8$-channel, acting as the precursor for reductive elimination occurring along competing routes, to afford the principal cyclodimer products (cf. Scheme 1). The thermodynamic stability and also the reactivity of the different configurations and stereoisomers of this complex play a critical role for the regulation of the cyclodimer product selectivity. Likewise, the octadienediyl–Ni$^{II}$ complex is an important intermediate in the reaction course of the C$_{12}$-channel, as it represents the precursor for butadiene insertion into the allyl–Ni$^{II}$ bond to yield the critical [Ni$^{II}$(dodecatrienediyl)] complex (cf. Scheme 2). The different configurations of the octadienediyl–Ni$^{II}$ complex are in equilibrium, while the several stereoisomeric forms are interconverted via *syn–anti* isomerization as well as enantioface conversion of the terminal allylic groups (cf. Section 4.3).

The relative stability and reactivity of the different octadienediyl–Ni$^{II}$ configurations is known to be decisively influenced by the ligand's properties. The bis($\eta^3$) species (**4a**, **4b**) represents the energetically preferred mode of the Ni$^{II}$-bis(allyl-anion) coordination, since the formal negative charge is delocalized over the allylic moieties. In the $\eta^3,\eta^1$ species, the formal negative charge on one of the allylic groups has to be localized on either the terminal unsubstituted C$^1$ (**2a**, **2b**), or on the terminal substituted C$^3$ (**3a**, **3b**), which is less favorable compared with the bis($\eta^3$) species. The localization of the negative charge, however, can be supported by the presence of an electron-releasing ligand L (cf. Section 5.2). Finally, the bis($\eta^1$) coordination mode is energetically highly disfavored, since it requires negative charge localization at both allylic moieties.

This qualitative picture is confirmed by the calculated thermodynamic stabilities of the most favorable stereoisomers of the different species **2a–7a** of the generic [Ni$^{II}$(octadienediyl)PH$_3$] complex (Fig. 3). The formal 16e$^-$ square-planar (SP) $\eta^3,\eta^1$(C$^1$) species **2a** and the formal 18e$^-$ square-pyramidal (SPY) bis($\eta^3$) species **4a** are the prevalent forms of the PR$_3$/P(OR)$_3$-stabilized octadienediyl–Ni$^{II}$ complex, that are predicted to occur in similar amounts for the generic catalyst. The localization of the negative charge on the terminally substituted C$^3$, **3a**, is seen to be energetically less favorable, relative to the charge accumulation on the C$^1$, as in **2a**. The bis($\eta^1$) species **5a–7a**, however, are at a higher energy, well separated from

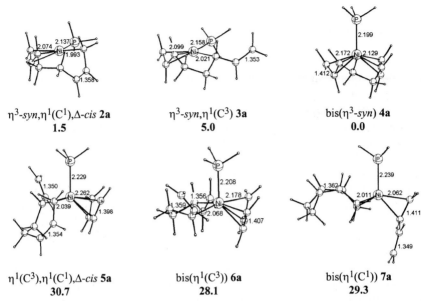

$\eta^3\text{-}syn,\eta^1(C^1),\Delta\text{-}cis$ **2a**
**1.5**

$\eta^3\text{-}syn,\eta^1(C^3)$ **3a**
**5.0**

bis($\eta^3\text{-}syn$) **4a**
**0.0**

$\eta^1(C^3),\eta^1(C^1),\Delta\text{-}cis$ **5a**
**30.7**

bis($\eta^1(C^3)$) **6a**
**28.1**

bis($\eta^1(C^1)$) **7a**
**29.3**

FIG. 3. Most favorable isomer for each of the several species **2a**–**7a** of the generic [$Ni^{II}$(octadienediyl)$PH_3$] complex, together with the relative thermodynamic stabilities ($\Delta G$ in kcal mol$^{-1}$).

the favorable species **2a** and **4a**, which indicates these species to be negligibly populated. The coordination of additional butadiene was found to be necessary for the coordinative stabilization of bis($\eta^1$) species. On the other hand, for bis($\eta^3$) and $\eta^3,\eta^1$ species, the weak donor butadiene is not able to compete efficiently for coordination with the allylic groups and the ligand L. The small enthalpic stabilization of butadiene complexation cannot compensate for the associated entropic costs. Therefore, coordination of additional butadiene is unlikely for **2a**, **3a**, **4a**, respectively.

A similar picture is revealed for the octadienediyl–$Ni^{II}$ complex involved in the $C_{12}$-cyclotrimer reaction channel.[11c] In the $\eta^3,\eta^1$ species, **2b**, **3b**, $\eta^2$-butadiene preferably occupies the fourth position around $Ni^{II}$, while the coordination of butadiene is unfavorable at the $\Delta G$ surface for the bis($\eta^3$) species **4b**. The predominant configurations are **2b** and **4b**, while bis($\eta^1$) species represent isomers lying at much higher energies. The bis($\eta^3$) species **4b** is favorable by 5.6 kcal mol$^{-1}$ ($\Delta G$) relative to **2b**, which is due to the limited ability of the weak donor butadiene to electronically stabilize the $\eta^3,\eta^1(C^1)$ coordination mode.

We shall elaborate further, in subsequent sections, on the role played by the bis($\eta^1$) species in the reaction course as possible reactive intermediates involved in allylic isomerization and/or reductive elimination.

## C. *Syn–Anti Isomerization and Enantioface Conversion of the Terminal Allylic Group in the Octadienediyl–Ni$^{II}$ Complex*

The interconversion between the stereoisomeric forms of the octadiene-diyl–Ni$^{II}$ complex plays a crucial role in the catalytic reaction course. Provided, for example, that the formation of the octadienediyl–Ni$^{II}$ complex via oxidative coupling and the subsequent processes of either reductive elimination (cf. Scheme 1) or butadiene insertion into the allyl–Ni$^{II}$ bond (cf. Scheme 2) involve different stereoisomers along the respective most feasible pathways, it becomes clear that the interconversion step is indispensable. Furthermore, the relative rates of interconversion and the rates for the subsequent elementary steps may have a pronounced influence on which of the cyclo-oligomer products are predominantly generated. In the case of a kinetically impeded interconversion, several of the stereochemical pathways of the reductive elimination or butadiene insertion steps would be disabled due to the negligible population of the corresponding precursor species, which is only one of the several possible mechanistic scenarios.

The interconversion between octadienediyl–Ni$^{II}$ stereoisomers may involve two different processes; namely the *syn–anti* isomerization and also the enantioface conversion of one or both terminal allylic groups (Fig. 4). The isomerization of the allylic group is connected with two aspects; firstly the interconversion of its *syn* and *anti* configuration and secondly the inversion of its enantioface.[36] On the other hand, the process of enantioface conversion is not accompanied by alternation of the allylic configuration.

Allylic isomerization has been demonstrated to preferably proceed via an $\eta^3$-$\pi \rightarrow \eta^1$-$\sigma$-$C^3$ allylic rearrangement followed by internal rotation of the vinyl group around the formal $C^2$–$C^3$ single bond (Fig. 4) from evidence provided by both experimental[36,37] and theoretical[38] studies. The several $\eta^1(C^3)$-octadienediyl–Ni$^{II}$ species can be envisioned as possible precursors; that are **3a**, **5a**, **6a**, for instance, for the C$_8$-channel (cf. Scheme 1). For both reaction channels, allylic isomerization is most likely to take place commencing from the $\eta^3,\eta^1(C^3)$ species, **3a** and **3b**, respectively, and occurring through formal 16e$^-$ rotational transition states TS$_{ISO}$[**3a**] and TS$_{ISO}$[**3b**], respectively, that constitute the internal vinyl group's rotation around the $C^2$–$C^3$ bond (Fig. 5). The conversion of the allylic enantioface was found to preferably proceed in the $\eta^3,\eta^1(C^1)$ species **2a** and **2b**, respectively, by inversion of the $\eta^1$-allylic group. It has been shown that incoming butadiene does not accelerate either allylic isomerization or enantioface conversion via coordinative stabilization of the corresponding transition state. Thus, these two elementary processes are unlikely to be assisted by butadiene. However, bis($\eta^1$) species, which in general are sparsely populated, have been shown to be not involved along any viable path for allylic isomerization.[11]

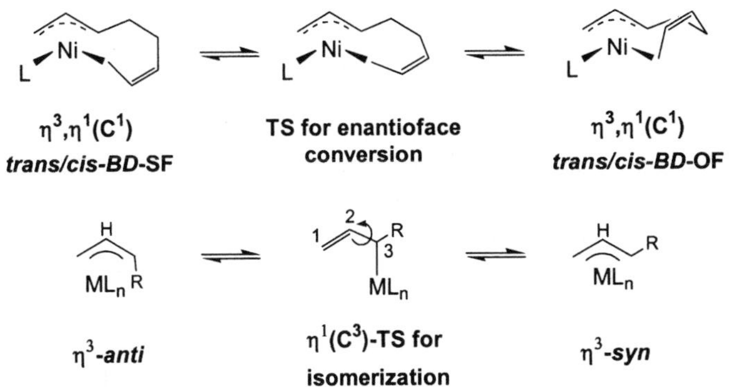

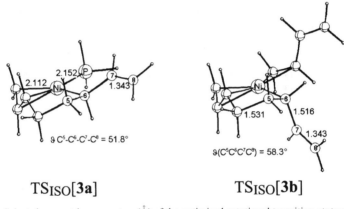

FIG. 4. Allylic enantioface conversion in $\eta^3,\eta^1(C^1)$ species (top), and allylic isomerization taking place via an $\eta^3,\eta^1(C^3)$-allylic intermediate (below).

TS$_\mathrm{ISO}$[**3a**]                    TS$_\mathrm{ISO}$[**3b**]

FIG. 5. Selected geometric parameters (Å) of the optimized rotational transition-state structures for allylic isomerization via the $\eta^3$-*syn*,$\eta^1(C^3)$-octadienediyl–Ni$^{II}$ TS$_\mathrm{ISO}$[**3a**] and TS$_\mathrm{ISO}$[**3b**], respectively.

It has not been possible to determine rates of allylic conversion processes by NMR for the octadienediyl–Ni$^{II}$ complex in stoichiometric cyclodimerization reactions, even at low temperatures (−30 to 25 °C), suggesting that these processes are too fast to be observed on the NMR timescale. Accordingly, allylic enantioface conversion is indicated to be a facile process, that is connected with an overall largest total barrier of 14.3 kcal mol$^{-1}$ ($\Delta G^{\ddagger}$, relative to the favorable $\eta^3$-*syn*,$\eta^1(C^1)$,$\Delta$-*cis* isomer of **2a**) for conversion occurring in **2a**, for example. Similarly, moderate total barriers for allylic isomerization involving $\eta^3$-*syn*,$\eta^1(C^3)$ isomers (i.e., conversion between $\eta^2$-*trans*/$\eta^2$-*cis*-butadiene and $\eta^2$-*trans*/$\eta^2$-*trans*-butadiene coupling products) have to be overcome along the C$_8$- and C$_{12}$-product channels, which amount to 9.4 and 15.3 kcal mol$^{-1}$ ($\Delta G^{\ddagger}$, relative to the favorable $\eta^3$-$syn,\eta^1(C^1),\Delta$-$cis$ isomer of **2a** and **2b**), respectively. On the other hand, allylic isomerization is predicted to be significantly slower commencing from the corresponding $\eta^3$-$anti,\eta^1(C^3)$ isomers (i.e., conversion between $\eta^2$-$trans/$$\eta^2$-$cis$-butadiene and $\eta^2$-$cis/\eta^2$-$cis$-butadiene coupling products) due to connected barriers that are $\sim$ 7–8 kcal mol$^{-1}$ ($\Delta G^{\ddagger}$) higher. Thermodynamic reasons are seen to be critical for the different total reactivity of the $\eta^3$-$syn,\eta^1(C^3)$ and the $\eta^3$-$anti,\eta^1(C^3)$ forms to undergo allylic isomerization. Both forms show comparable intrinsic reactivities, as indicated by similar intrinsic barriers of 7–8 kcal mol$^{-1}$ ($\Delta G^{\ddagger}_{int}$, relative to the respective precursors **3a** and **3b**, respectively). However, $\eta^3$-$anti,\eta^1(C^3)$ isomers of **3a** and **3b** are thermodynamically disfavored by more than 7.5 kcal mol$^{-1}$ ($\Delta G$) relative to the $\eta^3$-$syn,\eta^1(C^3)$ counterparts. The $\eta^3$-$syn,\eta^1(C^3)$ and $\eta^3$-$anti,\eta^1(C^3)$ isomers are connected by a facile $\eta^3$-$syn,\eta^1$-$anti \rightleftharpoons \eta^3$-$anti,\eta^1$-$syn$ conversion.[11]

The conversion processes of the terminal allylic groups of the octadienediyl–Ni$^{II}$ complex show very similar characteristics for the two reaction channels. The influence of electronic and steric factors on the allylic isomerization will be scrutinized in Section 5.3 and the overall role played by allylic conversion will be elucidated in the context of the entire reaction course (see Sections 6.1 and 6.2).

## D. *Butadiene Insertion Into the Allyl–Ni$^{II}$ Bond of the Octadienediyl–Ni$^{II}$ Complex*

The formation of the crucial [Ni$^{II}$(dodecatrienediyl)] complex of the $C_{12}$-channel, occurring via butadiene insertion into the terminal allyl–Ni$^{II}$ bond of the octadienediyl–Ni$^{II}$ complex, can be envisioned to commence from the $\eta^3,\eta^1$ species **2b**, **3b** or the bis($\eta^3$) species **4b**, respectively. In general, the most favorable transition states that are involved along the different conceivable insertion paths are characterized by a quasi-planar arrangement of the reactive moieties; namely, the terminal carbon of the allylic group, the nickel atom, and the double bond of the coordinated butadiene which will be inserted. A careful inspection of the several possible insertion paths revealed that both the bis($\eta^3$) species **4b-BD** (with an axial coordinated $\eta^2$-butadiene) and the $\eta^3,\eta^1(C^3)$ species **3b** are precluded from the energetically most favorable path for butadiene insertion.[11c] The $\eta^3,\eta^1(C^1)$ species **2b**, which is formed as the initial product of oxidative coupling, is the precursor for the most feasible insertion path. Commencing from **2b**, butadiene preferably inserts into the terminal $\eta^3$-allyl–Ni$^{II}$ bond via square-planar transition-state structures, that constitute the $\eta^2$-butadiene insertion into the $\eta^3$-allyl–Ni$^{II}$ bond (Fig. 6) while the alternative insertion into the

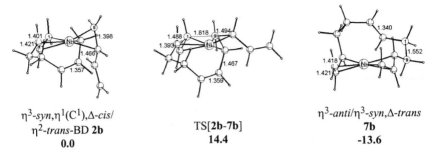

$\eta^3$-*syn*,$\eta^1$(C$^1$),$\Delta$-*cis*/                                           $\eta^3$-*anti*/$\eta^3$-*syn*,$\Delta$-*trans*
$\eta^2$-*trans*-BD **2b**            TS[**2b**-**7b**]                    **7b**
**0.0**                           **14.4**                         **-13.6**

FIG. 6. Selected geometric parameters (Å) of the optimized structures of the key species for the feasible $\eta^2$-*trans*-butadiene insertion into the $\eta^3$-*syn*-allyl–Ni$^{II}$ bond along **2b** → **7b**. Free energies ($\Delta G$, $\Delta G^\ddagger$ in kcal mol$^{-1}$) are given relative to the favorable $\eta^3$-*syn*,$\eta^1$(C$^1$),$\Delta$-*cis*/$\eta^2$-*trans*-butadiene stereoisomer of the precursor **2b**.

$\eta^1$(C$^1$)–Ni$^{II}$ bond is kinetically impeded. Among the several configurations of the [Ni$^{II}$(dodecatrienediyl)] product complex, the bis($\eta^3$),$\Delta$-dodecatrienediyl–Ni$^{II}$ species **7b** is thermodynamically favored (cf. Section 4.5). Notably, a similar preference of the $\eta^3$-$\pi$ vs. $\eta^1$-$\sigma$ coordination mode of the allyl–transition-metal bond has also been shown in computational investigations of the monomer insertion step in the allylnickel(II)-catalyzed[39a,39b] and the allyltitanium(III)-catalyzed[39c] polymerization of butadiene and for the $\alpha$-olefin insertion into the allyl–Zr$^{III}$ bond.[40]

Among the several stereoisomers of the precursor **2b**, the preferably generated $\eta^3$-*syn*,$\eta^1$(C$^1$),$\Delta$-*cis* isomer ($\eta^2$-*trans*/$\eta^2$-*cis*-butadiene coupling product) is also seen to be most reactive. Nearly identical total free-energies of activation (relative to the favorable $\eta^3$-*syn*,$\eta^1$(C$^1$),$\Delta$-*cis*/$\eta^2$-*trans*-butadiene isomers **2b**) of 14.0–15.1 kcal mol$^{-1}$ are connected with the $\eta^2$-*trans*- and $\eta^2$-*cis*-butadiene insertion into the $\eta^3$-*syn*-allyl–Ni$^{II}$ bond, affording $\eta^3$-*syn*/$\eta^3$-*anti*,$\Delta$-*trans* and bis($\eta^3$-*anti*),$\Delta$-*trans* isomers of **7b**, in a process that is highly exergonic by $-(13.6$–$16.4)$ kcal mol$^{-1}$. On the other hand, butadiene insertion into either the $\eta^3$-*anti*–Ni$^{II}$ bond of $\eta^3$-*anti*,$\eta^1$(C$^1$),$\Delta$-*cis* isomers of **2b** (coupling product of two $\eta^2$-*cis*-butadienes) or into the $\eta^3$-*syn*–Ni$^{II}$ bond of $\eta^3$-*syn*,$\eta^1$(C$^1$),$\Delta$-*trans* isomers of **2b** (coupling product of two $\eta^2$-*trans*-butadienes) is kinetically impeded due to distinctly higher activation barriers ($\Delta\Delta G^\ddagger > 7$ kcal mol$^{-1}$) for these pathways. Furthermore, these pathways are also disabled from thermodynamic considerations, since the corresponding $\eta^3$-*anti*,$\eta^1$(C$^1$),$\Delta$-*cis* and $\eta^3$-*syn*,$\eta^1$(C$^1$),$\Delta$-*trans* isomers of **2b** are likely to be sparsely populated. In the latter case, $\eta^3$-*syn*,$\eta^1$(C$^1$),$\Delta$-*trans* isomers are accessible from the predominant $\eta^3$-*syn*,$\eta^1$(C$^1$),$\Delta$-*cis* coupling isomer of **2b** via ready allylic conversion (cf. Section 4.3). These isomers, however, are likely to undergo facile reductive decoupling **2b** → **1'b** (cf. Section 4.1). On the other hand, the slow conversion between $\eta^2$-*trans*/$\eta^2$-*cis*-butadiene and $\eta^2$-*cis*/$\eta^2$-*cis*-butadiene coupling isomers of **2b** via allylic isomerization ($\Delta G^{\ddagger} = 16.0$–18.0 kcal mol$^{-1}$), relative to the more rapid competing butadiene insertion process, prevents the $\eta^3$-*anti*,$\eta^1$(C$^1$),$\Delta$-*cis* isomers from occurring in an appreciable concentration. This leads to important mechanistic consequences for the C$_{12}$-product channel. The complete branch for the generation of bis(allyl),$\Delta$-*cis*-dodecatrienediyl–Ni$^{II}$ forms is entirely suppressed: (i) by the unfavorable coupling of two $\eta^2$-*cis*-butadienes along **1′b** $\rightarrow$ **2b** together with a slow allylic conversion through $\eta^3$-*anti*,$\eta^1$(C$^3$) isomers of TS$_{ISO}$[**3b**], giving rise to a negligible amount of $\eta^3$-*anti*,$\eta^1$(C$^1$), $\Delta$-*cis* isomers of **2b** (coupling species of two $\eta^2$-*cis*-butadienes), and (ii) by the kinetically retarded insertion of butadiene into the $\eta^3$-*anti*-allyl–Ni$^{II}$ bond along **2b** $\rightarrow$ **7b**. Consequently, the *all-c*-CDT generation route, which would be accessible via formation of bis($\eta^3$),$\Delta$-*cis* isomers of **7b**, followed by facile allylic isomerization (if required) and subsequent reductive elimination involving bis($\eta^3$-*anti*),$\Delta$-*cis*-dodecatrienediyl–Ni$^{II}$ isomers, is entirely precluded.

Identical stereoisomers (coupling species of $\eta^2$-*trans*/$\eta^2$-*cis*-butadiene with opposite enantiofaces) participate along the most feasible pathways for oxidative coupling and butadiene insertion. Thus, allylic conversion in the octadienediyl–Ni$^{II}$ complex is not required along the C$_{12}$-channel. In agreement with experimental observation, octadienediyl–Ni$^{II}$ species are indicated as highly reactive intermediates that occur in very low stationary concentrations, since the generating oxidative coupling and the consuming insertion processes involve similar moderate activation barriers. Accordingly, octadienediyl–Ni$^{II}$ species are unlikely to be isolable either in the catalytic or in the stoichiometric cyclotrimerization process.

### E. *Allylic Isomerization in the Dodecatrienediyl–Ni$^{II}$ Complex*

Similar to the octadienediyl–Ni$^{II}$ complex, the $\eta^3,\eta^1$ configurations **5b**, **6b** and the bis($\eta^3$) configuration **7b** of the [Ni$^{II}$(dodecatrienediyl)] complex are also in a dynamic equilibrium. The bis($\eta^3$),$\Delta$-*cis*/*trans* species **7b** is prevalent, while **5b**, **6b** are thermodynamically disfavored, since these species are not efficiently stabilized by the coordinated olefinic double bond. The formation of the [Ni$^{II}$(dodecatrienediyl)] complex along **2b** $\rightarrow$ **7b** is driven by the highest thermodynamic force among all of the critical elementary steps of the C$_{12}$-channel. Therefore, **7b** is likely to serve as a thermodynamic sink, which is supported by the isolation of a bis($\eta^3$-*anti*),$\Delta$-*trans*-dodecatrienediyl–Ni$^{II}$ compound as the reactive intermediate in the stoichiometric cyclotrimerization.[9a,10] Among the four possible stereoisomers of the bis($\eta^3$-*anti*),$\Delta$-*trans* intermediate, the two isolated stereoisomers with *trans* oriented $\eta^3$-*anti* allylic groups, which have been confirmed by NMR,[20,21] are predicted to be the thermodynamically favorable dodecatrienediyl–Ni$^{II}$ species.[41] Moreover, these isomers are likely to be involved in the catalytic reaction course of the C$_{12}$-channel, since they are formed along feasible stereochemical pathways for $\eta^2$-*cis*-butadiene insertion into the $\eta^3$-*syn*–Ni$^{II}$ bond of the reactive $\eta^2$-*trans*/$\eta^2$-*cis* coupling species **2b** (cf. Scheme 5 in Section 6.2).

Commencing from the preferably generated bis($\eta^3$-*anti*),$\Delta$-*trans* and $\eta^3$-*anti*/$\eta^3$-*syn*,$\Delta$-*trans* isomers along **2b** → **7b** (cf. Section 4.4), allylic isomerization is most likely to occur via a facile $\eta^3$-$\pi$ → $\eta^1$(C$^3$) rearrangement and subsequent passage through an $\eta^3$,$\eta^1$(C$^3$),$\Delta$-*trans*-dodecatrienediyl–Ni$^{II}$ rotational transition state, TS$_{ISO}$[**6b**] (Fig. 7) that is stabilized by the coordinated *trans* double bond. The bis(*anti*),$\Delta$-*trans* ⇌ *anti*/*syn*,$\Delta$-*trans* conversion is connected with an total barrier (i.e., relative to the most favorable bis($\eta^3$-*anti*),$\Delta$-*trans* isomer of **7b**) of 15.0–20.0 kcal mol$^{-1}$ ($\Delta G^\ddagger$), and the total barrier for *anti*/*syn*,$\Delta$-*trans*- ⇌ bis(*syn*),$\Delta$-*trans* isomerization amounts to 13.0–16.5 kcal mol$^{-1}$ ($\Delta G^\ddagger$). The variation of the isomerization barrier is found to be primarily due to the different ability of the coordinated *trans* double bond to stabilize TS$_{ISO}$[**6b**] for individual stereoisomers.

The isomerization barrier of 15.0–20.0 kcal mol$^{-1}$ ($\Delta G^\ddagger$) can be considered to be large enough to allow isolation and characterization of bis($\eta^3$-*anti*),$\Delta$-*trans*-dodecatrienediyl–Ni$^{II}$ stereoisomers of **7b**[41] as reactive intermediates in the stoichiometric cyclotrimerization process. Furthermore, the *trans* orientation of the two allylic groups gives rise to an insurmountable barrier for reductive elimination for these cases, which prevents these species from readily leaving the thermodynamic sink via a facile reductive elimination. The isolated intermediates clearly constitute dead-end

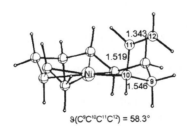

$\vartheta$(C$^9$C$^{10}$C$^{11}$C$^{12}$) = 58.3°

## TS$_{ISO}$[**6b**]

FIG. 7. Selected geometric parameters (Å) of the optimized rotational transition-state structure for *anti*/*syn*,$\Delta$-*trans* ⇌ bis(*syn*),$\Delta$-*trans* conversion via the $\eta^3$-*syn*,$\eta^1$(C$^3$),$\Delta$-*trans*-dodecatrienediyl–Ni$^{II}$ TS$_{ISO}$[**6b**].

points within the catalytic cycle and they first have to undergo allylic isomerization in order to make feasible pathways for reductive elimination accessible, which is consistent with the observation of the stoichiometric reaction.[22]

Overall, allylic isomerization in the dodecatrienediyl–Ni[II] complex is predicted to require a distinctly lower barrier than for reductive elimination ($\Delta\Delta G^{\ddagger} > 5.5\,\text{kcal}\,\text{mol}^{-1}$, see Section 4.6). This leads to the conclusion, that isomerization should be significantly more facile than the subsequent reductive elimination, which is confirmed by NMR investigations of the stoichiometric reaction.[22] Consequently, the several configurations and stereoisomers of the bis(allyl),$\Delta$-*trans*-dodecatrienediyl–Ni[II] forms of **5b**–**7b** are in a kinetically mobile, pre-established equilibrium, with **7b** as the prevalent form. The various bis($\eta^3$-allyl),$\Delta$-*trans* stereoisomers of **7b** are found to be close in energy, while bis(allyl),$\Delta$-*cis* forms are shown to be negligibly populated (cf. Section 4.4) and therefore play no role within the catalytic reaction course.

Experiments have demonstrated that the stoichiometric cyclotrimerization becomes accelerated by the presence of donor phosphines (i.e., PMe$_3$, PEt$_3$, PPh$_3$) and also by excess butadiene.[9a] However, the rotational transition-state structure TS$_{ISO}$[**6b**] is found to be not stabilized in enthalpy by coordination of butadiene. Therefore, incoming butadiene does not serve to facilitate allylic isomerization and will not assist this process. Accordingly, reductive elimination is indicated to be accelerated by excess butadiene, which will be examined in the next section.

## F.  *Reductive Elimination under Ring Closure*

The cyclo-oligomer products are formed in final reductive elimination steps commencing from the octadienediyl–Ni[II] and dodecatrienediyl–Ni[II] complexes for the C$_8$- and C$_{12}$-cyclo-oligomer production channels, respectively. Reductive elimination is accompanied with a formal electron redistribution between the nickel and the organyl moieties, which will be analyzed in Section 5.4.

Reductive elimination, in general, involves $\eta^3$-allylic species along the most feasible paths, while bis($\eta^1$) species do not participate along any viable reaction paths.[11] In the proximity of the respective transition states, the bis($\eta^1$) species display a high tendency to adopt the favorable $\eta^3$-allyl–Ni[II] coordination via a facile $\eta^1 \rightarrow \eta^3$ rearrangement of one or both of the terminal allylic groups. Although the $\eta^3,\eta^1$-coordination can be stabilized by attached donor ligands, the $\eta^3$-$\pi$-allyl coordination is clearly demonstrated to be the reactive mode of the bis(allyl-anion)–Ni[II] coordination (cf. butadiene insertion, Section 4.4). In general, bis($\eta^1$)

species are sparsely populated and are also not involved along any viable path for either allylic isomerization or reductive elimination. Overall, this leads to the conclusion that the bis($\eta^1$) species play no role in the reaction cycles of the nickel-catalyzed cyclo-oligomerization of 1,3-butadiene.

## 1. Reductive Elimination under Ring Closure to Occur in the Octadienediyl–Ni$^{II}$ Complex

The generation of the three principal C$_8$-cyclodimers occurs via competing routes for reductive elimination commencing from different forms of the [Ni$^{II}$(octadienediyl)L] complex. VCH is formed directly from the $\eta^3,\eta^1(C^1)$ species **2a** along **2a** $\rightarrow$ **8a**, and the bis($\eta^3$) species **4a** acts as the precursor for both the generation of cis-1,2-DVCB along **4a** $\rightarrow$ **9a** and the **4a** $\rightarrow$ **10a** cis,cis-COD production path. The other conceivable paths outlined in Scheme 1 have been shown to be kinetically impeded.[11a] From the [Ni$^0$($\eta^4$-cyclodimer)L] products **8a–10a**, the cyclodimers are liberated in subsequent consecutive substitution steps with butadiene, that regenerates the active catalyst **1a** in an overall exergonic process.

Following the **2a** $\rightarrow$ **8a** route, VCH is formed via the establishment of a C–C $\sigma$-bond between the terminal substituted $\eta^3$-allylic carbon, C$^3$, and the $\eta^1$-allylic carbon, C$^8$, in the square-planar TS[**2a–8a**] (Fig. 8) which occurs at a distance of $\sim 2.0$—2.1 Å for the emerging bond. In the product-like TS[**2a–8a**], the $\eta^3$-allylic group is partly converted into a vinyl group and VCH is essentially pre-formed. Similar total barriers (i.e., relative to the most favorable $\eta^3$-syn,$\eta^1(C^1)$,$\Delta$-cis isomer of **2a**, cf. Fig. 2) are connected with the several stereochemical pathways along **2a** $\rightarrow$ **8a**, that amount to 23 kcal mol$^{-1}$ ($\Delta G^{\ddagger}$) for the most feasible pathway, involving coupling species of two $\eta^2$-cis-butadiene with identical enantiofaces. Formation of VCH is driven by the strongest thermodynamic force among the three competing cyclodimer generating routes, with an exergonicity of $-12.5$ kcal mol$^{-1}$. Although conceivable, it has been shown that the VCH generating route involving formal 16e$^-$ species is not facilitated by coordination of incoming butadiene.

Formation of cis-1,2-DVCB and cis,cis-COD commences through the formation of a $\sigma$-bond between the terminal substituted carbons, C$^3$, C$^6$, and the terminal unsubstituted carbons, C$^1$, C$^8$, of the two $\eta^3$-allylic groups along **4a** $\rightarrow$ **9a** and **4a** $\rightarrow$ **10a**, respectively (Fig. 8). The transition states TS[**4a–9a**] and TS[**4a–10a**] occur at a distance of $\sim 1.9$ and $\sim 2.1$ Å for the newly formed C–C bond and decay into **9a** and **10a**, respectively, where the cyclodimers are each coordinated to Ni$^0$ by two olefinic double bonds. The several stereochemical pathways are connected with activation barriers that differ significantly. Moderate barriers have to be overcome for **4a** $\rightarrow$ **9a**

## VCH generating route

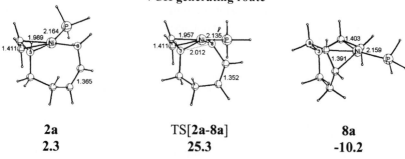

**2a**
**2.3**

**TS[2a-8a]**
**25.3**

**8a**
**-10.2**

### *cis*-1,2-DVCB generating route

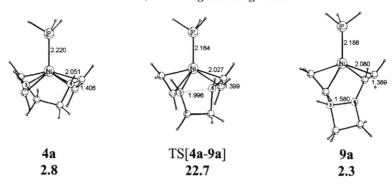

**4a**
**2.8**

**TS[4a-9a]**
**22.7**

**9a**
**2.3**

### *cis,cis*-COD generating route

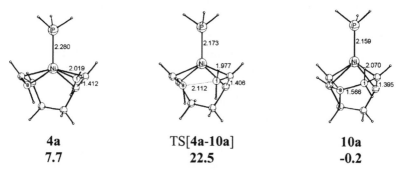

**4a**
**7.7**

**TS[4a-10a]**
**22.5**

**10a**
**-0.2**

FIG. 8. Selected geometric parameters (Å) of the optimized structures of the key species for reductive elimination via the most feasible stereochemical pathway for the competing routes affording VCH, *cis*-1,2-DVCB, and *cis,cis*-COD, respectively, for the generic catalyst along $2a \rightarrow 8a$, $4a \rightarrow 9a$, and $4a \rightarrow 10a$. Free energies ($\Delta G$, $\Delta G^{\ddagger}$ in kcal mol$^{-1}$) are given relative to the favorable bis($\eta^3$-*syn*) stereoisomer of **4a**.

along *cis*-1,2-DVCB generating pathways that involve coupling species of either two *cis*-butadienes or two *trans*-butadienes with opposite enantiofaces and of *trans*/*cis*-butadiene with identical enantiofaces, while the corresponding *trans*-1,2-DVCB pathways are kinetically disfavored by substantially higher barriers ($\Delta\Delta G^{\ddagger} > 10.0 \, \text{kcal mol}^{-1}$). This behavior can be rationalized by simple molecular orbital arguments. Although strict symmetry constraints are not involved in any of the pathways, the pathways including coupling species of either *cis*/*cis*-butadiene or *trans*/*trans*-butadiene with identical and opposite enantiofaces can formally be considered to occur with the preservation of $C_2$- and $C_s$-symmetry, respectively (cf. Fig. 1). Inspection of the frontier orbitals along the reaction reveals that the $C_s$-symmetrical *cis*-1,2-DVCB pathways are characterized by the preservation of the orbital symmetry (thus indicating a symmetry allowed process),[42] while the $C_2$-symmetrical *trans*-1,2-DVCB pathways appear to be symmetry forbidden, since the symmetry of occupied and vacant orbitals change along the reaction, which is similar to previous findings.[18] The same orbital symmetry arguments hold true for the formation of COD along **4a** → **10a**. In addition to these electronic factors, reductive elimination with coupling isomers of *cis*/*cis*-butadiene with identical enantiofaces requires a high barrier via **4a** → **10a** due to the *trans* orientation of the two allylic groups (cf. Fig. 1), and the formation of *trans*,*trans*-COD is accompanied with insurmountable high barriers of ∼ 60 kcal mol$^{-1}$ ($\Delta G^{\ddagger}$), owing to severe steric strain involved in the corresponding TS[**4a**–**10a**] and **10a** structures.

cis-1,2-DVCB is preferably generated along **4a** → **9a** due to both kinetic and thermodynamic reasons, and the formation of the thermodynamically favorable *cis*,*cis*-COD involves also the lowest kinetic barrier along **4a** → **10a**. Very similar total barriers (i.e., relative to the most favorable bis($\eta^3$-*syn*) isomer of **4a**) of 22.7 and 22.5 kcal mol$^{-1}$ ($\Delta G^{\ddagger}$) have to be overcome along the most feasible stereochemical pathways for the *cis*-1,2-DVCB and *cis*,*cis*-COD generating routes, that involve *cis*/*trans*-butadiene and *cis*/*cis*-butadiene coupling species with identical and opposite enantiofaces, respectively. Furthermore, reductive elimination along the routes commencing from **4a** is indicated to be reversible, since *cis*-1,2-DVCB and *cis*,*cis*-COD are formed in an endergonic and slightly exergonic process, respectively. The thermodynamic instability of **9a**, however, indicates a facile reverse oxidative addition under C–C bond cleavage. As a consequence, the intermediately formed, but thermodynamic less stable *cis*-1,2-DVCB product **9a** is likely to undergo a facile rearrangement into **10a** under thermodynamic control via **9a** → **4a** → **10a**, which is consistent with experimental observation.[6a,19] Accordingly, *cis*,*cis*-COD is predicted to be the predominantly formed cyclodimer along the reductive elimination routes that start from **4a** as the precursor.

## 2. Reductive Elimination under Ring Closure to Occur in the Dodecatrienediyl–Ni$^{II}$ Complex

The thermodynamically favorable bis($\eta^3$),$\Delta$-*cis*/*trans* configuration **7b** of the [Ni$^{II}$(dodecatrienediyl)] complex also represents the catalytically active species for reductive elimination. The new C–C $\sigma$-bond is preferably established between the terminal unsubstituted carbons on two $\eta^3$-allylic groups (Fig. 9) giving rise to the formal 16e$^-$ [Ni$^0$(CDT)] product **8b**, where CDT is coordinated to nickel by its three olefinic double bonds.

The several isomers of CDT (cf. Chart 2) are generated along competing paths for reductive elimination, which are schematically depicted in Scheme 3. For a general overview of possible CDT generation routes, the paths commencing from bis($\eta^3$),$\Delta$-*cis* isomers of **7b** are also included in Scheme 3, although the bis($\eta^3$),$\Delta$-*cis* forms are practically not present in the catalytic reaction course (cf. Section 4.4). Bis($\eta^3$-*syn*),$\Delta$-*trans* and bis($\eta^3$-*anti*),$\Delta$-*cis* isomers of **7b** are the precursors for the generation of *all-t*-CDT and *all-c*-CDT, respectively. On the other hand, two different paths can be envisioned for the formation of *c,c,t*-CDT and *c,t,t*-CDT. These include either bis($\eta^3$-allyl),$\Delta$-*trans* (from butadiene insertion into the $\eta^3$-*syn*–Ni$^{II}$ bond via **2b** $\rightarrow$ **7b**) or bis($\eta^3$-allyl),$\Delta$-*cis* (from *trans*-butadiene insertion into the $\eta^3$-*anti*–Ni$^{II}$ bond via **2b** $\rightarrow$ **7b** followed by facile allylic isomerization via TS$_{ISO}$[**6b**]) precursor species **7b**, having either identical or different configurations of the two $\eta^3$-allylic groups. The complete branch for the formation of the bis($\eta^3$-allyl),$\Delta$-*cis* forms of **7b**, however, is entirely suppressed (cf. Section 4.4) and only bis($\eta^3$-allyl),$\Delta$-*trans* isomers are present in a sufficient concentration. The precursors for the *c,t,t*-CDT and *c,c,t*-CDT generating paths, namely the $\eta^3$-*syn*,$\eta^3$-*anti*,$\Delta$-*trans* and bis($\eta^3$-*anti*),$\Delta$-*trans* isomers of **7b**, respectively, are formed in a direct fashion via **2b** $\rightarrow$ **7b** (cf. Section 4.4). Furthermore, it follows from Scheme 3, that a facile isomerization of one or both allylic groups of the directly formed isomers via TS$_{ISO}$[**6b**] is required, for making the *all-t*-CDT generating path accessible, since the direct formation of bis($\eta^3$-*syn*),$\Delta$-*trans* precursors along **2b** $\rightarrow$ **7b** is kinetically impeded (cf. Section 4.4). All the bis($\eta^3$-allyl),$\Delta$-*trans* precursors for competing paths for reductive elimination affording *c,c,t*-CDT, *c,t,t*-CDT and *all-t*-CDT should be populated to a similar amount, since they are found to be very close in energy.

The transition state TS[**7b**–**8b**] that occurs at a distance of $\sim$ 1.9–2.1 Å of the emerging C–C $\sigma$-bond can adopt two different conformations. A square-planar transition state is crossed along the path for formation of *all-t*-CDT, in which the *trans* double bond is not coordinated to nickel. On the other hand, the *c,c,t*-CDT and *c,t,t*-CDT generating paths involve a square-pyramidal transition state, in which the coordinated *trans* double bond is

## *c,c,t*-CDT generating path

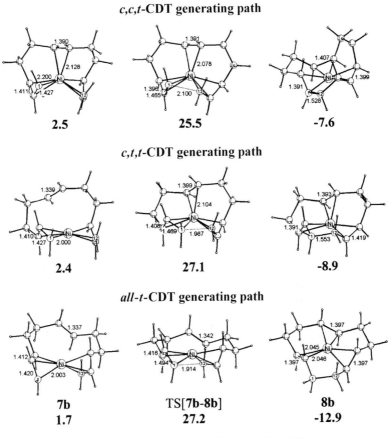

**2.5**                 **25.5**                 **-7.6**

## *c,t,t*-CDT generating path

**2.4**                 **27.1**                 **-8.9**

## *all-t*-CDT generating path

**7b**            **TS[7b-8b]**            **8b**
**1.7**               **27.2**               **-12.9**

### *all-t*-CDT generating path + assisting BD

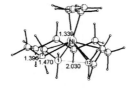

**TS[7b-8b]-BD**
**23.0**

FIG. 9. Selected geometric parameters (Å) of the optimized structures of the key species for reductive elimination via the most feasible stereochemical pathway for the competing paths affording *c,c,t*-CDT, *c,t,t*-CDT and *all-t*-CDT, respectively, along **7b** → **8b**. Free energies ($\Delta G$, $\Delta G^{\ddagger}$ in kcal mol$^{-1}$) are given relative to the favorable bis($\eta^{3}$-*anti*),$\Delta$-*trans* stereoisomer of the precursor **7b**[41] and {**7b** + C$_4$H$_6$} for the butadiene assisted *all-t*-CDT path.

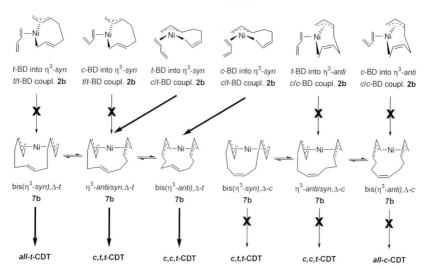

t-BD into η³-syn          c-BD into η³-syn          t-BD into η³-syn          c-BD into η³-syn          t-BD into η³-anti         c-BD into η³-anti
t/t-BD coupl. **2b**      t/t-BD coupl. **2b**      c/t-BD coupl. **2b**      c/t-BD coupl. **2b**      c/c-BD coupl. **2b**      c/c-BD coupl. **2b**

bis(η³-*syn*),Δ-*t*       η³-*anti/syn*,Δ-*t*       bis(η³-*anti*),Δ-*t*      bis(η³-*syn*),Δ-*c*       η³-*anti/syn*,Δ-*c*       bis(η³-*anti*),Δ-*c*
**7b**                    **7b**                    **7b**                    **7b**                    **7b**                    **7b**

*all-t*-CDT               *c,t,t*-CDT               *c,c,t*-CDT               *c,t,t*-CDT               *c,c,t*-CDT               *all-c*-CDT

SCHEME 3. Competing paths for formation of *all-t*-CDT, *c,t,t*-CDT, *c,c,t*-CDT and *all-c*-CDT commencing from the initial coupling product species **2b**. Butadiene insertion along **2b** → **7b** preferably takes place into the η³-allyl–Ni$^{II}$ bond of **2b**. (NB Only one of the four possible stereoisomers is displayed for each of the given species **2b** and **7b**, respectively.)

situated in an axial site (Fig. 9). As a consequence, incoming butadiene influences the three competing paths in a different way. Overall, the precursor **7b** shows no notable tendency for coordination of additional butadiene, since only weakly bonded adducts could be located, that are found to be not stabilized even in enthalpy. On the other hand, incoming butadiene is able to compete for coordination with the *trans* double bond in the square-pyramidal transition state, which, however, cannot compensate for the energy costs of the deformation of the $C_{12}$-chain. Therefore, the *c,c,t*-CDT and *c,t,t*-CDT paths are clearly indicated to be not assisted by additional butadiene. In contrast, the square-planar transition state of the *all-t*-CDT production path, which shows a notable η³-allyl → vinyl conversion of both allylic groups, is considerably stabilized by coordination of new butadiene on the empty axial site (TS[**7b–8b**]-**BD**, Fig. 9). Donor phosphines, like PMe₃, are also seen to coordinatively stabilize the square-planar transition state.[11c] The butadiene adduct of the [Ni⁰(*all-t*-CDT)] product **8b-BD**, however, is stabilized to only a minor extent in enthalpy, which is not unexpected for the weak donor butadiene. Thus, incoming butadiene participates along the *all-t*-CDT path by stabilizing the transition state, but the monomer is unlikely to assist the process at the very early and the very late stages.

Overall, the reductive elimination paths along **7b** → **8b** affording *c,c,t*-CDT and *c,t,t*-CDT, respectively, are not assisted by either incoming butadiene or the presence of $PR_3$ ligands. On the other hand, butadiene as well as donor ligands are likely serve to kinetically facilitate the *all-t*-CDT production path, which rationalizes the experimental results for the stoichiometric reactions[9a] (cf. last paragraph of Section 4.5). Among the competing paths for reductive elimination, formation of *all-t*-CDT is seen to be the most favorable due to the overall lowest total kinetic barrier (i.e., relative to the favorable bis($\eta^3$-*anti*),$\Delta$-*trans* isomer of **7b**[41] + butadiene) of $\sim 23 \, \text{kcal mol}^{-1}$ ($\Delta G^{\ddagger}$), giving rise to the thermodynamically most stable $[\text{Ni}^0(\textit{all-t}\text{-CDT})]$ compound of all product species **8a**, in a process that is $-12.9 \, \text{kcal mol}^{-1}$ exergonic (cf. Fig. 9). *All-t*-CDT is liberated through subsequent, consecutive substitution steps with butadiene in an overall exothermic process, which regenerates the active catalyst **1′b**. The *c,c,t*-CDT and *c,t,t*-CDT paths, however, are connected with higher total barriers of 25.5 and $27.1 \, \text{kcal mol}^{-1}$, respectively. This indicates that the *all-t*-CDT route is the most facile of the competing CDT production paths, while formation of *c,c,t*-CDT and *c,t,t*-CDT is less feasible.

## V

## INFLUENCE OF ELECTRONIC AND STERIC FACTORS ON INDIVIDUAL ELEMENTARY STEPS OF THE C$_8$-CYCLO-OLIGOMER GENERATING REACTION CHANNEL

So far we have been able to predict the most feasible pathway for each of the critical elementary steps of the two reaction channels and to characterize them by locating the involved key species. The $C_8$-channel has been initially explored for the generic $[\text{Ni}^0(\eta^2\text{-butadiene})_2\text{PH}_3]$ catalyst. The role of electronic and steric factors for the thermodynamic and kinetic aspects of individual elementary processes shall now be elaborated further for this channel,[11b] that represents a pre-requisite for a detailed understanding of the regulation of the cyclodimer product selectivity. Here, we will concentrate entirely on the most likely of the several stereochemical pathways for each process. A complete collection of all the stereochemical pathways can be found elsewhere.[11b] It is worth noting, that the preference for a certain path and stereochemical pathway of an individual elementary step does not change upon going from the generic catalyst to the real catalysts **I–VI**. Hence, the investigation of the generic catalyst serves to provide a principle mechanistic insight, which will be enhanced further by exploring the real catalysts **I–VI**. The catalysts **V** with L = P(OMe)$_3$, and **VI** with L = PBu$_3^t$ are known to predominantly catalyze the formation of $C_{12}$-cyclo-oligomers (cf. Section 8).[3,8] Nevertheless, catalysts **V** and **VI** are included in the exploration of the influence of electronic and steric factors, aimed at corroborating the drawn conclusions, although these catalysts will not be discussed in the context of the $C_8$-channel.

## A. *Oxidative Coupling*

Formally, oxidative coupling is accompanied by electron redistribution between the two [NiL] and [$C_8H_{12}$] moieties, such that the oxidation number of nickel increases by two ($Ni^0 \rightarrow Ni^{II}$) during this process. A high lying orbital of primarily nickel 3d character, the occupation of which changes along the process, is seen to support the **1a** $\rightarrow$ **2a** step (Fig. 10). Accordingly, σ-donor ligands are found to facilitate the oxidative coupling kinetically (cf. $\Delta\Delta E_{el}^{\ddagger}$ in Table I) relative to the generic catalyst, in two ways. On the one hand, the antibonding $3d_\sigma$–$sp_\sigma$ Ni–L interaction destabilizes the high lying occupied orbital in **1a** (Fig. 10) and on the other hand, σ-donor ligands tend to alleviate the formal electron deficiency on nickel in TS[**1a–2a**]. Both effects contribute to a reduction of the intrinsic activation barrier. Moreover, electron-donating ligands act to stabilize **2a** (see Section 5.2) and therefore, increasing σ-donor strength favors the process thermodynamically as well (cf. $\Delta\Delta E_{el}$ in Table I).

With regard to the steric factors, it follows from the $\Delta\Delta E_{st}^{\ddagger}$ contribution that the coupling of two $\eta^2$-butadienes becomes impeded for ligands that are sterically bulky. This is understandable, since the steric interactions are likely to increase during the process that commences from the trigonal planar **1a** and ends up at **2a**, where the ligand L resides in a square-planar conformation together with the $\eta^3$- and the $\eta^1(C^1)$-allylic groups.

Overall, steric and electronic factors, which are seen to be small, are found to work in opposite directions and, to some degree, cancel each other out. Consequently, the intrinsic free activation barriers and reaction free energies ($\Delta G_{int}^{\ddagger}$, $\Delta G_{int}$), respectively, span a small range for catalysts **I–IV** and differ by less than $1.0 \, kcal \, mol^{-1}$. Thus, oxidative coupling represents the one process (beside allylic isomerization, cf. Section 5.3) among all the critical elementary steps of the $C_8$-cyclodimer channel, that is least influenced by electronic and steric factors.

## B. *Thermodynamic Stability of the Several Species of the Octadienediyl–Ni$^{II}$ Complex*

The ligand's influence on the **2a** $\rightleftharpoons$ **4a** equilibrium is examined for the thermodynamically favorable stereoisomers for each one of the two species, namely the $\eta^3$-*syn*,$\eta^1(C^1)$,$\Delta$-*cis* isomer of **2a** and the bis($\eta^3$-*syn*) isomer of **4a**.

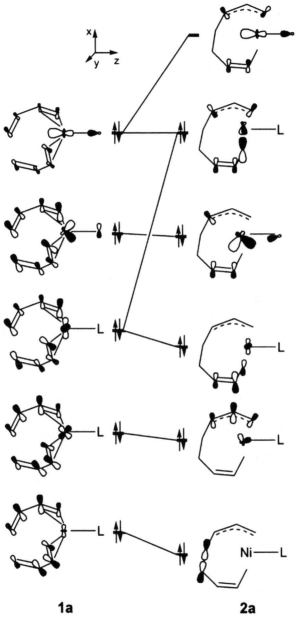

FIG. 10. Schematic correlation diagram for oxidative coupling along **1a** → **2a**, that is focused on the principally involved orbitals.

TABLE I

Estimated Electronic and Steric Contributions to Intrinsic Activation Barriers and Reaction Energies for Oxidative Coupling via the Most Feasible Pathway along $1a \rightarrow 2a^a$

|  | I | II | III | IV | V | VI |
|---|---|---|---|---|---|---|
| Catalyst | $L = PMe_3$ | $L = PPh_3$ | $L = PPr_3^i$ | $L = P(OPh)_3$ | $L = P(OMe)_3$ | $L = PBu_3^t$ |
| $\Delta\Delta E_{el}^{\ddagger}/\Delta\Delta E_{st}^{\ddagger\,b}$ | −0.25/0.02 | −0.13/0.41 | −0.40/0.51 | −0.10/0.38 | −0.05/0.03 | −0.35/0.82 |
| $\Delta\Delta E_{el}/\Delta\Delta E_{st}^{\,c}$ | −0.72/0.02 | 0.23/0.05 | −1.03/0.30 | −0.35/0.15 | −0.12/0.02 | −1.89/−0.85 |

[a] *Positive*/negative sign indicates *increased*/decreased relative barriers and reaction energies, respectively, relative to the generic [Ni$^0$($\eta^2$-butadiene)$_2$PH$_3$] catalyst. For computational details see Reference 11b.
[b] $\Delta\Delta E_{el}^{\ddagger}/\Delta\Delta E_{st}^{\ddagger}$-estimate of electronic and steric contributions (in kcal mol$^{-1}$) to the activation barriers relative to the parent generic catalyst.
[c] $\Delta\Delta E_{el}/\Delta\Delta E_{st}$-estimate of electronic and steric contributions (in kcal mol$^{-1}$) to the reaction energies relative to the parent generic catalyst.

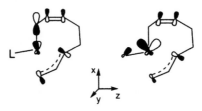

**Chart 3**. Important orbital interactions acting to stabilize the $\eta^3,\eta^1(C^1)$ form **2a** of the [Ni$^{II}$(octadienediyl)L] complex.

The relative stabilities of **2a** and **4a** are known to be influenced by the ligand's properties. As far as electronic factors are concerned, the localization of the negative charge on the $\eta^1$-allylic moiety is assisted by a σ-bonding interaction between the $C^1$-allylic π-orbital with a suitably polarized 3d-hybrid on nickel, which also involves a small contribution from the 2π-orbital of the $\eta^3$-allyl moiety (Chart 3, left side). Furthermore, the $\eta^3,\eta^1$ coordination mode is supported by a bonding interaction between the formal $C^2$–$C^3$ double bond with the properly polarized Ni $d_{xz}$ orbital, that benefits from a mixing of the ligand's L sp$_\sigma$ donor orbital (Chart 3, right side). The contribution of ligand donor orbital is seen to increase for electron-releasing ligands. Thus, an enhanced donor ability of the ancillary ligand tends to stabilize the $\eta^3,\eta^1(C^1)$ mode electronically.

This is confirmed by the estimated energetic contributions of electronic and steric factors ($\Delta\Delta E_{el}$ and $\Delta\Delta E_{st}$ in Table II) to the relative stability of **2a** and **4a**, which follows a regular trend. The increase in the ligand's donating ability correlates with the relative stabilization of **2a**, and is largest for the strong σ-donors PPr$_3^i$ and PBu$_3^t$. On the other hand, **4a** becomes

TABLE II

ESTIMATED ELECTRONIC AND STERIC CONTRIBUTIONS TO THE THERMODYNAMIC STABILITIES OF
THE FAVORABLE ISOMERS OF **2a** AND **4a** OF THE [Ni$^{II}$(OCTADIENEDIYL)L] COMPLEX$^a$

| Catalyst | I<br>L = PMe$_3$ | II<br>L = PPh$_3$ | III<br>L = PPr$_3^i$ | IV<br>L = P(OPh)$_3$ | V<br>L = P(OMe)$_3$ | VI<br>L = PBu$_3^t$ |
|---|---|---|---|---|---|---|
| $\Delta\Delta E_{el}/\Delta\Delta E_{st}$$^b$ | 1.74/0.01 | 0.58/0.30 | 3.93/0.63 | −0.45/0.46 | −0.42/0.03 | 3.69/−1.17 |

$^a$Difference in the thermodynamic stability between the bis($\eta^3$-*syn*) isomer of **4a** and the $\eta^3$-*syn*,$\eta^1$(C$^1$),$\Delta$-*cis* isomer of **2a** of the [Ni$^{II}$(octadienediyl)L] complex; *positive*/negative sign indicates *higher*/lower stability of **2a** relative to **4a**, with the generic [Ni$^{II}$(octadienediyl)PH$_3$] complex serves as the reference. For computational details see Reference 11b.
$^b\Delta\Delta E_{el}/\Delta\Delta E_{st}$-estimate of electronic and steric contributions (in kcal mol$^{-1}$) to the relative thermodynamic stabilities.

favored relative to **2a** for the π-acceptor ligands P(OPh)$_3$ and P(OMe)$_3$. Moderate steric pressure introduced by the ligand is predicted to slightly favor **2a** over **4a** by ∼0.5 kcal mol$^{-1}$. Severe steric bulk, however, acts to destabilize the square-planar species **2a** relative to the square-pyramidal species **4a**, as observed for PBu$_3^t$.

Overall, the **2a** ⇌ **4a** equilibrium is seen to be predominantly determined by electronic factors, with steric interactions having a less pronounced influence. The predicted position of the **2a** ⇌ **4a** equilibrium is in excellent agreement with experiment (cf. Section 2).[14] For the strong σ-donor PPr$_3^i$ as well as for the moderate σ-donor PMe$_3$ **2a** is favored relative to **4a** by 3.3 and 0.5 kcal mol$^{-1}$ ($\Delta G$), thus the [Ni$^{II}$(octadienediyl)L] complex is predicted to occur predominantly in the $\eta^3$,$\eta^1$(C$^1$) form **2a**. On the other hand, the bis($\eta^3$) form **4a** is preferred by 0.4 and 1.1 kcal mol$^{-1}$ ($\Delta G$) in the case of the weak σ-donor PPh$_3$ as well as the π-acceptor P(OPh)$_3$, and is therefore predicted to have the highest thermodynamic population.

## C. Allylic Isomerization and Allylic Enantioface Conversion

The conversion processes of the terminal allylic groups in the [Ni$^{II}$(octadienediyl)L] complex are seen to be influenced to only a minor extent by electronic and steric factors. Sterics are found to have a negligible effect on the intrinsic barriers for allylic enantioface conversion occurring in **2a** and for allylic isomerization via TS$_{ISO}$[3a], due to the very similar structure of the respective transition states and the corresponding direct precursor species. The σ-donor or π-acceptor ability of the ligand L also plays a minor role on the reactivity of the $\eta^3$,$\eta^1$-mode of the Ni$^{II}$-bis(allyl-anion) coordination to undergo allylic conversion processes. Accordingly, the total barriers ($\Delta G^{\ddagger}$, relative to the favorable $\eta^3$-*syn*,$\eta^1$(C$^1$),$\Delta$-*cis* isomer of **2a**) for allylic isomerization and enantioface conversion fall within a narrow range and differ by less than $1\,\mathrm{kcal\,mol^{-1}}$ for individual catalysts. For enantioface conversion, a total free-energy barrier of $\sim 10.5\,\mathrm{kcal\,mol^{-1}}$ has to be overcome, while the overall highest activation free energy for allylic isomerization amounts to $\sim 18.7\,\mathrm{kcal\,mol^{-1}}$.

The allylic conversion processes clearly represent the most feasible ones among all the critical elementary steps along the $C_8$-channel, involving the $[Ni^{II}(octadienediyl)L]$ complex. Consequently, the several forms as well as the different stereoisomers are in a dynamic, pre-established equilibrium, that can likely be assumed as always being attained.

## D. *Reductive Elimination*

Reductive elimination under C–C bond formation involves a formal electron redistribution between the [NiL] and $[C_8H_{12}]$ fragments, giving rise to a reduction of the oxidation number on nickel by two, namely $Ni^{II} \rightarrow Ni^0$. A low-lying acceptor d-orbital on nickel, that is able to mix efficiently into the $2\pi$-orbital of the octadienediyl fragment, is indicated to kinetically facilitate the process by stabilizing the transition state (Fig. 11). $\pi$-Acceptor ligands decrease the energetic gap between these two orbitals, giving rise to a more efficient mixing, and therefore serve to accelerate the reductive elimination by diminishing the activation barrier.

The estimated electronic contributions to the activation barriers ($\Delta\Delta E_{\mathrm{el}}^{\ddagger}$) for the VCH generating route confirm the supposition that $\pi$-acceptor ligands act to stabilize TS[**2a**–**8a**] relative to **2a**, giving rise to a lowering of the intrinsic activation barrier, that correlates inversely with the ligand's acceptor strength (Table III). On the other hand, $\sigma$-donor ligands raise the activation energy uniformly by $\sim 1\,\mathrm{kcal\,mol^{-1}}$, when compared to the generic catalyst. This leads to the conclusion, that the kinetic barrier along **2a** $\rightarrow$ **8a**, with regard to the electronic influence, is indicated to be predominantly determined by the ancillary ligand's $\pi$-acceptor ability. Steric pressure on the ligand is seen to reduce the activation barrier as well ($\Delta\Delta E_{\mathrm{st}}^{\ddagger}$, Table III), which is understandable, since this process goes along with a formal reduction of the coordination number on nickel from 4 to 3 along **2a** $\rightarrow$ **8a**. Moderate steric bulk ($L = PMe_3$, $PPh_3$, $P(OMe)_3$) has a minor effect and for this case the elimination barrier is predicted to be mainly determined by electronic factors. With increase of the steric pressure, the steric effect becomes prevalent, which is most remarkable for the bulky, space-demanding $PBu_3^t$. Similar to the trends obtained for the kinetic barrier, the formation of VCH is also favored thermodynamically by the increase of the $\pi$-acceptor ability of the ligand. The dominant electronic effect for the stabilization of **8a** is the [VCH] $\rightarrow$ [NiL] donation, along with the support due to steric bulk on the ligand ($\Delta\Delta E_{\mathrm{el}}$, $\Delta\Delta E_{\mathrm{st}}$, Table III).

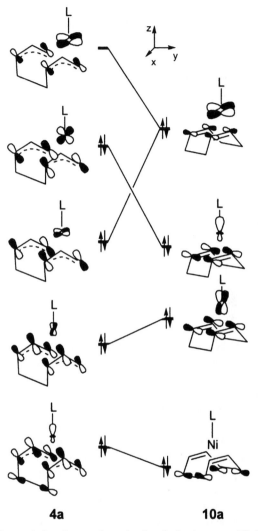

**4a**                    **10a**

FIG. 11. Schematic correlation diagram for reductive elimination exemplified for the **4a** → **10a** route, that is focused on the principally involved orbitals.

Among catalysts **I–IV**, which predominantly catalyze the generation of cyclodimer products, the overall lowest intrinsic free-energy barrier of 20.5 kcal mol$^{-1}$ ($\Delta G_{int}^{\ddagger}$) for **2a** → **8a** appears for catalyst **IV** with L = P(OPh)$_3$, where both electronic and steric factors are seen to assist the formation of VCH to a similar amount. Reductive elimination involves a higher intrinsic barrier ($\Delta G_{int}^{\ddagger}$) for catalysts bearing moderately bulky, donor phosphines,

TABLE III

ESTIMATED ELECTRONIC AND STERIC CONTRIBUTIONS TO INTRINSIC ACTIVATION BARRIERS AND
REACTION ENERGIES FOR REDUCTIVE ELIMINATION AFFORDING VCH VIA THE MOST FEASIBLE
PATHWAY ALONG $2a \rightarrow 8a^a$

| Catalyst | I<br>$L = PMe_3$ | II<br>$L = PPh_3$ | III<br>$L = PPr_3^i$ | IV<br>$L = P(OPh)_3$ | V<br>$L = P(OMe)_3$ | VI<br>$L = PBu_3^t$ |
|---|---|---|---|---|---|---|
| $\Delta\Delta E_{el}^{\ddagger}/\Delta\Delta E_{st}^{\ddagger\,b}$ | 1.13/−0.02 | 0.95/−0.25 | 1.11/−1.91 | −1.04/−0.91 | −1.15/−0.05 | 0.84/−6.88 |
| $\Delta\Delta E_{el}/\Delta\Delta E_{st}^{\,c}$ | 1.58/−0.03 | 1.10/−1.08 | 1.18/−1.92 | −0.55/−1.00 | −0.45/−0.07 | 1.00/−7.30 |

$^{a,b,c}$See Table I.

TABLE IV

ESTIMATED ELECTRONIC AND STERIC CONTRIBUTIONS TO INTRINSIC ACTIVATION BARRIERS AND
REACTION ENERGIES FOR REDUCTIVE ELIMINATION AFFORDING *CIS,CIS*-COD VIA THE MOST
FEASIBLE PATHWAY ALONG $4a \rightarrow 10a^a$

| Catalyst | I<br>$L = PMe_3$ | II<br>$L = PPh_3$ | III<br>$L = PPr_3^i$ | IV<br>$L = P(OPh)_3$ | V<br>$L = P(OMe)_3$ | VI<br>$L = PBu_3^t$ |
|---|---|---|---|---|---|---|
| $\Delta\Delta E_{el}^{\ddagger}/\Delta\Delta E_{st}^{\ddagger\,b}$ | −0.85/0.01 | −0.65/−0.12 | −0.73/−1.43 | −1.84/0.09 | −1.53/0.02 | −0.93/−2.61 |
| $\Delta\Delta E_{el}/\Delta\Delta E_{st}^{\,c}$ | −1.39/−0.02 | −1.94/−0.12 | −3.06/−1.42 | −2.81/−0.99 | −2.41/−0.06 | −2.52/−5.91 |

$^{a,b,c}$See Table I.

that amounts to 23.7 and 23.3 kcal mol$^{-1}$ for **I** with L = PMe$_3$ and **II** with L = PPh$_3$, respectively, owing to the limited π-acceptor ability of these ligands. The kinetic hindrance caused by the electronic destabilization of the transition state can, however, be compensated by introducing steric bulk on the ligand. Accordingly, the intrinsic barrier is lowered to 21.8 kcal mol$^{-1}$ ($\Delta G_{int}^{\ddagger}$) for **III** bearing the bulky, donor PPr$_3^i$.

Reductive elimination commencing from the bis($\eta^3$) precursor **4a**, affording *cis,cis*-COD as the predominant product (cf. Section 4.6.1) is influenced in a similar fashion, but to a different extent, by electronic and steric factors as predicted for the VCH generating route. Electron-donating ligands serve to reduce the intrinsic activation barrier uniformly by ~0.8 kcal mol$^{-1}$, when compared with the generic catalyst ($\Delta\Delta E_{el}^{\ddagger}$, Table IV). The ligand's π-acceptor strength is indicated to represent the critical electronic factor for reductive elimination. This process is facilitated by increasing the ligand's π-acceptor ability, since the barriers are predicted to be reduced further for the π-acceptors P(OMe)$_3$ and P(OPh)$_3$. On the other hand, space demanding ligands act to stabilize the transition state ($\Delta\Delta E_{st}^{\ddagger}$, Table IV) and thus assist the **4a** → **10a** process as well.

The catalysts **III** with L = PPr$_3^i$ and **IV** with L = P(OPh)$_3$ show the overall lowest and very similar intrinsic barriers ($\Delta G_{int}^{\ddagger}$) of 12.6 and 12.4 kcal mol$^{-1}$, respectively, for the *cis,cis*-COD route, which however are determined by different factors. The moderate barriers are the result of the stabilization of TS[4a–10a] relative to the bis($\eta^3$-*anti*) precursor 4a, which for III is mainly due to the reduced steric interaction, while for IV the $\pi$-acceptor ability of P(OPh)$_3$ is clearly the decisive factor, with sterics playing a minor role. Higher intrinsic elimination barriers ($\Delta G_{int}^{\ddagger}$) of 13.7 and 13.8 kcal mol$^{-1}$ have to be overcome for I and II (L = PMe$_3$ and PPh$_3$, respectively), since steric factors here have a minor influence on the barrier.

It is worth noting, that the mechanistic conclusions for the competing 4a → 9a and 4a → 10a routes drawn for the generic catalyst are corroborated for the real catalysts I–IV. The formation of *cis*-1,2-DVCB and *cis,cis*-COD is connected with very similar total activation barriers (i.e., relative to the favorable bis($\eta^3$-*syn*) isomer of 4a) for each of the individual catalysts. Furthermore, *cis,cis*-COD is clearly seen to be the thermodynamically preferred product of the two cyclodimers. The difference in the thermodynamic stability between the [Ni$^0$($\eta^4$-cyclodimer)L] products 9a and 10a is most remarkable for IV with L = P(OPh)$_3$ and amounts to 6.7 kcal mol$^{-1}$ ($\Delta G$). This confirms the conclusion (cf. Section 4.6.1), that *cis,cis*-COD is generated as the predominant product along the reductive elimination routes that commence from the bis($\eta^3$) precursor 4a.

# VI

# THEORETICALLY REFINED CATALYTIC CYCLES FOR THE NICKEL-CATALYZED CYCLO-OLIGOMERIZATION REACTIONS OF 1,3-BUTADIENE

In the preceding sections, all critical elementary steps have been scrutinized for the C$_8$- and C$_{12}$-cyclo-oligomer generating reaction channels. Furthermore, the roles of electronic and steric factors have been elucidated for individual elementary processes of the C$_8$-channel. On the basis of the original mechanistic proposal of Wilke *et al.* and on recent computational mechanistic investigations,[11] we now present a refined mechanistic view of the nickel-catalyzed cyclo-oligomerization of 1,3-butadiene. For this purpose condensed free-energy schemes are provided for the complete cycles of the C$_8$- (Scheme 4) and C$_{12}$-channel (Scheme 5), consisting of the most feasible routes for the important elementary reaction processes.

## A. C$_8$-Cyclo-Oligomer Generating Reaction Channel

The formal 16e$^-$ trigonal planar [Ni$^0$($\eta^2$-butadiene)$_2$L] species 1a is the thermodynamically favorable form of the active catalyst, with the

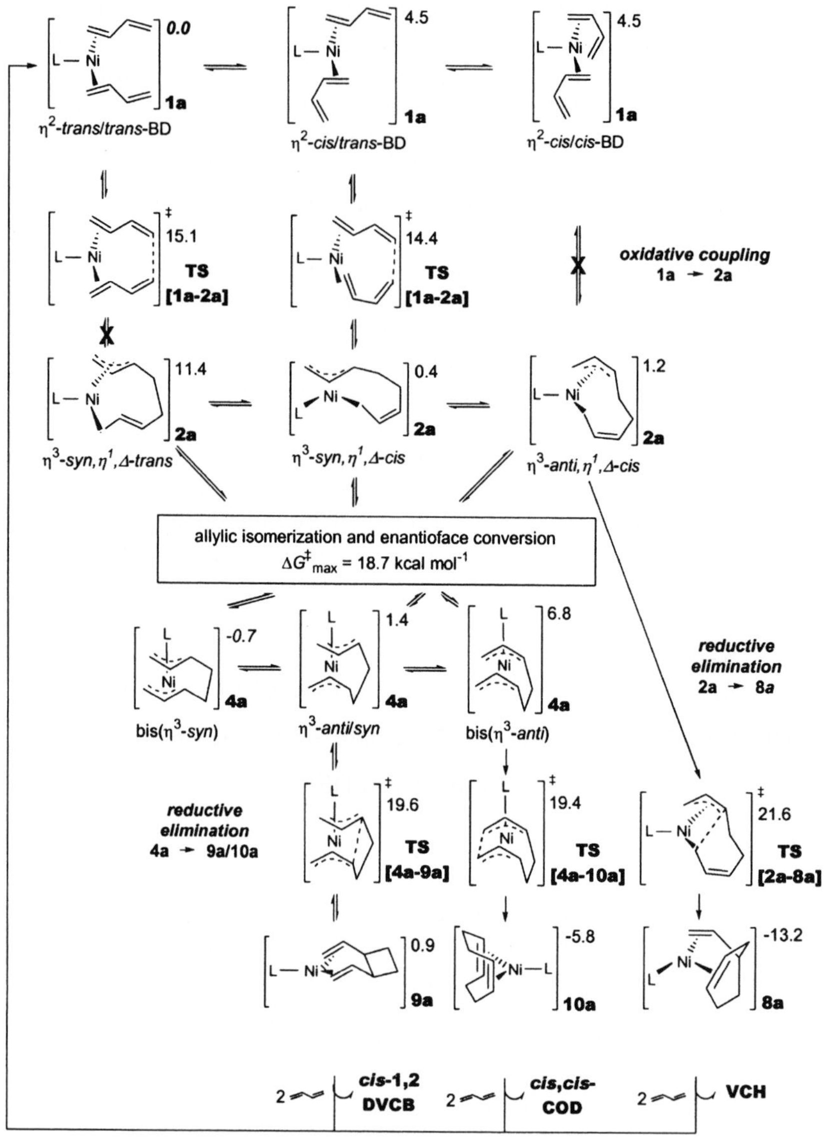

SCHEME 4. Condensed free-energy profile (kcal mol$^{-1}$) of the complete catalytic cycle of the C$_8$-reaction channel of the nickel-catalyzed cyclo-oligomerization of 1,3-butadiene for catalyst **IV** with L = P(OPh)$_3$. The favorable [Ni$^0$($\eta^2$-*trans*-butadiene)$_2$L] isomer of the active catalyst **1a** was chosen as reference and the activation barriers for individual steps are given relative to the favorable stereoisomer of the respective precursor (given in italics; **4a** for both allylic conversion and reductive elimination).

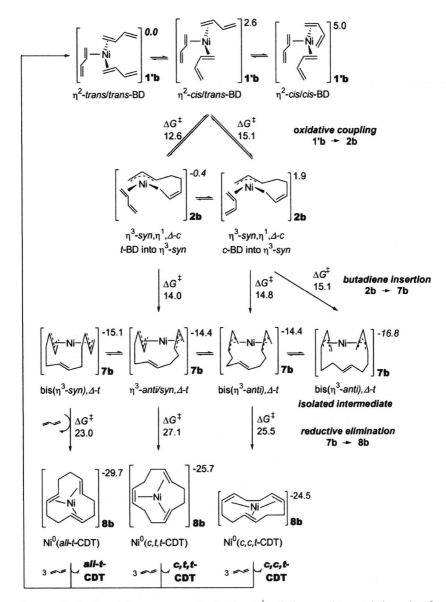

**SCHEME 5.** Condensed free-energy profile (kcal mol$^{-1}$) of the complete catalytic cycle of the C$_{12}$-reaction channel of the nickel-catalyzed cyclo-oligomerization of 1,3-butadiene, focused on viable routes for individual elementary steps. The favorable [Ni$^0$($\eta^2$-*trans*-butadiene)$_3$] isomer of the active catalyst **1'b** was chosen as reference and the activation barriers for individual steps are given relative to the favorable stereoisomer of the respective precursor (given in italics).

bis($\eta^2$-*trans*) isomers being most stable, and also represents the catalytically active form that undergoes oxidative coupling along **1a** → **2a**. Oxidative coupling is seen to occur preferably via establishing of a C–C $\sigma$-bond between the terminal noncoordinated carbons of the two $\eta^2$-butadienes, affording the $\eta^3,\eta^1(C^1)$-octadienediyl–Ni$^{II}$ species **2a** as the initial coupling product. The most feasible pathway occurs via coupling of $\eta^2$-*trans*/$\eta^2$-*cis*-butadiene involving a moderate kinetic barrier, which also gives rise to the thermodynamically favored $\eta^3$-*syn*,$\eta^1(C^1)$,$\Delta$-*cis* isomer of **2a** in a thermoneutral process (Scheme 4). This indicates that the oxidative coupling is reversible. The coupling of two $\eta^2$-*cis*-butadienes is impeded due to the associated higher barriers, while the direct formation of bis-$\eta^2$-*trans* coupling species **2a**, although kinetically manageable, is disabled by a significantly more facile reverse **2a** → **1a** process due to highly strained, thermodynamically unfavorable coupling species **2a** in this case.

Among the several configurations of the crucial [Ni$^{II}$(octadienediyl)L] complex, all of which are in equilibrium, the $\eta^3,\eta^1(C^1)$ species **2a** and the bis($\eta^3$) species **4a** are predicted to be prevalent. The $\sigma$-donor/$\pi$-acceptor ability of the ancillary ligand is shown to predominantly determine the position of the kinetically mobile **2a** ⇌ **4a** equilibrium. The conversion of the terminal allylic groups via allylic isomerization and/or allylic enantioface conversion are indicated to be the most facile of all the elementary processes that involve the [Ni$^{II}$(octadienediyl)L] complex. Consequently, the several octadienediyl–Ni$^{II}$ configurations and their stereoisomers are likely to be in a dynamic pre-established equilibrium, that can be assumed to be always present.

Bis($\eta^1$)-octadienediyl–Ni$^{II}$ species are shown: (i) to be thermodynamically highly unfavorable, thus indicating them to be sparsely populated, and (ii) not to be involved as reactive intermediates along any viable path either for allylic isomerization or for reductive elimination. This leads to the conclusion, that bis($\eta^1$) species play no role within the catalytic reaction cycle.

Oxidative coupling via **1a** → **2a** and the reductive elimination routes, that commence from **4a** as the precursor, involve different stereoisomers along the most feasible pathway. Accordingly, the conversions of the terminal allylic groups of the [Ni$^{II}$(octadienediyl)L] complex represent indispensable elementary processes.

The formation of the three principal C$_8$-cyclo-oligomers occurs via reductive elimination along three competing reaction routes, which preferably proceeds through direct paths involving the prevalent octadienediyl–Ni$^{II}$ species **2a** and **4a**. VCH is generated along **2a** → **8a**, commencing from the $\eta^3$-*anti*,$\eta^1(C^1)$,$\Delta$-*cis* isomer of **2a**, and $\eta^3$-*syn*,$\eta^3$-*anti* and bis($\eta^3$-*anti*) isomers of **4a** serve as the precursor for the formation of *cis*-1,2-DVCB along **4a** → **9a** and the **4a** → **10a** *cis*,*cis*-COD production route, respectively.

Along the routes that start from **4a**, *cis*-1,2-DVCB and *cis,cis*-COD are the stereoisomers that are exclusively formed due to both kinetic and thermodynamic considerations. Reductive elimination is rate-determining in the catalytic cycle, since this step is connected with the highest kinetic barriers among all the critical elementary steps (Scheme 4). VCH is the thermodynamically preferred of the three principal cyclodimers. Accordingly, its formation is driven by the strongest thermodynamic force in an exergonic process, which indicates that the VCH generating route along **2a** → **8a** is irreversible. The production routes for *cis*-1,2-DVCB and *cis,cis*-COD have very similar activation barriers, but the corresponding elimination products **9a** and **10a** (as well as the free cyclodimers) display distinctly different stabilities. The thermodynamically preferred *cis,cis*-COD is formed in a slightly exergonic process, while the thermodynamically less stable intermediately formed *cis*-1,2-DVCB is predicted to readily undergo conversion under thermodynamic control into *cis,cis*-COD along **9a** → **4a** → **10a**. Consequently, VCH and *cis,cis*-COD are the prevalent products of the [Ni$^0$L]-catalyzed cyclo-oligomerization of 1,3-butadiene.

The cyclodimers are liberated from the respective elimination products **8a** and **10a** via successive substitution processes with incoming butadiene, that regenerates the active catalyst **1a** in an overall exergonic process. For the rate determining reductive elimination step of the C$_8$-channel free-energy activation barriers of 20.1–24.1 kcal mol$^{-1}$ are predicted for catalysts **I–IV**, that are in excellent agreement with experimental estimates.[43] Thus, moderate reaction conditions are required for the catalytic cyclodimerization of 1,3-butadiene.[6]

## B. $C_{12}$-Cyclo-Oligomer Generating Reaction Channel

Among the several possible forms, formal 16e$^-$ trigonal planar [Ni$^0$($\eta^2$-butadiene)$_3$] compound **1′b**, with the tris($\eta^2$-*trans*-butadiene) isomers being most favorable, are indicated to be the prevalent species of the active catalyst. **1′b** also represents the precursor for the favorable route for oxidative addition under C–C bond formation, that, in analogy to the C$_8$-channel, occurs between the terminal noncoordinated carbons of two reactive $\eta^2$-butadiene moieties and is assisted by an ancillary butadiene in the $\eta^2$-mode. Similarly to the C$_8$-channel, the coupling preferably takes place between $\eta^2$-*trans* and $\eta^2$-*cis*-butadiene due to both kinetic and thermodynamic reasons, affording the $\eta^3$-*syn*,$\eta^1$(C$^1$),$\Delta$-*cis* isomer of **2b** as the initial coupling product in a thermoneutral reaction. Oxidative coupling via the most feasible pathway along **1′b** → **2b** is indicated to be a reversible process that requires a moderate activation barrier (Scheme 5).

Alternative pathways for coupling of two *cis*-butadiene or two *trans*-butadiene moieties, however, are unfeasible on considerations similar to those discussed for the $C_8$-channel (cf. Section 6.1).

The dominant dodecatrienediyl–$Ni^{II}$ production path occurs through the insertion of butadiene into the $\eta^3$-allyl–$Ni^{II}$ bond of the initial coupling product **2b**. The most feasible pathway involves butadiene insertion into the $\eta^3$-*syn*–$Ni^{II}$ bond of the predominantly formed $\eta^3$-*syn*,$\eta^1(C^1)$,$\Delta$-*cis* stereoisomer of **2b**. Similar, moderate activation barriers are associated with butadiene insertions from its s-*cis* and s-*trans* configuration, giving rise to bis($\eta^3$-*anti*),$\Delta$-*trans* and $\eta^3$-*syn*/$\eta^3$-*anti*,$\Delta$-*trans* isomers of **7b**. Isomerization of the terminal allylic groups of the octadienediyl–$Ni^{II}$ complex, that is most likely to proceed via the $\eta^3$,$\eta^1(C^3)$ $TS_{ISO}$[**3b**], is not a necessary process in the catalytic reaction course, since identical stereoisomers are involved along the most feasible pathways for oxidative coupling and subsequent butadiene insertion. Butadiene is inserted irreversibly into the $\eta^3$-allyl–$Ni^{II}$ bond along **2b** → **7b** in a highly exergonic process (Scheme 5). Accordingly, the thermodynamically favorable bis($\eta^3$) species **7b** of the [$Ni^{II}$(dodecatrienediyl)] complex serves as a thermodynamic sink.

The complete branch for formation of bis(allyl),$\Delta$-*cis*-dodecatrienediyl–$Ni^{II}$ forms is shown to be disabled, because of: (i) the unfavorable coupling of two *cis*-butadienes along **1′b** → **2b** together with a slow isomerization via $\eta^3$-*anti*,$\eta^1(C^3)$ isomers of $TS_{ISO}$[**3b**], which prevents a sufficient concentration of $\eta^3$-*anti*,$\eta^1(C^1)$,$\Delta$-*cis* precursors **2b**, and (ii) owing to a kinetically impeded butadiene insertion into the $\eta^3$-*anti*-allyl–$Ni^{II}$ bond along **2b** → **7b**. Consequently, the *all-c*-CDT production route is entirely precluded.

The bis($\eta^3$-allyl),$\Delta$-*trans*-dodecatrienediyl–$Ni^{II}$ form of **7b** is the direct precursor for reductive elimination, which has been shown to preferably occur along with the formation of a C–C $\sigma$-bond between the terminal carbons of two $\eta^3$-allylic groups. Bis($\eta^1$) species of the octadienediyl–$Ni^{II}$ and dodecatrienediyl–$Ni^{II}$ complexes are clearly demonstrated to play no role in the catalytic reaction cycles of the nickel-catalyzed cyclooligomerization of 1,3-butadiene, either in the $C_8$- or the $C_{12}$-production channel. Commencing from the thermodynamically highly favored species **7b**, reductive elimination is connected with the overall largest kinetic barrier among all critical elementary steps and gives rise to the [$Ni^0$(CDT)] product **8b** in an exergonic irreversible process (Scheme 5). Thus, reductive elimination along **7b** → **8b** is predicted to be rate-controlling.

Of the four principal isomers of CDT, the three accessible CDT-isomers are generated along competing **7b** → **8b** paths, with the bis($\eta^3$-*anti*),$\Delta$-*trans*, $\eta^3$-*syn*/$\eta^3$-*anti*,$\Delta$-*trans*, and bis($\eta^3$-*syn*),$\Delta$-*trans* isomers of **7b** acting as the precursors for the production of *c,c,t*-CDT, *c,t,t*-CDT, and *all-t*-CDT, respectively. The first two isomers of **7b** are formed in a direct way through feasible pathways for **2b** → **7b**, while the bis($\eta^3$-*syn*),$\Delta$-*trans* precursor for the *all-t*-CDT path, although not directly generated, is readily available from these isomers via facile allylic isomerization through the $\eta^3,\eta^1(C^3)$-dodecatrienediyl–Ni$^{II}$ TS$_{ISO}$[**6b**]. The thermodynamically most stable bis($\eta^3$-*anti*),$\Delta$-*trans* stereoisomers of **7b** with *trans* oriented allylic groups,[41] that have been located as reactive intermediates in stoichiometric reactions, are formed under catalytic reaction conditions via feasible pathways for **2b** → **7b**. These stereoisomers, however, constitute dead-end points within the catalytic cycle and they first have to undergo allylic isomerization in order to make feasible pathways for reductive elimination accessible.

The *c,c,t*-CDT and *c,t,t*-CDT production paths are shown to be not assisted by incoming butadiene, while the square-planar transition state involved along the *all-t*-CDT path is significantly stabilized by an axial coordination of butadiene. Hence, the *all-t*-CDT route becomes the most facile of the three CDT production paths with a free-energy barrier for reductive elimination of $\sim 23\,\mathrm{kcal\,mol^{-1}}$, that perfectly corresponds with experimental estimates.[44] Accordingly, the production of $C_{12}$-cyclo-oligomers requires moderate reaction conditions,[9] although **7b** represents a thermodynamic sink within the catalytic cycle.

The cyclotrimer products are liberated in subsequent, consecutive substitution steps with new butadiene, which is an exothermic reaction ($\Delta H$) for expulsion of *all-t*-CDT by three *trans*-butadienes along **8b** → **1'b**. This process, however, is endergonic by $\sim 7\,\mathrm{kcal\,mol^{-1}}$ ($\Delta G$) after entropic costs are taken into account. Therefore, **7b** and **8b** (stabilized by donors) are indicated to be likely candidates for isolable intermediates of the catalytic process, while the active catalyst **1'b**, the intermediate species **2b**, in particular, and other species will not be present in a sufficient concentration, since they are either too reactive or thermodynamically too unfavorable, for experimental characterization. Overall, the cyclotrimerization process is driven by a strong thermodynamic force with an exothermicity of $-44.6\,\mathrm{kcal\,mol^{-1}}$ ($\Delta H$ for the process without a catalyst) for the fusion of three *trans*-butadiene to afford the favorable *all-t*-CDT.

# VII

# REGULATION OF THE CYCLO-OLIGOMER PRODUCT SELECTIVITY

On the basis of the free-energy profiles presented so far for the refined catalytic reaction cycles for the $C_8$- and $C_{12}$-cyclo-oligomer production cycles, we now rationalize the critical factors that are decisive for the regulation of the product selectivity for the two reaction channels.

## A. Selectivity Control for the Reaction Channel Affording $C_8$-Cyclo-Oligomer Products

The important mechanistic aspects discussed so far are: firstly, the several configurations of the [Ni$^{II}$(octadienediyl)L] complex, with **2a** and **4a** being the prevalent species, and their various stereoisomeric forms occurring in a kinetically mobile pre-established equilibrium, prior to the rate-determining reductive elimination step. Secondly, **2a** and **4a** represent the precursors for the competing routes for reductive elimination affording VCH along **2a** → **8a** and *cis,cis*-COD along **4a** → **10a** as the principal products of the $C_8$-cyclo-oligomer channel.

For this typical Curtin–Hammett situation,[45] the selectivity of the cyclodimer formation is, as a consequence, entirely determined kinetically by the ratio of the entire reductive elimination barriers (i.e., their difference in free-energy of activation, $\Delta\Delta G^{\ddagger}$) for the competing VCH and *cis,cis*-COD generating routes. Two aspects are of interest for elucidating the selectivity control. On the one hand, the catalytic activity as well as the selectivity is regulated by the concentration of the active precursors **2a** and **4a**, which is determined by the position of the pre-established, dynamic equilibrium between these two species. This describes the thermodynamically related aspect. The thermodynamic population of the $\eta^3,\eta^1(C^1)$ and bis($\eta^3$) species of the [Ni$^{II}$(octadienediyl)L] complex has been shown to be primarily determined by the electronic properties of the ancillary ligand L, with steric factors playing a minor role (cf. Section 5.2). The bis($\eta^3$) species **4a** is the prevalent species for $\pi$-acceptor ligands as well as for weak $\sigma$-donors, while the **2a** ⇌ **4a** equilibrium becomes displaced to the left upon increasing the ligand's $\sigma$-donor strength. The [Ni$^{II}$(octadienediyl)L] complex exists predominantly in the $\eta^3,\eta^1(C^1)$ configuration **2a** for strong $\sigma$-donors. On the other hand, the intrinsic reactivity of **2a** and **4a** is a further important point for the selectivity control. The reductive elimination is predicted to be facilitated kinetically by the increase of the ligand's $\pi$-acceptor ability as well as by bulky, space demanding ligands L. In the absence of high steric pressure, the electronic influence is prevalent, while for ligands that are sterically bulky the steric factor becomes dominant (cf. Section 5.4). The competing VCH and *cis,cis*-COD production routes, however, are not influenced in a uniform way by the electronic and steric properties of the ancillary ligand, L. Among the two routes, the VCH route is seen to be affected by electronic and steric factors to a larger extent.

Heimbach *et al.*[8] conducted a careful experimental investigation of the influence of the ancillary ligand on the cyclodimer product distribution. Strong $\pi$-acceptors that are sufficiently space-demanding were found to catalyze the formation of *cis,cis*-COD almost exclusively. The VCH portion at first becomes enlarged with the increase in the ligand's σ-donor strength, passing through a maximum with a VCH: *cis,cis*-COD product ratio of ∼ 50:50, and decreases afterwards for strong σ-donor ligands, with *cis,cis*-COD becoming predominant. From their statistical analysis Heimbach *et al.* concluded,[8] that electronic factors are entirely decisive for the cyclodimer product distribution and steric factors overall have no pronounced influence.

In agreement with experiment (VCH: *cis,cis*-COD product ratio of 24:74 for **II** and of 9:91 for **IV**),[3c,6a,8b,8c] the catalysts **II** and **IV** bearing the weak σ-donor PPh$_3$ and the π-acceptor P(OPh)$_3$, respectively, are predicted to predominantly catalyze the formation of *cis,cis*-COD, since **4a** → **10a** is indicated to be more facile than **2a** → **8a** due to a barrier that is 2.8 (**II**) and 2.2 kcal mol$^{-1}$ (**IV**) ($\Delta\Delta G^{\ddagger}$) lower for the *cis,cis*-COD route.[11b] The pronounced preference of 2.8 kcal mol$^{-1}$, in advance of the *cis,cis*-COD route, that is predicted for PPh$_3$, seems to be slightly overestimated, since **2a** is likely to be populated to a certain extent for weak σ-donors (cf. Section 5.2), that would give rise to a moderate VCH portion of ∼ 24%.[3c,6a,8b,8c] For catalyst **IV** *cis,cis*-COD should almost exclusively be generated, while the **2a** → **8a** route affording VCH is likely to be essentially suppressed. Both thermodynamic and kinetic considerations are seen to favor the *cis,cis*-COD generating route. π-Acceptor ligands are shown: (i) to displace the pre-established **2a** ⇌ **4a** equilibrium strongly far to the right (cf. Section 5.2) and thus give rise to a high thermodynamic population of the precursor **4a**, and (ii) to reduce the kinetic barrier (cf. Section 5.4). Consequently, the increase in the *cis,cis*-COD selectivity is accompanied by an increase of the catalytic reactivity.[3c,6a] The catalytic activity together with the *cis,cis*-COD selectivity can be enhanced further by π-acceptor ligands that are sterically bulky (cf. Section 5.4), due to kinetic reasons (that leads to a further reduction of the activation barrier) and to a lesser extent also due to thermodynamic reasons (that increase of the concentration of **4a**). This is consistent with the experimental observation that the highest reaction rate together with the highest yield of *cis,cis*-COD is obtained for the bulky P(OC$_6$H$_4$–*o*-Ph)$_3$ ligand (96%).[3c,6a,8a,8b,43]

As shown in Section 5.4, the reductive elimination barrier becomes higher with increase of the ligand's σ-donor strength. The activation barrier, however, is not uniformly raised for the two competing routes. The VCH generating route is indicated to be retarded in a more pronounced way by electronic factors when compared with the route for *cis,cis*-COD formation (cf. Section 5.4). This handicap, however, is compensated for by the displacement of the pre-established **2a** ⇌ **4a** equilibrium to the left for strong σ-donors, thus leading to an enhanced thermodynamic population of **2a**. Consequently, the portion of VCH becomes enlarged up to a certain amount that is connected with a decrease in the catalytic activity. Both aspects are consistent with experiment.[8b,8c] The subtle balance between these two effects is likely to be responsible for the fact that the maximal yield of VCH never exceeds 55%.[3c,6a,8b,8c]

For catalysts **I** and **III** with the stronger σ-donors PMe$_3$ and PPr$_3^i$ the similar activation barriers for the formation of VCH and *cis,cis*-COD ($\Delta\Delta G^{\ddagger} = 0.7$ and $0.3\,\text{kcal mol}^{-1}$ with a slight preference for the *cis,cis*-COD route[11b]) indicate that the two routes have a comparable probability to be passed through. Accordingly, the *cis,cis*-COD portion becomes reduced to approximately 50% for these catalysts, which is in excellent agreement with experiment (VCH: *cis,cis*-COD product ratio of 35:56 for the catalyst with L = PEt$_3$, that is similar to **I** and of 34:57 for **III**).[3c,8b,8c,46]

## B. *Selectivity Control for the Reaction Channel Affording C$_{12}$-Cyclo-Oligomer Products*

It follows from the free-energy profile (cf. Scheme 5) that the production of C$_{12}$-cyclo-oligomers can be formally considered to occur in two phases. In the first stage, the crucial dodecatrienediyl–Ni$^{II}$ complex is formed commencing from the active catalyst **1′b** through facile oxidative coupling and butadiene insertion processes, with the intermediately generated octadienediyl–Ni$^{II}$ complex clearly seen to be highly reactive. In a second part, the thermodynamically favored bis($\eta^3$),Δ-dodecatrienediyl–Ni$^{II}$ species **7b**, that constitutes a thermodynamic sink in the catalytic cycle, undergoes reductive elimination, affording three of the four possible CDT stereoisomers along competing paths via **7b** → **8b**. The several stereoisomers of **7b** must be considered to exist in a dynamic, pre-established equilibrium prior to the rate-controlling reductive elimination step. Furthermore, the respective bis($\eta^3$-*syn*),Δ-*trans*, bis($\eta^3$-*anti*),Δ-*trans*, and $\eta^3$-*syn*/$\eta^3$-*anti*,Δ-*trans* precursors **7b** for the competing *all-t*-CDT, *c,c,t*-CDT, and *c,t,t*-CDT generating paths are very close in energy, thus indicating that they occur in comparable concentrations. Incoming butadiene does not serve to displace the equilibrium between the precursors.

Accordingly, the product selectivity is entirely regulated kinetically by the difference of the entire reductive elimination barriers ($\Delta\Delta G^{\ddagger}$)[45] for the three competing paths. The *all-t*-CDT path, that is assisted by incoming butadiene, is indicated to be the most facile of the three competing paths, since it involves the lowest activation barriers overall. Thus, *all-t*-CDT is predicted to be generated as the predominant C$_{12}$-cyclo-oligomer, together with significantly smaller amounts of *c,c,t*-CDT and *c,t,t*-CDT, since their formation is kinetically retarded by elimination barriers that are more than

2.5 kcal mol$^{-1}$ ($\Delta\Delta G^{\ddagger}$) higher (cf. Scheme 5). The fourth possible *all-c*-CDT product will never be formed in the nickel-catalyzed cyclo-oligomerization of 1,3-butadiene, since the complete branch for generation of bis($\eta^3$),$\Delta$-*cis* isomers of **7b** is completely suppressed due to both thermodynamic and kinetic considerations (cf. Section 4.4).

## VIII

## REGULATION OF THE C$_8$:C$_{12}$-CYCLO-OLIGOMER PRODUCT DISTRIBUTION

So far we have analyzed the catalytic structure–activity relationships for the regulation of the product selectivity for the cyclo-oligomerization reaction to proceed either via the C$_8$- or the C$_{12}$-product channel. The catalytic cyclo-oligomerization, however, is known to not afford exclusively either C$_8$- or C$_{12}$-cyclo-oligomer products, but a mixture of both, with the predominant product depending on the active catalyst.[6,8,9] As outlined in the outset, PR$_3$/P(OR)$_3$-stabilized nickel complexes catalyze, in most cases, the formation of cyclodimers as the prevalent product, while so-called 'bare' nickel catalysts predominantly yield cyclotrimer products. For these two major types of nickel catalysts, the critical factors that are responsible for directing the cyclo-oligomerization process into one of the two competing C$_8$- and C$_{12}$-cyclo-oligomer production channels will now be elucidated.

The initial coupling $\eta^3,\eta^1$(C$^1$)-octadienediyl–Ni$^{II}$ compound is likely to represent the critical species that connects the alternative reaction channels. The concentration of the respective species **2a** and **2b** and their reactivity when undergoing subsequent butadiene insertion or reductive elimination, which is accompanied by intramolecular C–C bond formation, represent important factors for the regulation of the C$_8$:C$_{12}$ product ratio.

### A. Cyclo-Oligomerization with Zerovalent PR$_3$/P(OR)$_3$-Stabilized Nickel Complexes as the Catalyst

Scheme 6 displays the interplay of the two reaction channels for this type of nickel catalyst. The electronic and steric properties of the ancillary ligand are shown to have a pronounced influence on the kinetic and thermodynamic aspects of the VCH (via **2a** → **8a**) and *cis,cis*-COD (via **4a** → **10a**) generating routes (cf. Section 5.4) and will also affect the substitution process of the PR$_3$/P(OR)$_3$ ligand by incoming butadiene in the octadienediyl–Ni$^{II}$ complex (*vide infra*). The species **2a** and **2b** are likely to

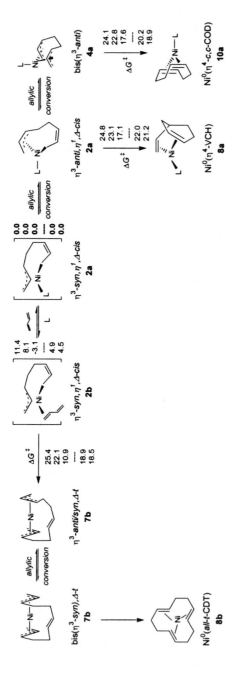

**SCHEME 6.** Interplay of the C₈- and C₁₂-production channels for the cyclo-oligomerization of 1,3-butadiene with zerovalent PR₃/P(OR)₃-stabilized nickel complexes as the catalyst. Free energies ($\Delta G$, $\Delta G^{\ddagger}$ in kcal mol$^{-1}$) are given relative to the favorable $\eta^3$-$syn$,$\eta^1$(C$^1$),$\Delta$-$cis$ isomer of **2a** for catalysts bearing strong $\sigma$-donor ligands; namely **I** (L = PMe₃), **III** (L = PPr$^i_3$), **VI** (L = PBu$^i_3$), and $\pi$-acceptor ligands; namely **V** (L = P(OMe)₃), **IV** (L = P(OPh)₃), respectively.

216

be in a pre-established equilibrium, since a facile substitution can reasonably be supposed to occur.[47] Thus, on account of the Curtin–Hammett principle,[45] the difference in the entire kinetic barriers ($\Delta\Delta G^{\ddagger}$) for reductive elimination (along **2a** → **8a** or **4a** → **10a**) and butadiene insertion (along **2b** → **7b**) entirely discriminates which of the two reaction channels is passed through.

For $PR_3/P(OR)_3$-stabilized nickel complexes, there are two borderline cases known from the experimental investigation of Heimbach *et al.*,[8a] which, unlike the usual behavior, redirect the cyclo-oligomerization reaction into the $C_{12}$-cyclo-oligomer production channel. Catalysts bearing either strong σ-donor ligands that must also introduce severe steric pressure (e.g., $PBu^tPr^i_2$) or sterically compact π-acceptors (like $P(OMe)_3$) are known to yield CDT as the predominant product. From a statistical analysis it was concluded,[8a,8c] that the $C_8 : C_{12}$-cyclo-oligomer product ratio is mainly determined by steric factors (75%) with electronic factors are less important.

These two cases will now be analyzed in the context of Scheme 6. Strong σ-donors serve to strengthen the Ni–L bond in **2a**, thus leading the substitution equilibrium between **2a** and **2b** strongly far on the side of **2a**. For catalyst **I** with the sterically compact strong σ-donor $PMe_3$ the energetic gap between **2a** and **2b** amounts to 11.4 kcal mol$^{-1}$ ($\Delta G$) in advance of **2a**. Accordingly, for this case, the $C_{12}$-production channel is not accessible due to the sparsely populated species **2b**, giving rise to a total barrier **2a** → **7b** (that includes the thermodynamic gap between **2a** and **2b** together with the overall lowest barrier of 14.0 kcal mol$^{-1}$ for **2b** → **7b**, Section 4.6.1) that is higher than the barrier for the competitive **2a** → **8a** VCH and **4a** → **10a** *cis,cis*-COD generating routes. Steric pressure introduced by the ligand tends to support both channels, but to a different extent. On the one hand, increasing steric bulk weakens the Ni–L bond in **2a**, thus displacing the substitution equilibrium toward **2b**. On the other hand, steric bulk has been shown to kinetically facilitate the reductive elimination along **2a** → **8a** and **4a** → **10a** (cf. Section 5.4). Between the two effects, the first is indicated to be more prominent. The gap between **2a** and **2b** decreases from 11.4 kcal mol$^{-1}$ ($\Delta G$) for catalysts **I** (L = $PMe_3$) to 8.1 kcal mol$^{-1}$ for **III** (L = $PPr^i_3$) and to $-3.1$ kcal mol$^{-1}$ for **VI** (L = $PBu^t_3$), where **2b** is suggested to be prevalent, which is more pronounced than the decrease of the elimination barriers for $C_8$-cyclodimer formation (cf. Scheme 6). Severe steric pressure on σ-donor ligands primarily serves to enhance the thermodynamic population of **2b** and therefore acts in this direction to make the $C_{12}$-cyclo-oligomer production cycle more likely to be entered. Overall, catalyst **I** with the sterically compact $PMe_3$ should preferably catalyze the formation of cyclodimers, since this channel is favorable by 0.6–1.3 kcal mol$^{-1}$ ($\Delta\Delta G^{\ddagger}$), while larger portions of CDT can be expected for catalysts **III** and **VI** bearing the space-demanding $PPr_3^i$, $PBu_3^t$. This is indicated by $\Delta\Delta G^{\ddagger}$ values of 0.7–1.0 and 6.2–6.7 kcal mol$^{-1}$, respectively, in favor of the C$_{12}$-channel (cf. Scheme 6). For σ-donor ligands that introduce severe steric pressure, CDT is clearly indicated to be the predominantly formed product. Experimental measurements provided the following C$_8$ : C$_{12}$ product ratios for the catalysts investigated:[8a] for catalyst **I** (L = PMe$_3$, which can be considered as an appropriate model for the experimentally examined catalyst with L = PEt$_3$) 65 : 29, for **III** (L = PPr$_3^i$) 69 : 24, and for **VI** (L = PBu$_3^t$, for which the experimentally investigated catalyst with L = PBu$^t$Pr$_2^i$ is most similar) 46 : 50. Although the computed $\Delta\Delta G^{\ddagger}$ values do not exactly reproduce the experimentally observed C$_8$ : C$_{12}$-cyclo-oligomer product ratios,[8a] the predicted trends are regular and allow a detailed understanding and a consistent rationalization of the experimental results.

For catalysts bearing ligands that are π-acceptors, the situation is different. Here, the substitution equilibrium does not strongly favor **2a** as for the moderate bulky σ-donors, because of the reduced stability of the Ni–L bond due to electronic reasons. This already gives rise to a certain population of **2b** without the steric factors coming into play. Thus, the two reaction channels are likely to be kinetically comparable for sterically compact π-acceptors. Although increasing steric bulk on π-acceptors also serves to favor both channels, the role played by steric factors here is quite different from that reported for ligands that are strong σ-donors. Firstly, the thermodynamic gap between **2a** and **2b** of 4.9 kcal mol$^{-1}$ ($\Delta G$) for catalyst **V** (L = P(OMe)$_3$) is reduced to 4.5 kcal mol$^{-1}$ for catalyst **IV** with L = P(OPh)$_3$, which is mainly due to steric factors. Secondly, and indicated to be the more pronounced effect, moderate steric bulk on the ligand together with its increasing π-acceptor strength act to lower the activation energy along the favored *cis,cis*-COD route from 20.2 kcal mol$^{-1}$ ($\Delta G^{\ddagger}$ for **2a** → **10a**) for **V** to 18.9 kcal mol$^{-1}$ ($\Delta G^{\ddagger}$) for **IV**. Furthermore, severe steric pressure introduced by the ligand is shown to facilitate the *cis,cis*-COD route to a considerable extent, as exemplified for PBu$_3^t$ (cf. Section 5.4). Accordingly, the dominant role of steric bulk for π-acceptor ligands is to kinetically accelerate the C$_8$-channel. This is confirmed by the estimated difference of the discriminating **4a** → **10a** and **2b** → **7b** barriers ($\Delta\Delta G^{\ddagger}$) for the two channels. For catalyst **V**, formation of CDT is predicted to be favored by 1.3 kcal mol$^{-1}$ ($\Delta\Delta G^{\ddagger}$), while both channels are indicated to be kinetically equivalent ($\Delta\Delta G^{\ddagger} = 0.4$ kcal mol$^{-1}$) for **IV** (cf. Scheme 6). Experimental measurements provided C$_8$ : C$_{12}$ product ratios of 38 : 60 for catalyst **V** (L = P(OMe)$_3$) and 87 : 12 for **IV** (L = P(OPh)$_3$).[8a] Similar to the findings for the selectivity control for catalysts that carry σ-donor ligands, the $\Delta\Delta G^{\ddagger}$ estimates do not exactly match the experimental selectivities, but the enlightened regular trends for the contribution of electronic and steric effects lead to a consistent understanding and interpretation of the experiment.

### B. Cyclo-Oligomerization with Zerovalent 'Bare' Nickel Complexes as the Catalyst

This type of catalyst is characterized by a weakly coordinating ligand sphere around the nickel that can be readily displaced, in a facile substitution, by incoming butadiene. Therefore, the $[Ni^0(\eta^2$ butadiene)$_3]$ compound **1'b** is the active catalyst species and **2b** represents the crucial species to enter into the two competing reaction channels (Scheme 7). Similar to the catalysts discussed in the previous section, the precursors **2b** and **4b-BD** for the critical reductive elimination and butadiene insertion steps are in a dynamic pre-established equilibrium. Thus, the Curtin–Hammett principle[45] can be applied to elucidate the $C_8 : C_{12}$ product selectivity, which accordingly is entirely determined by the absolute reactivity of the octadienediyl–$Ni^{II}$ complex to undergo the competing butadiene insertion (making the $C_{12}$-channel accessible via **2b** → **7b**) and reductive elimination processes (making the $C_8$-channel accessible via **2b** → **8'b** and/or **4b-BD** → **10'b**).

Passing through the $C_8$-production cycle involves an overall lowest total barrier of 17.2 and 16.8 kcal mol$^{-1}$ ($\Delta G^{\ddagger}$, relative to the **2b**) for the VCH and *cis,cis*-COD generating routes, respectively (cf. Scheme 7). On the other hand, butadiene insertion along **2b** → **7b** requires a free-energy of activation of 14.0 kcal mol$^{-1}$ (cf. Section 4.4). Accordingly, the $C_{12}$-production channel is favored, while the alternative channel for generation of cyclodimers is indicated to be kinetically impeded by higher elimination barriers of 2.8–3.2 kcal mol$^{-1}$ ($\Delta\Delta G^{\ddagger}$). In agreement with experiment,[9a] cyclotrimer products, with *all-t*-CDT as the prevalent isomer, are predicted to be the major products of the cyclo-oligomerization of 1,3-butadiene catalyzed by zerovalent 'bare' nickel compounds, and cyclodimers, with VCH and *cis,cis*-COD should occur in similar amounts, formed only in minor portions.

## IX

## CONCLUSION

In this account we have presented a consistent and theoretically well-founded view of the catalytic structure–reactivity relationships for the nickel-catalyzed cyclo-oligomerization of 1,3-butadiene, which represents one of the key reactions in homogeneous catalysis that, furthermore, has

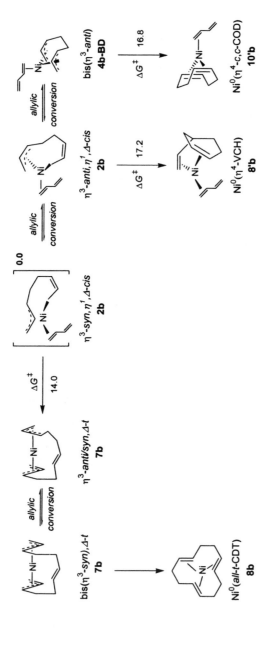

SCHEME 7. Interplay of the $C_8$- and $C_{12}$-production channels for the cyclo-oligomerization of 1,3-butadiene with zerovalent 'bare' nickel complexes as the catalyst. Free energies ($\Delta G$, $\Delta G^{\ddagger}$ in kcal mol$^{-1}$) are given relative to the favorable $\eta^3$-$syn$,$\eta^1$(C$^1$),$\Delta$-$cis$ isomer of 2b.

been explored in great detail by experiment. This account clearly illustrates the level reached by theoretical methods for the exploration of catalytic structure–reactivity relationships and gives a demonstrated example of how computational modeling can substantially contribute to the elucidation of reaction mechanisms.

On the basis of the original proposal by Wilke *et al.* the important elementary steps of the $C_8$- and $C_{12}$-cyclo-oligomer generating reaction channels have been critically scrutinized by means of an accurate quantum-chemical method. The first mechanism suggested by Wilke *et al.* was confirmed in essential details, but supplemented by novel insights into how the catalytic cyclo-oligomerization reaction operates. We have been able: (1) to predict the most feasible reaction path for each of the critical elementary steps of the $C_8$- and $C_{12}$-cyclo-oligomer production channels by scrutinizing several conceivable paths, (2) to characterize each elementary step pinpointing the crucial structural, electronic and energetic (including both thermodynamic and kinetic) features, (3) to elucidate the role of electronic and steric factors for individual elementary processes, and (4) to rationalize on what considerations a particular path is preferred or handicapped among the several plausible paths for a given process. This leads us: (1) to propose refined catalytic reaction cycles for the $C_8$- and $C_{12}$-production channels, (2) to identify and enlighten the determining factors for the regulation of the selectivity of the $C_8$- and $C_{12}$-cyclo-oligomer formation, and (3) to derive the catalytic structure–reactivity relationships for the nickel-catalyzed cyclo-oligomerization of 1,3-butadiene.

#### ACKNOWLEDGEMENTS

I would like to thank Professor Dr. Tom Ziegler (University of Calgary) for providing the opportunity to conduct important parts of this investigation during a stay in his research group and for his generous support and advice. Furthermore, I am grateful to Professor Dr. Rudolf Taube for his ongoing interest in this research, which was a continual stimulus. Excellent service by the computer centers URZ Halle and URZ Magdeburg is gratefully acknowledged. Finally, I thank the Deutsche Forschungsgemeinschaft (DFG) for financial support by a Habilitandenstipendium.

#### REFERENCES

(1) (a) Jolly, P. W.; Wilke, G. The oligomerization and co-oligomerization of butadiene and substituted 1,3-dienes *The Organic Chemistry of Nickel. Vol. 2, Organic Synthesis,* Academic Press, New York, 1975. pp. 133–212. (b) Jolly, P. W. Nickel-catalyzed oligomerization of 1,3-dienes related reactions, (Wilkinson, G., Stone, F. G. A., Abel, E. W. Eds.), *Comprehensive Organometallic Chemistry,* Vol. 8, Pergamon, New York, 1982. pp. 671–711.

(2) (a) Wilke, G. *Angew. Chem.* **1957**, *69*, 397. (b) Breil, H.; Heimbach, P.; Kröner, M.; Müller, H.; Wilke, G. *Makromol. Chem.* **1963**, *69*, 18. (c) Baker, R. *Chem. Rev.* **1973**, *73*, 487.

(3) (a) Wilke, G. *Angew. Chem. Int. Ed. Engl.* **1963**, *2*, 105. (b) Wilke, G. *Angew. Chem. Int. Ed. Engl.* **1988**, *27*, 185. (c) Wilke, G.; Eckerle, A. Cyclooligomerizations and cyclo-co-oligomerizations of 1,3-dienes, (Cornils, B., Herrmann, W. A. Eds.) *Applied Homogeneous Catalysis with Organometallic Complexes*, VCH, Weinheim, Germany, 1996. pp. 358–373.

(4) Reed, H. W. B. *J. Chem. Soc.* **1954**, 1931.

(5) (a) Wilke, G. *Pure Appl. Chem.* **1978**, *50*, 677. (b) Wilke, G. *J. Organomet. Chem.* **1980**, *200*, 349.

(6) (a) Brenner, W.; Heimbach, P.; Hey, H.; Müller, E. W.; Wilke, G. *Liebigs Ann. Chem.* **1967**, *727*, 161. (b) The catalytic cyclodimerization has been performed under the following reaction conditions: 80 °C, normal pressure, and nickel:ligand ratio 1:1.

(7) (a) A fourth cyclodimer of butadiene, the 1-methylene-2-vinylcyclopentane (MVCP), is formed in the presence of secondary alcohols. The reaction channel that affords this product, however, is not investigated in the present study. (b) Furukawa, J.; Kiji, J.; Konishi, H.; Yamamoto, K. *Makromol. Chem.* **1973**, *174*, 65.

(8) (a) Heimbach, P.; Kluth, J.; Schenkluhn, H.; Weimann, B. *Angew. Chem. Int. Ed. Engl.* **1980**, *19*, 569. (b) Heimbach, P.; Kluth, J.; Schenkluhn, H.; Weimann, B. *Angew. Chem. Int. Ed. Engl.* **1980**, *19*, 570. (c) Kluth J. Ph.D. Thesis, University of Essen-Gesamthochschule, 1980.

(9) (a) Bogdanovic, B.; Heimbach, P.; Kröner, M.; Wilke, G. *Liebigs Ann. Chem.* **1967**, *727*, 143. (b) The catalytic cyclotrimerization has been performed under the following conditions: 40–80 °C in liquid butadiene under pressure.

(10) Wilke, G.; Kröner, M.; Bogdanovic, B. *Angew. Chem.* **1961**, *71*, 755.

(11) (a) Tobisch, S.; Ziegler, T. *J. Am. Chem. Soc.* **2002**, *124*, 4881. (b) Tobisch, S.Ziegler, T. *J. Am. Chem. Soc.* **2002**, *124*, 13,290. (c) Tobisch S. *Chem. Eur. J.* **2003**, *9*, accepted.

(12) Heimbach, P.; Jolly, P. W.; Wilke, G. *Adv. Organomet. Chem.* **1970**, *8*, 29.

(13) (a) Mango, F. D. *Adv. Catal.* **1969**, *20*, 291. (b) Mango, F. D. *Tetrahedron Lett.* **1969**, 4813.

(14) Benn, R.; Büssemeier, B.; Holle, S.; Jolly, P. W.; Mynott, R.; Tkatchenko, I.; Wilke, G. *J. Organomet. Chem.* **1985**, *279*, 63.

(15) (a) Jolly, P. W.; Tkatchenko, I.; Wilke, G. *Angew. Chem. Int. Ed. Engl.* **1971**, *10*, 329. (b) Brown, J. M.; Golding, B. T.; Smith, M. J. *Chem. Commun.* **1971**, 1240. (c) Barnett, B.; Büssemeier, B.; Heimbach, P.; Jolly, P. W.; Krüger, C.; Tkatchenko, I.; Wilke, G. *Tetrahedron Lett.* **1972**, 1457. (d) Büssemeier B. Ph.D. Thesis, University of Bochum, 1973. (e) Jolly, P. W.; Mynott, R.; Salz, R. *J. Organomet. Chem.* **1980**, *184*, C49.

(16) (a) Heimbach, P.; Hey, H. *Angew. Chem. Int. Ed. Engl.* **1970**, *9*, 528. (b) Buchholz, H.; Heimbach, P.; Hey, H. J.; Selbeck, H.; Wiese, W. *Coord. Chem. Rev.* **1972**, *8*, 129. (c) Heimbach, P. *Angew. Chem. Int. Ed. Engl.* **1973**, *12*, 975.

(17) (a) Yamamoto, A.; Morifuji, K.Ikeda, S.Saito, T.Uchida, Y.Misono, A. *J. Am. Chem. Soc.* **1968**, *90*, 1878. (b) Graham, C. R.Stephenson, L. M. *J. Am. Chem. Soc.* **1977**, *99*, 7098.

(18) Heimbach, P.; Traunmüller, R. Chemie der Metall-Olefin-Komplexe, Verlag Chemie GmbH, Weinheim, 1970.

(19) Heimbach, P.; Brenner, W. *Angew. Chem. Int. Ed. Engl.* **1967**, *6*, 800.

(20) Jolly, P. W.; Mynott, R. *Adv. Organomet. Chem.* **1981**, *19*, 257.

(21) Henc, B.; Jolly, P. W.; Salz, R.; Wilke, G.; Benn, R.; Hoffmann, E. G.; Mynott, R.; Schroth, G.; Seevogel, K.; Sekutowski, J. C.; Krüger, C. *J. Organomet. Chem.* **1980**, *191*, 425.

(22) Henc, B.; Jolly, P. W.; Salz, R.; Stobbe, S.; Wilke, G.; Benn, R.; Mynott, R.; Seevogel, K.; Goddard, R.; Krüger, C. *J. Organomet. Chem.* **1980**, *191*, 449.

(23) Wilke, G.; Müller, E. W.; Kröner, M. *Angew. Chem.* **1961**, *73*, 33.

(24) (a) Wilke, G. *Angew. Chem.* **1960**, *72*, 581. (b) Dietrich, H.; Schmidt, H. *Naturwiss.* **1965**, *52*, 301. (c) Brauer, D. J.; Krüger, C. *J. Organomet. Chem.* **1972**, *44*, 397. (d) Jonas, K.; Heimbach, P.; Wilke, G. *Angew. Chem. Int. Ed. Engl.* **1968**, *7*, 949.

(25) Hoffmann, E. G.; Jolly, P. W.; Küsters, A.; Mynott, R.; Wilke, G. *Z. Naturforsch.* **1976**, *B31*, 1712.

(26) (a) Tolman, C. A. *J. Am. Chem. Soc.* **1970**, *92*, 2953. (b) Tolman, C. A. *Chem. Rev.* **1977**, *77*, 313.

(27) (a) Gonzalez-Banco, O.; Branchadell, V. *Organometallics* **1997**, *16*, 5556. (b) Senn, H. M.; Deubel, D. V.; Bloechl, P. E.; Togni, A.; Frenking, G. *J. Mol. Struct. (THEOCHEM)* **2000**, *506*, 233.

(28) (a) Jolly, C. A.; Chan, F.; Marynick, D. S. *Chem. Phys. Lett.* **1990**, *174*, 320. (b) Müller, B.; Reinhold, J. *Chem. Phys. Lett.* **1992**, *196*, 363. (c) Pacchioni, G.; Bagus, P. S. *Inorg. Chem.* **1992**, *31*, 4391. (d) Dias, P. B.; Minas de Piedade, M. E.; Martinho Simoes, J. A. *Coord. Chem. Rev.* **1994**, *135/136*, 737. (e) Howard, S. T.; Foreman, J. P.; Edwards, P. G. *Can. J. Chem.* **1997**, *75*, 60.

(29) (a) A complete collection of all the stereochemical pathways for the crucial elementary processes of the $C_8$- and $C_{12}$-cyclo-oligomer reaction channels can be found elsewhere (Reference 11). (b) Detailed descriptions of the employed computational methodology are provided elsewhere (Reference 11). (c)For details of the computational procedure for estimating the energetic contribution of electronic and steric effects of the real $PR_3/P(OR)_3$ ligands, the reader is referred to Reference 11b.

(30) For an overview see the recent special issue Computational Transition Metal Chemistry. *Chem. Rev.* **2000**, *100*, 351–818.

(31) (a) Dirac, P. A. M. *Proc. Cambridge Philos. Soc.* **1930**, *26*, 376. (b) Slater, J. C. *Phys. Rev.* **1951**, *81*, 385. (c) Vosko, S. H.; Wilk, L.; Nussiar, M. *Can. J. Phys.* **1980**, *58*, 1200. (d) Becke, A. D. *Phys. Rev.* **1988**, *A38*, 3098. (e) Perdew, J. P. *Phys. Rev.* **1986**, *B33*, 8822. (f) Perdew, J. P. *Phys. Rev.* **1986**, *B34*, 7406.

(32) (a) Bernardi, F.; Bottoni, A.Calcinari, M.Rossi, I.Robb, M. A. *J. Phys. Chem.* **1997**, *101*, 6310. (b) Jensen, V. R.Børve, K. *J. Comput. Chem.* **1998**, *19*, 947.

(33) (a) Warshel, A.; Levitt, M. *J. Mol. Biol.* **1976**, *103*, 227. (b) Singh, U. C.; Kollman, P. A. *J. Comput. Chem.* **1986**, *7*, 718. (c) Field, M.; Bash, P. A.; Karplus, M. *J. Comput. Chem.* **1990**, *11*, 700. (d) Maseras, F.; Morokuma, K. *J. Comput. Chem.* **1995**, *16*, 1170. (e) Woo, T. K.; Cavallo, L.; Ziegler, T. *Theor. Chem. Acc.* **1998**, *100*, 307.

(34) (a) Maseras, F., (Cundari, T. R. Ed.) *Computational Organometallic Chemistry*, Marcel Dekker, Basel, 2001. pp. 159–183. (b) Deng, L.; Woo, T. K.; Margl, P. M.; Ziegler, T. *J. Am. Chem. Soc.* **1997**, *119*, 6177. (c) Svensson, M.; Humbel, S.; Froese, R. D. J.; Matsubara, T.; Sieber, S.; Morokuma, K. *J. Phys. Chem.* **1996**, *100*, 19,357.

(35) (a) Full MM3(96) parameter set including π-systems. (b) Allinger, N. L.; Yuh, Y. H.; Lii, J. H. *J. Am. Chem. Soc.* **1989**, *111*, 8551. (c) Lii, J. H.; Allinger, N. L. *J. Am. Chem. Soc.* **1989**, *111*, 8566. (d) Lii, J. H.; Allinger, N. L. *J. Am. Chem. Soc.* **1989**, *111*, 8576. (e) Allinger, N. L.; Geise, H. J.; Pyckhout, W.; Paquette, L. A.; Gallucci, J. C. *J. Am. Chem. Soc.* **1989**, *111*, 1106. (f) Allinger, N. L.; Li, F.; Yan, L. *J. Comput. Chem.* **1990**, *11*, 848. (g) Allinger, N. L.; Li, F.; Yan, L.; Tai, J. C. *J. Comput. Chem.* **1990**, *11*, 868.

(36) Lukas, J.; van Leeuwen, P.W.N.M.; Volger, H. C.; Kouwenhoven, A. P. *J. Organomet. Chem.* **1973**, *47*, 153.

(37) (a) Faller, J. W.; Thomsen, M. E.; Mattina, M. J. *J. Am. Chem. Soc.* **1971**, *93*, 2642. (b) Vrieze, K. Fluxional allyl complexes, (Jackman, L. M., Cotton, F. A. Eds.) *Nuclear Magnetic Resonance Spectroscopy*. Academic Press, New York, **1975**, pp. 441–487.

(38) Tobisch, S.; Taube, R. *Organometallics* **1998**, *18*, 3045.

(39) (a) Tobisch, S. *Acc. Chem. Res.* **2002**, *35*, 96. (b) Tobisch, S. *Chem. Eur. J.* **2002**, *8*, 4756. (c) Tobisch S. To be submitted for publication.

(40) (a) Margl, P. M.; Woo, T. K.; Ziegler, T. *Organometallics* **1998**, *17*, 4997. (b) Lieber, S.; Prosenc, M. H.; Brintzinger, H. H. *Organometallics* **2000**, *19*, 377.

(41) The two stereoisomers of the bis($\eta^3$-*anti*),$\Delta$-*trans* intermediate **7b** confirmed by NMR (References 20, 21) are distinguished by the *trans* double bond coordination of opposite enantiofaces. Different from experiment, the major stereoisomer (established by NMR) with a parallel orientation of the *trans* double bond is predicted to be 0.6 kcal mol$^{-1}$ ($\Delta G$) less favorable than the minor stereoisomer with a nonparallel double bond arrangement.

(42) (a) Hoffmann, R.; Woodward, R. B. *Acc. Chem. Res.* **1968**, *1*, 17. (b) Hoffmann, R.; Woodward, R. B. *Angew. Chem. Int. Ed. Engl.* **1969**, *8*, 781.

(43) We have estimated the experimental free-energy of activation for the bulky $\pi$-acceptor P(OC$_6$H$_4$–*o*-Ph)$_3$, that almost exclusively catalyzes the formation of *cis,cis*-COD (96% selectivity), by the following crude approximation: with a TOF of 780 g butadiene (g Ni)$^{-1}$ h$^{-1}$ at 353 K (Reference 3c) one obtains an effective rate constant $k \sim 0.23$ s$^{-1}$ and $\Delta G^{\ddagger} \sim 21.8$ kcal mol$^{-1}$, by applying the Eyring equation with $k = 2.08 \times 10^{10}$ T exp($-\Delta G^{\ddagger}$/RT). This value agrees very well with the calculated barrier of 20.1 kcal mol$^{-1}$ ($\Delta G^{\ddagger}$) for catalyst **IV** (L = P(OPh)$_3$) for the rate-determining reductive elimination along **4a** $\rightarrow$ **10a**, relative to the overall most stable species of the catalytic cycle, namely the bis($\eta^3$-*syn*) species **4a**.

(44) The overall free-energy of activation for the [Ni$^0$]-catalyzed cyclotrimerization of 1,3-butadiene, with *all-t*-CDT formed as the predominant product ($\sim$90% selectivity), has been estimated from available experimental data by the following crude approximation: with a TOF of 4/75 g butadiene (g Ni)$^{-1}$ h$^{-1}$ at 313/353 K (Reference 9a), one obtains effective rate constants $k \sim 1.21 \times 10^{-3}/2.26 \times 10^{-2}$ s$^{-1}$ and $\Delta G^{\ddagger} \sim 22.5/23.4$ kcal mol$^{-1}$, respectively, by applying the Eyring equation with $k = 2.08 \times 10^{10}$ T exp($-\Delta G^{\ddagger}$/RT). These values agree very well with the computational estimated total barrier of $\sim$23 kcal mol$^{-1}$ ($\Delta G^{\ddagger}$) for the most feasible stereochemical pathway of the *all-t*-CDT generating route.

(45) (a) Seemann, J. I. *Chem. Rev.* **1983**, *83*, 83. (b) Seemann, J. I. *J. Chem. Educ.* **1985**, *63*, 42.

(46) A *cis,cis*-COD yield of 56–57% was observed for catalysts with the strong $\sigma$-donors PEt$_3$ and PPr$_3^i$ (References 8b, 8c). Catalyst **I** with L = PMe$_3$ can be considered as a suitable model for the experimentally examined catalyst with L = PEt$_3$.

(47) (a) Basolo, F., Pearson, R. G. Eds. Mechanisms in Inorganic Chemistry, G. Thieme Verlag, Stuttgart, Germany, 1973. (b) Cross, R. J. *Chem. Soc. Rev.* **1985**, *14*, 197. (c) Tobe, M. L., (Wilkinson, G.Gillard, R. D., McCleverty, J. A. Eds.), *Comprehensive Coordination Chemistry*, Vol. 1, Pergamon Press, New York, 1987. pp. 81–384. (d) Cross, R. J. *Adv. Inorg. Chem.* **1989**, *34*, 219.

# Group 13/15 Organometallic Compounds—Synthesis, Structure, Reactivity and Potential Applications

## STEPHAN SCHULZ*

*Institut für Anorganische Chemie der Universität Bonn,
Gerhard-Domagk-Str. 1, D-53121 Bonn, Germany*

# I

# INTRODUCTION

Group 13/15 compounds have a long-standing history in inorganic chemistry and have been known for almost two centuries. First reports on such compounds go back to 1809, when Gay Lussac synthesized $F_3B \leftarrow NH_3$,[1] the historical prototype of a Lewis acid–base adduct, by reaction of $BF_3$ and $NH_3$. Since this initial study, numerous Lewis acid–base adducts of boranes, alanes, gallanes and indanes $MX_3$, $MH_3$ and $MR_3$ (M = B, Al, Ga, In; X = F, Cl, Br, I; R = alkyl, aryl) of the type $R_3M \leftarrow ER'_3$ (E = N, P, As) (Type A) have been synthesized and

*Phone: Int +0228 735326, Fax: Int +0228 735327; E-mail: sschulz@uni-bonn.de

225

ADVANCES IN ORGANOMETALLIC CHEMISTRY
VOLUME 49 ISSN 0065-3055/DOI 10.1016/S0065-3055(03)49006-0

©2003 Elsevier Science (USA)
All rights reserved.

structurally characterized both in the solid state and in the gas phase, demonstrating the general tendency of electron-deficient, Lewis acidic (*electron pair acceptor*) group 13 trihalides, trihydrides and triorganyls to react with electron-rich Lewis bases (*electron pair donor*) under Lewis acid–base adduct formation. In addition, hypercoordinated adducts $MX_3(ER'_3)_2$ (Type B) containing five-coordinated group 13 metal centers have also been studied in detail (Fig. 1).

The second major class of group 13/15 organometallic compounds that has been investigated in detail is represented by monomeric, heterocyclic and cage-like compounds of the type $[R_2MER'_2]_x$ (Types C, D, E) and $[RMER']_x$ ($x \geq 2$) (Types F, G, H) featuring regular σ-bonds consisting of *2-electron-2-center-bonds* between group 13 and group 15 elements. Epoch-making studies in this field have been performed by Stock and Poland, who synthesized borazine $B_3N_3H_6$, sometimes referred to as "*inorganic benzene*" due to its electronic and structural analogy to $C_6H_6$, and by Wiberg, who demonstrated the complete range of group 13/15 chemistry by investigating the reaction of $AlH_3$ and $NH_3$ in detail,[2] and others (Scheme 1). However,

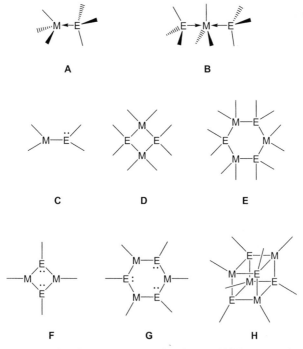

FIG. 1. Preferred structure geometries of group 13/15 compounds.

besides the synthesis of such important new compounds, the development of new synthetic strategies and new preparation techniques such as the *Schlenck technique* and the *vacuum line technique*, allowing the storage, handling and manipulation of extremely air- and moisture-sensitive pyrophoric compounds, were the most striking breakthroughs of their studies. These techniques are still essential for every preparative chemist working today in any field of modern inorganic and organometallic chemistry or material sciences.

According to their electronic structure, compounds of the type $[R_2MER'_2]_x$ typically form four- or six-membered heterocycles ($x = 2$, 3). Monomeric compounds ($x = 1$), which are of particular interest due to their bonding properties ($\pi$-bonding parts?), usually require sterically demanding substituents (*kinetic stabilization*) at both the group 13 and group 15 elements. Their synthesis and solid state structures have been studied in detail in the past decade, in particular, by Power *et al.*

Compounds of the type $[RMER']_x$ have been investigated to a far lesser extent. They also tend to oligomerize, preferably leading to the formation of dimeric, trimeric and tetrameric compounds of the type $[RMER']_x$ ($x = 2-4$; M = Al, Ga, E = N, P, As). Particularly, six-membered heterocycles $[RMER']_3$ have been the subject of detailed preparative and theoretical studies since they are isoelectronic to borazine $B_3N_3H_6$. Higher oligomers $[RMER']_x$ ($4 < x \leq 16$) could also be obtained, especially by reaction of Lewis base-stabilized $AlH_3$ with primary amines $RNH_2$ at elevated temperatures, leading to the elimination of two equivalents of dihydrogen.[3] In contrast, monomeric compounds of the type RMER' featuring a M=E double bond remained unknown until Power and Roesky *et al.* reported on the synthesis and solid state structure of the first iminogallane RGaNR' in 2001.[4] Very recently, von Hänisch and Hampe synthesized $\{Li(thf)_3\}_2Ga_2\{AsSi(i\text{-}Pr)_3\}_4$ containing two Ga=As double bonds.[5] Kinetical stabilization by use of sterically

$$3/2\ B_2H_6 + 3\ NH_3 \xrightarrow[-6\ H_2]{} [HBNH]_3 \qquad (1)$$

$$AlH_3 + NH_3 \xrightarrow{-80\ °C} AlH_3{\leftarrow}NH_3 \xrightarrow[-H_2]{>\,-50\ °C} [H_2AlNH_2]_x$$

$$[H_2AlNH_2]_x \xrightarrow[-H_2]{>\,30\ °C} [HAlNH]_x \xrightarrow[-H_2]{>\,300\ °C} AlN \qquad (2)$$

SCHEME 1. Synthesis of borazine and reaction of $AlH_3$ and $NH_3$.

$$GaR_3 + AsH_3 \xrightarrow[-3\ RH]{600\ -\ 750\ °C} GaAs \qquad (3)$$

$$R = Me,\ Et$$

SCHEME 2. Manasevit's initial study on the synthesis of GaAs films by MOCVD process.

encumbered organic substituents again plays the key role for their successful synthesis.

Besides such fundamental studies concerning the synthesis, structure and reactivity of group 13/15 compounds, which are basically of pure academic interest, materials sciences undoubtedly had the most important impact on group 13/15 chemistry within the last two decades. Binary group 13/15 materials, typically referred to as III–V materials, are semiconducting materials with several applications in opto- and micro-electronic devices.[6] Usually, thin films of such materials are needed for the desired applications. The MOCVD process (*metal organic chemical vapor deposition*), that was introduced by Manasevit in 1968,[7] has become the most advantageous industrial process for the synthesis of thin films of such materials.[8] It makes use of metalorganic precursors such as group 13 trialkyls. Manasevit's initial study described the deposition of GaAs thin films by subsequent thermolysis of $GaEt_3$ and $AsH_3$ at elevated temperatures on a particular substrate (Scheme 2).

Later on, Cowley, Jones and others extended this *two-source concept* to precursor compounds containing both elements of the desired III–V material connected by a stable chemical bond in a single molecule, so-called *single source precursors*. Consequently, this concept had a tremendous impact on group 13/15 chemistry in the following years, forcing in particular the synthesis of Lewis acid–base adducts $R_3M \leftarrow ER'_3$ and heterocycles of the type $[R_2MER'_2]_x$, which have been demonstrated to be useful precursors for the deposition of a variety of III–V materials.[9] However, the majority of such studies focused on the synthesis of group 13/15 compounds containing the lighter elements of group 15, N, P and As. In contrast, the synthesis of Sb- and Bi-containing compounds of these types played a Cindarella-like role in group 13/15 chemistry for a long time. In an attempt to gain some deeper knowledge of these particular classes of compounds, we and Wells *et al.* independently started in 1996 detailed investigations on the syntheses, structures and properties of Sb- and Bi-containing Lewis acid–base adducts and heterocycles. The results obtained since then are the subject of this review. In addition, selected Lewis acid–base adducts, heterocycles and cages of the lighter elements of group 15, N, P and As, will be presented.

## II

## LEWIS ACID–BASE ADDUCTS

### A. Adducts of the Type $R_3M \leftarrow ER'_3$—General Trends

#### 1. Nature of Dative Bonding

The reaction between a Lewis acid $R_3M$ and a Lewis base $ER'_3$ is of fundamental interest in main group chemistry. Synthetic and computational chemists have investigated the influence of both the Lewis acid and the base on the solid state structure and the thermodynamic stability of the corresponding adduct, that is usually expressed in terms of the dissociation enthalpy $D_e$. This led to a sophisticated understanding of the nature of dative bonding interactions. In particular, reactions of boranes, alanes and gallanes $MR_3$ with amines and phosphines $ER'_3$, typically leading to adducts of the type $R_3M \leftarrow ER'_3$, have been studied.[10]

Donor–acceptor bonds (*dative bonds*) are usually classified as covalent bonds. When compared to "regular" covalent bonds, "dative" covalent bonds undergo different dissociation pathways as can be exemplified by comparing the bonding properties of ethane, $C_2H_6$, with the isoelectronic amine–borane, $H_3B \leftarrow NH_3$. Homolytic rupture of the C–C bond yields electrically neutral methyl radicals $H_3C^{\bullet}$ while the amine–borane adduct gives the radical ions $H_3N^{\bullet +}$ and $H_3B^{\bullet -}$. In contrast, heterolytic rupture of the dative $B \leftarrow N$ bond yields neutral $H_3N$ and $BH_3$ whereas ethane forms the ions $H_3C^+$ and $H_3C^-$. Molecular orbital calculations[11] showed that the formation of the dative bond between $BH_3$ and $NH_3$ is accompanied by a transfer of only 0.20 $e^-$, which of course is significantly lower than 1.00 $e^-$ as would be the case for perfect electron sharing. Consequently, the dative electron pair remains closer to the N center. In addition, regular and dative covalent bonds differ by their bonding energies and bond distances. Dative bonds are much weaker (by the factor $3^{12}$) and significantly longer compared to regular covalent bonds as can be seen for instance by comparing the dissociation enthalpies of $H_3B \leftarrow NH_3$ ($D_e = 31.1 \pm 1.0$ kcal/mol)[13] and $C_2H_6$ ($D_e = 89.8 \pm 0.5$ kcal/mol)[14] and the bond distances (B ← N: 165.8 pm; C–C: 153.3 pm).[15]

#### 2. Thermodynamic Stability of $R_3M \leftarrow ER'_3$ Adducts

Theoretical calculations and experimental studies provided a detailed understanding of the parameters that determine the thermodynamic stabilities of $R_3M \leftarrow ER'_3$ adducts.[16] It was demonstrated that the "strength" of

TABLE I

DISSOCIATION ENTHALPIES $D_e$ [kcal/mol] OF GROUP 13/15 LEWIS ACID–BASE ADDUCTS

| M = | $Me_3M \leftarrow NMe_3$ | $Me_3M \leftarrow PMe_3$ | $Ph_3M \leftarrow NC_5H_5$ |
|-----|------|------|----|
| B  | 17.6 | 16.5 | 24 |
| Al | 30.0 | 21.0 | 36 |
| Ga | 21.0 | 18   | 30 |
| In | 19.9 | 17.1 | 24 |

TABLE II

BOND ANGLES [°] OF ANALOGOUSLY SUBSTITUTED TRIORGANOPENTELES

|       | E = N | E = P | E = As | E = Sb | E = Bi |
|-------|-------|-------|--------|--------|--------|
| $EH_3$  | 106.8 | 93.5 | 92.0 | 91.5, 91.3 | 90.5 |
| $EMe_3$ | 110.6 | 98.9 | 96.1 | 94.2 | 96.7 |
| $EPh_3$ | 119.6 | 102.8 | 99.7 | 96.6, 96.3 | 93.9 |

Lewis acids and bases, which play the key roles for the thermodynamic stability of the resulting adducts,[17] generally agree to the following trends, as can also be seen in Table I:

Lewis acidity : $R_3Al > R_3Ga > R_3In \approx R_3B > R_3Tl$

Lewis basicity : $R_3N > R_3P > R_3As > R_3Sb > R_3Bi$

However, these trends, which cannot provide an *absolute scale* for Lewis acidity and basicity, do not necessarily allow predictions of what kind of adducts will be formed.[18] Gallane $GaH_3$ for instance forms more stable adducts with phosphines than with amines. Consequently, reactions between $H_3Ga \leftarrow NMe_3$ and tertiary phosphines $PR'_3$ yield the corresponding phosphine adducts. In sharp contrast, $AlH_3$ forms significantly more stable adducts with amines than phosphines.[19]

The trend described for the group 15 compounds corresponds to the increasing *s*-character of the electron *lone pair*,[20] that is already reflected by the decreasing sum of the bond angles of analogously substituted tri-organopenteles $EH_3$,[21] $EMe_3$[22] and $EPh_3$,[23] respectively, as shown in Table II.

It was also found that both the Lewis acidity and basicity depend not only on specific electronic properties of the central group 13 and group 15 elements, but significantly on substituent effects.[24] Electron-withdrawing substituents increase the Lewis acidity but decrease the Lewis basicity, whereas electron-donating substituents decrease the Lewis acidity and

TABLE III

DISSOCIATION ENTHALPIES $D_e$ [kcal/mol] OF SELECTED BORANE–, ALANE– AND GALLANE–AMINE ADDUCTS

|  | $BF_3$ | $BCl_3$ | $BBr_3$ | $BH_3$ | $BMe_3$ |
|---|---|---|---|---|---|
| $Me_3N$ | 26.6 | 30.5 | a | 31.5 | 17.6 |
| $Et_3N$ | 26.5 | a | a | 35.2 | 10.0 |
| $C_5H_5N$ | 24.8 | 30.8 | 32.0 | 28.8 | 17.0 |
|  | $AlCl_3$ | $AlBr_3$ | $AlI_3$ | $AlH_3$ | $AlMe_3$ |
| $H_3N$ | 41.0 | 41.0 | 33.0 | a | 27.6 |
| $Et_3N$ | a | 44.8 | a | a | 26.5 |
| $C_5H_5N$ | 45.7 | 44.7 | 43.9 | a | 27.6 |
|  | $GaCl_3$ | $GaBr_3$ | $GaI_3$ | $GaH_3$ | $GaMe_3$ |
| $Et_3N$ | 34.4 | 10.0 | 1.7 | a | a |
| $C_5H_5N$ | 35.2 | 33.7 | 28.7 | a | a |

[a]No data available.

increase the Lewis basicity. The following trends have been established experimentally and by computational calculation for the strengths of Lewis acids $MR_3$ (M = Al, Ga, In) and Lewis bases $ER_3$ (E = N–Bi), as is also shown in Table III:[17]

Lewis acidity: $MF_3 > MCl_3 > MBr_3 > MI_3 > MH_3 > MMe_3 > MEt_3$
$> M(^tBu)_3$

Lewis basicity: $EF_3 < ECl_3 < EBr_3 < EI_3 < EH_3 < EMe_3 < EEt_3$
$< E(^tBu)_3$

It should also be stated that borane trihalides exhibit the opposite trend:

Lewis acidity : $MF_3 < MCl_3 < MBr_3 < MI_3$[25]

Steric interactions between bulky substituents such as $t$-Bu, leading to larger C–E–C bond angles, obviously affect the Lewis basicity caused by the increased $p$-character of the electron *lone pair*. However, the strength of the Lewis acid–base interaction within an adduct as expressed by its dissociation enthalpy does not necessarily reflect the Lewis acidity and basicity of the pure fragments, because steric (repulsive) interactions between the substituents bound to both central elements may play a contradictory role. In particular, adducts containing small group 13/15 elements are very sensitive to such interactions as was shown for amine–borane and –alane adducts

TABLE IV

DISSOCIATION ENTHALPIES $D_e$ [kcal/mol] OF SELECTED BORANE– AND ALANE–AMINE ADDUCTS

| Acceptor | $NH_3$ | $MeNH_2$ | $Me_2NH$ | $NMe_3$ | $NEt_3$ |
|----------|--------|----------|----------|---------|---------|
| $H_3B$ | $31.1 \pm 1.0$ | $35.0 \pm 0.8$ | $36.4 \pm 1.0$ | $34.8 \pm 0.5$ | 35.2 |
| $Me_3B$ | $13.8 \pm 0.3$ | $17.6 \pm 0.2$ | $19.3 \pm 1.0$ | $17.6 \pm 0.2$ | 10.0 |
| $Me_3Al$ | $27.6 \pm 0.3$ | $30.7 \pm 0.3$ | $30.8 \pm 0.3$ | $30.0 \pm 0.2$ | $26.5 \pm 0.2$ |

of the type $R_3M \leftarrow N(H)_x R_{3-x}$ ($x = 0$–3). According to the electronic properties of the substituents (positive inductive-effect of Et > Me > H), the Lewis basicity of the amines is expected to take the following order:

Lewis basicity : $NEt_3 > NMe_3 > HNMe_2 > H_2NMe > H_3N$

However, only $BH_3$ adducts display the expected tendency with $H_3B \leftarrow NEt_3$ being the most stable adduct. In contrast, $Me_3B$ and $Me_3Al$ adducts do not exhibit the maximum thermodynamic stability with $Et_3N$ but with $Me_2NH$, as can be seen from Table IV.[13,17] Theoretical calculations of Jarid and coworkers on borane and alane adducts with amines and phosphines clearly confirmed these experimental findings (Table IV).[26]

Unfortunately, experimental data on the thermodynamic stability of group 13–stibine and –bismuthine adducts are very rare. The only enthalpy of formation has been reported for $Br_3Al \leftarrow SbBr_3$ ($4.3 \pm 0.6 \, kJ/mol$).[27] A detailed study by Coates in the early fifties, who investigated the reaction between $Me_3Ga$ and $EMe_3$ (E = N, P, As, Sb, Bi), showed that only $Me_3N$, $Me_3P$, $Me_3As$ and $Me_3Sb$ gave the expected adducts $Me_3Ga \leftarrow EMe_3$, whereas $Me_3Bi$ did not react.[28] The thermodynamic stability of the adducts was found to decrease from $Me_3Ga \leftarrow NMe_3$ ($D_e = 21.0 \, kcal/mol$) and $Me_3Ga \leftarrow PMe_3$ ($D_e = 18 \, kcal/mol$) to $Me_3Ga \leftarrow AsMe_3$ ($D_e = 10 \, kcal/mol$), while $Me_3Ga \leftarrow SbMe_3$ was unstable in the gas phase. Studies by Mills and coworkers on the reaction of $EMe_3$ (E = P, As, Sb) with boron trihalides $BX_3$, diborane $B_2H_6$ and trimethylborane $BMe_3$ yielded analogous results, clearly proving the decreasing stability of these adducts with increasing atomic number of the group 15 element.[29]

### 3. General Structural Trends Observed for $R_3M \leftarrow ER'_3$ Adducts

Adduct formation between group 13 compounds $R_3M$ and group 15 compounds $ER'_3$ affect the geometry of both compounds. According to the VSEPR model, the C–M–C bond angles of the Lewis acid $R_3M$ should decrease (from $120°$ toward tetrahedral) and the M–C bond distances

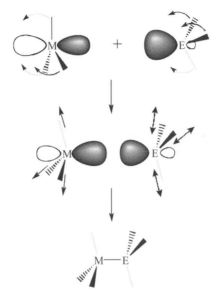

M = Group 13 Lewis Acid
E = Group 15 Lewis Base

Fig. 2. Structural rearrangements within the Lewis acid and base according to the VSEPR
model.

slightly increase upon adduct formation, whereas the E–C bond lengths of
the Lewis base should decrease and the C–E–C bond angles increase. These
predictions reflect the decreasing spatial requirements of an electron lone
pair (*ELP*), donated electron bonding pair (*DEBP*) and a normal electron
bonding pair (*NEBP*), which take the following order (Fig. 2):

*ELP > DEBP > NEBP*

The structural trends according to the VSEPR model are in contrast to
Gutmann's rules, which predict that adduct formation should generally lead
to an elongation of *both the M–C and the E–C bond lengths* and a *decrease* of
the C–E–C bond angles.[30] It was found that amines such as $NMe_3$ in borane
and alane adducts follow Gutmann's rules, whereas Lewis bases such as
$PMe_3$ and $AsMe_3$ tend to follow the VSEPR model.[13] Haaland suggested
steric interactions to be responsible for these contradictory observations.
Steric repulsion between a Lewis acid and a base fragment becomes less
effective with increasing atomic radii of the central atoms and with
decreasing size of the substituents. Consequently, the VSEPR model seems

to be valid for adducts containing small substituents such as hydrogen and methyl as well as large central atoms, whereas sterically more demanding substituents lead to repulsive interactions, forcing decreased C–E–C bond angles and increased E–C bond distances according to Gutmann's rule.

The second structural parameter of major interest is the central M–E bond distance. Dative bonds are significantly elongated compared to normal covalent bonds as mentioned previously. For instance, dative Al–N bonds range from 194 ($Cl_3Al \leftarrow NMe_3$ 194.9 pm;[31] $Cl_3Al \leftarrow NH_2t$-Bu 193.6 pm[32]) to 210 pm ($Me_3Al \leftarrow NMe_3$), whereas regular covalent Al–N bond distances vary between 180 and 195 pm. However, this particular structural parameter significantly depends on the method used for its determination. Differences between gas phase and single crystal X-ray diffraction studies of 6 (210 vs 204.5 pm for $Me_3Al \leftarrow NMe_3$;[31] 206.3 vs. 200.4 pm for $Me_3Al \leftarrow NH_3$[33]) up to 80 pm (167.2 vs 156.4 pm for $H_3B \leftarrow NH_3$;[34] 200.1 vs 163.0 for $F_3B \leftarrow NCMe$;[35] 247.3 vs. 163.8 for $F_3B \leftarrow NCH$[36]) have been reported. They are most likely caused by the presence of dipolar interactions between the molecules in the solid state.

Even if the trends observed for the Al–N bond distances (if not stated otherwise, all values reported in this paper refer to solid state data) in $Cl_3Al \leftarrow NMe_3$ (194.9 pm), $H_3Al \leftarrow NMe_3$ (201.8 pm)[37] and $Me_3Al \leftarrow NMe_3$ 204.5 pm reflect their decreasing thermodynamic stability, it has been demonstrated both experimentally and by theoretical calculations for several types of borane adducts that there is no correlation between the central M–E bond length and the dissociation energy: shorter M–E bond distances do not necessarily correlate with stronger bonded adducts.[38] This result can be explained by hybridization effects of the electron *lone pair*. An increase in *s*-character leads to a more compact $sp^n$ donor hybrid-orbital typically giving shorter bonds. Simultaneously, the donor–acceptor interaction (HOMO–LUMO) becomes weaker due to a larger energy difference between the interacting orbitals, resulting in less stable adducts.[18]

While group 13/15 adducts containing the lighter elements of group 15, N and P, have been intensely studied, investigations concerning the solid state structures of stibine and bismuthine adducts were much rarer.[39] Only $Br_3Al \leftarrow SbBr_3$ has been structurally characterized and found to be a molecular adduct in the gas phase (*electron diffraction*) but ionic ($[SbBr_2][AlBr_4]$) in the solid state (*single crystal X-ray diffraction*). Consequently, we became interested in the synthesis of stibine and bismuthine adducts in an attempt to gain further insights into the structures and thermodynamic stability of such adducts. At the same time, Wells *et al.* also started to investigate stibine adducts. Since then, a large number of such adducts have been synthesized and structurally characterized by single crystal X-ray diffraction. In addition, their thermodynamic stability has been studied by NMR spectroscopy. Theoretical calculations revealed the influence of both the central group 13/15 elements and their substituents on the structure and stability of such adducts.

## B. *Stibine and Bismuthine Adducts of the Type* $R_3M \leftarrow ER'_3$

### 1. Synthesis and Spectroscopic Characterization

Wells *et al.* first reported on the synthesis of group 13 stibine adducts in 1997. Reactions of boron trihalides $BX_3$ with $Sb(SiMe_3)_3$ yielded the corresponding borane–stibine adducts of the type $X_3B \leftarrow Sb(SiMe_3)_3$ (X = Cl, Br, I).[40] In the following years, they consequently expanded their studies on analogous reactions of trialkylgallanes and -indanes $R_3M$.[41] At the same time, we focused on reactions of dialkylchloroalanes $R_2AlCl$ as well as trialkylalanes and -gallanes both with $Sb(SiMe_3)_3$ and trialkylstibines $SbR_3$.[42] Very recently, we extended our investigations to reactions of trialkylalanes and -gallanes with triorganobismuthines $BiR'_3$ (R' = SiMe$_3$, *i*-Pr), yielding for the first time stable group 13 bismuthine adducts.[43] Mass spectroscopic studies on these stibine and bismuthine adducts indicated them to be very weak in the gas phase. Therefore, most of these adducts were carefully investigated by multinuclear NMR spectroscopy in solution. In addition, single crystal X-ray diffraction studies for the first time gave detailed insights into their solid state structures (Scheme 3).

$^1$H NMR spectra of stibine adducts $R_3M \leftarrow SbR'_3$ (M = Al, Ga) typically showed α-H resonances due to the organic ligands bound to Al and Ga shifted to lower field and those of the ligands bound to Sb shifted to higher field compared to the pure trialkyls. The same trends have been previously reported for amine and phosphine adducts such as $Me_3In \leftarrow NR_3$, $Me_3Al \leftarrow PR_3$ and $R_3Ga \leftarrow PR'_3$ (R = Me, Et).[44] These studies also demonstrated that the amount of the lowfield shift $\Delta\delta$ of the M–R proton resonance in the adduct $R_3M \leftarrow ER'_3$ is a sensitive parameter for the strength of the acid–base interaction, allowing an estimation of the

$$MR_3 + SbR'_3 \longrightarrow R_3M \leftarrow SbR'_3 \quad (4)$$

M = Al, Ga, In

$$R_2AlCl + SbR'_3 \longrightarrow R_2(Cl)Al \leftarrow SbR'_3 \quad (5)$$

$$MR_3 + BiR'_3 \longrightarrow R_3M \leftarrow BiR'_3 \quad (6)$$

M = Al, Ga

SCHEME 3. Synthesis of group 13-stibine and bismuthine adducts.

*relative stability* of $R_3M \leftarrow ER'_3$ adducts containing a *constant* Lewis acid $MR_3$ and different Lewis bases $ER'_3$ in solution. We observed for the stibine adducts $R_3M \leftarrow SbR'_3$ a strong correlation between $\Delta\delta$ and the steric requirements of the organic substituents R and R'. Adducts of Lewis acids containing small ligands such as $MMe_3$ or $MEt_3$ show the largest downfield shift of the $\alpha$-H resonance with sterically demanding Lewis bases such as $Sb(t\text{-}Bu)_3$ or $Sb(SiMe_3)_3$, containing large and electropositive substituents. In contrast, adducts of sterically encumbered Lewis acids such as $t\text{-}Bu_3Al$ and $t\text{-}Bu_3Ga$ show the biggest downfield shift with sterically less hindered Lewis bases such as $SbEt_3$ and $Sb(n\text{-}Pr)_3$. In contrast Lewis bases such as $Sb(t\text{-}Bu)_3$ and $Sb(SiMe_3)_3$ lead to $\Delta\delta$ values near zero, indicating excessive dissociation of the adduct in solution. Prior to our studies, repulsive ligand interactions were known largely to influence the stability of adducts containing small central atoms such as amine–borane adducts (see also Table IV). Obviously, they also play an important role in the stability of stibine adducts, which is rather unexpected, since steric interactions should become less important with increasing atomic radius of the central elements.

In contrast to the trends observed for stibine adducts, $^1$H NMR spectra of bismuthine adducts $R_3M \leftarrow BiR'_3$ show almost the same chemical shifts as the pure trialkylalanes and -gallanes, indicating an extensive dissociation in solution at ambient temperature. Reliable information on the thermodynamic stability of bismuthine adducts were obtained by use of temperature-dependent $^1$H NMR spectroscopy. The dissociation enthalpies $D_e$ of $^tBu_3Al \leftarrow BiR'_3$ (R' = $SiMe_3$, $i$-Pr) were determined to be 6.3 and 6.9 kcal/mol, respectively.[43b] In contrast, $Me_3Al \leftarrow BiR'_3$ (R' = $SiMe_3$, $i$-Pr) and $Et_3Al \leftarrow BiR_3$ (R = $SiMe_3$, $i$-Pr) were found to be fully dissociated in solution.[45] Interestingly, $Et_3Al \leftarrow Bi(SiMe_3)_3$ is a stable adduct in its pure form, as was shown by single crystal X-ray diffraction. These experimental findings strongly underline the very weak Lewis acid–base interaction in group 13-bismuthine adducts. Most likely, weak dipolar interactions between $Et_3Al \leftarrow Bi(SiMe_3)_3$ molecules are responsible for its stabilization in the solid state as was stated previously for amine–borane adducts such as $H_3B \leftarrow NH_3$ and $Me_3B \leftarrow NH_3$.[34b,46]

Computational calculations were performed in order to quantify the role of the substituents on the adduct stability of stibine and bismuthine adducts. These studies were completed by calculations of the corresponding phosphine and arsine adducts in order to reveal the influence of the group 15 element, as shown in Table V.[47]

Both the $AlH_3$ and $AlMe_3$ adduct families as well as the $Et_3Al \leftarrow E(SiMe_3)_3$ adducts show the predicted stability trends. The calculated dissociation energies within a homologous series of $H_3Al$ and $Me_3Al$ adducts constantly decrease with increasing atomic number of the central

TABLE V

CALCULATED DISSOCIATION ENTHALPIES $D_e$ [kcal/mol] OF ALANE ADDUCTS

| | E = P | E = As | E = Sb | E = Bi |
|---|---|---|---|---|
| $H_3Al \leftarrow EH_3$ | 13.5 | 9.5 | 6.4 | 2.9 |
| $H_3Al \leftarrow EEt_3$ | 21.3 | 16.6 | 12.6 | 9.2 |
| $H_3Al \leftarrow E(i\text{-}Pr)_3$ | 22.3 | 17.7 | 13.6 | 10.3 |
| $Me_3Al \leftarrow EH_3$ | 7.4 | 4.3 | 2.0 | 0.4 |
| $Me_3Al \leftarrow EEt_3$ | 12.6 | 8.8 | 5.7 | 3.6 |
| $Me_3Al \leftarrow E(i\text{-}Pr)_3$ | 13.6 | 9.9 | 6.7 | 4.4 |
| $t\text{-}Bu_3Al \leftarrow EH_3$ | 7.6 | 4.4 | 1.7 | 0.1 |
| $t\text{-}Bu_3Al \leftarrow EEt_3$ | 7.6 | 6.4 | 5.8 | 2.6 |
| $t\text{-}Bu_3Al \leftarrow E(i\text{-}Pr)_3$ | 2.6 | 4.0 | 4.1 | 3.0 |
| $Et_3Al \leftarrow E(SiMe_3)_3$ | 13.8 | 8.8 | 5.1 | 2.2 |

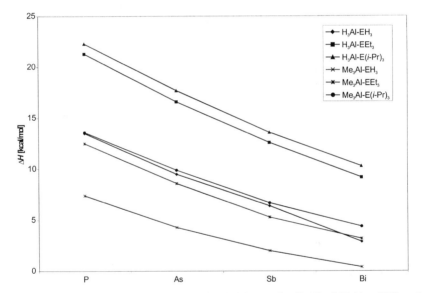

FIG. 3. B3LYP/SDD optimized dissociation enthalpies $D_e$ [kcal/mol] of $H_3Al \leftarrow ER'_3$ and $Me_3Al \leftarrow ER'_3$ adducts ($R' =$ H, Et, $i$-Pr) revealing the role of the pentele center on the adduct stability.

pentele as shown in Fig. 3. The dissociation energies $D_e$ of $Me_3Al$–trialkyl-pentele adducts range from 13 (Al–P) to 4 kcal/mol (Al–Bi), while the more Lewis acidic $AlH_3$ adducts are significantly more stable ($D_e$: 22 (Al–P) to 10 (Al–Bi) kcal/mol).

In accordance with the increasing Lewis basicity from $EH_3$ to $E(i$-Pr$)_3$, $EH_3$-adducts in each series exhibit the lowest dissociation energies as shown in Fig. 4.

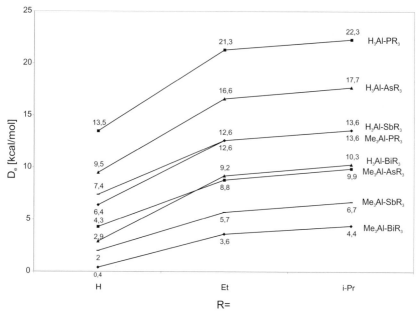

FIG. 4. B3LYP/SDD optimized dissociation enthalpies $D_e$ [kcal/mol] of $H_3Al \leftarrow ER'_3$ and $Me_3Al \leftarrow ER'_3$ adducts (R' = H, Et, $i$-Pr) revealing the role of the pentele substituents on the adduct stability.

In contrast, the results obtained for the $t$-Bu$_3$Al adduct groups are less obvious. The calculated dissociation energies $D_e$ within the $t$-Bu$_3$Al $\leftarrow$ EH$_3$ (P: 7.6; As: 4.4; Sb: 1.7; Bi: 0.1 kcal/mol) and $t$-Bu$_3$Al $\leftarrow$ EEt$_3$ (P: 7.6; As: 6.4; Sb: 5.8; Bi: 2.6 kcal/mol) adduct groups constantly decrease with increasing atomic number of the central pentele, whereas the $t$-Bu$_3$Al $\leftarrow$ E($i$-Pr)$_3$ adducts do not display a steady trend. In this group, the phosphine adduct ($D_e = 2.6$ kcal/mol) is even less stable than the bismuthine adduct ($D_e = 3.0$ kcal/mol) and both the arsine ($D_e = 4.0$ kcal/mol) and stibine ($D_e = 4.1$ kcal/mol) adducts are the most stable compounds. In addition, the triethylphosphine, -arsine and -stibine adducts $t$-Bu$_3$Al $\leftarrow$ EEt$_3$ were found to be more stable than the corresponding EH$_3$ and E($i$-Pr)$_3$ adducts. On the other hand, the bismuthine adducts show the expected trend with $t$-Bu$_3$Al $\leftarrow$ BiH$_3$ ($D_e = 0.1$ kcal/mol) being less stable than $t$-Bu$_3$Al $\leftarrow$ BiEt$_3$ ($D_e = 2.6$ kcal/mol) and $t$-Bu$_3$Al $\leftarrow$ Bi($i$-Pr)$_3$ ($D_e = 3.0$ kcal/mol), as shown in Fig. 5.

The observed tendencies for the $^t$Bu$_3$Al $\leftarrow$ ER'$_3$ adducts clearly reflect the influence of repulsive steric interactions between the large $t$-Bu and $i$-Pr substituents which become less important with increasing atomic radius of the central pentele. Such interactions overcompensate attractive dipolar

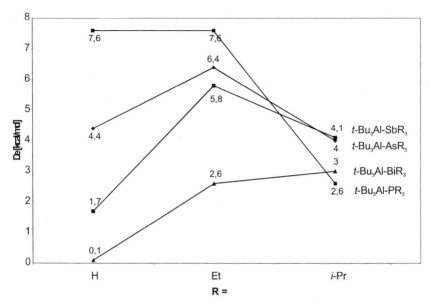

FIG. 5. B3LYP/SDD optimized dissociation enthalpies $D_e$ [kcal/mol] of $t$-Bu$_3$Al $\leftarrow$ ER$_3'$ adducts (R' = H, Et, $i$-Pr) revealing the role of the pentele substituents on the adduct stability.

interactions between the alane and the Lewis base, significantly reducing the stabilities of $t$-Bu$_3$Al $\leftarrow$ P($i$-Pr)$_3$ and $t$-Bu$_3$Al $\leftarrow$ As($i$-Pr)$_3$. However, such repulsive interactions appear to be more important in the gas phase than in solution. It was found by temperature dependent NMR studies that the stability of $t$-Bu$_3$Al $\leftarrow$ E($i$-Pr)$_3$ adducts in solution steadily decreases from the phosphine to the bismuthine adduct ($D_e$: 12.2, 9.9, 7.8, 6.9 kcal/mol). The observed trend properly reflects the decreasing Lewis basicity of E($i$-Pr)$_3$ with increasing atomic number of the group 15 element.

## 2. Solid State Structures

To date, 15 stibine R$_3$M $\leftarrow$ Sb(R')$_3$ (M = B, Al, Ga, In) and six bismuthine adducts R$_3$M $\leftarrow$ Bi(R')$_3$ (M = Al, Ga) have been structurally characterized by single crystal X-ray diffraction. Their M–E, M–X and E–X bond lengths as well as X–M–X and X–E–X bond angles are summarized in Tables VI, VII and VIII. Figures 6–9 show the solid state structures of four representative adducts.

### Stibine Adducts

R$_3$M $\leftarrow$ SbR$_3'$ adducts typically display distorted tetrahedral coordination geometries around the central atoms. The substituents R and R' are

TABLE VI

Selected Bonding Parameters [pm, °] of Alane Stibine Adducts $R_3Al \leftarrow SbR'_3$

| Adduct | M–E | M–X (av.) | E–X (av.) | $\sum$X–E–X | $\sum$X–M–X |
|---|---|---|---|---|---|
| | $R_3Al \leftarrow Sb(SiMe_3)_3$ | | | | |
| R = Et **1** | 284.1(1) | 198.4 | 256.0 | 310.8 | 347.3 |
| R = i-Bu **2** | 284.8(1) | 199.5 | 256.0 | 312.2 | 350.5 |
| | $R_2AlCl \leftarrow Sb(SiMe_3)_3$ | | | | |
| R = t-Bu **3**[a] | 282.1; 279.8 | 199.1; 199.4 | 258.6; 258.0 | 312.6; 309.1 | 339.6; 341.5 |
| | $R_3Al \leftarrow SbR'_3$ | | | | |
| R = Me; R′ = t-Bu **4** | 283.4(1) | 196.7 | 220.5 | 319.1 | 347.2 |
| R = Et; R′ = t-Bu **5** | 287.3(1) | 198.1 | 221.0 | 317.8 | 343.7 |
| R = t-Bu; R′ = Et **6** | 284.5(1) | 202.7 | 214.7 | 301.5 | 346.9 |
| R = t-Bu; R′ = i-Pr **7** | 292.7(1) | 203.0 | 218.2 | 294.1 | 348.7 |

[a]Two independent molecules within the asymmetric unit.

TABLE VII

Selected Bonding Parameters [pm, °] of Borane, Gallane and Indane Stibine Adducts $R_3M \leftarrow SbR'_3$

| Adduct | M–E | M–X (av.) | E–X (av.) | $\sum$X–E–X | $\sum$X–M–X |
|---|---|---|---|---|---|
| | $R_3M \leftarrow Sb(SiMe_3)_3$ | | | | |
| M = B; R = Cl **8** | 225.9(3) | 185.5 | 256.8 | 327.7 | 328.5 |
| M = B; R = Br **9** | 226.8(2) | 201.8 | 257.2 | 327.2 | 329.9 |
| M = B; R = I **10** | 225.7(8) | 224.2 | 258.1 | 325.6 | 334.3 |
| M = Ga; R = Et **11**[a] | 284.6(5); | 200.7; | 256.5; | 308.8; | 348.3; |
| | 285.4(1) | 199.5 | 255.9 | 309.3 | 349.5 |
| M = Ga; R = t-Bu **12** | 302.7(2) | 201.3 | 255.6 | 302.0 | 349.9 |
| M = In; R = Me_3SiCH_2 **13** | 300.8(1) | 220.8 | 255.4 | 312.8 | 353.1 |
| | $t\text{-}Bu_3Ga \leftarrow SbR'_3$ | | | | |
| R′ = Et **14** | 284.8(1) | 204.4 | 215.1 | 292.8 | 349.3 |
| R′ = i-Pr **15** | 296.2(1) | 204.2 | 218.4 | 300.5 | 347.6 |

[a]Two structure determination at different temperatures.

arranged in a staggered conformation in relation to one another. Surprisingly, the B–Sb bond lengths of the borane adducts $X_3B \leftarrow Sb(SiMe_3)_3$,[40] which range from 225.7 to 226.8 pm, are not elongated compared to the sum of the covalent radii ($\sum r_{cov}$(BSb): 223 pm). This is most likely caused by the electron withdrawing effect of the halogen atoms, leading to an increased Lewis acidity of the boranes. However, the B–Sb distances, which only differ by about 1 pm, do not reflect the different Lewis acidities of the boron trihalides. Comparable results were obtained for Me$_3$N–adducts of BF$_3$,

TABLE VIII
SELECTED BONDING PARAMETERS [pm, °] OF BISMUTHINE ADDUCTS $R_3M \leftarrow BiR'_3$

| Adduct | M–E | M–X (av.) | E–X (av.) | $\sum$X–E–X | $\sum$X–M–X |
|--------|-----|-----------|-----------|-------------|-------------|
| | | $Et_3M \leftarrow Bi(SiMe_3)_3$ | | | |
| M = Al **16** | 292.1(2) | 197.8 | 263.2 | 305.7 | 350.8 |
| M = Ga **17** | 296.6(1) | 199.0 | 263.5 | 303.5 | 353.9 |
| | | $t\text{-}Bu_3M \leftarrow Bi(i\text{-}Pr)_3$ | | | |
| M = Al **18** | 308.8(1) | 201.8 | 229.5 | 286.5 | 350.4 |
| M = Ga **19** | 313.5(1) | 203.0 | 229.2 | 286.1 | 352.1 |
| | | $t\text{-}Bu_3M \leftarrow BiEt_3$ | | | |
| M = Al **20** | 294.0(1) | 202.1 | 224.6 | 288.3 | 351.5 |
| M = Ga **21** | 296.6(1) | 203.7 | 225.5 | 285.9 | 352.3 |

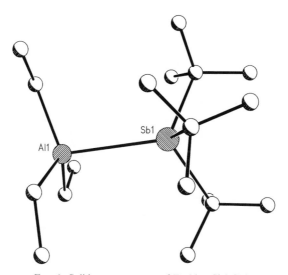

FIG. 6. Solid state structure of $Et_3Al \leftarrow Sb(t\text{-}Bu)_3$.

$BCl_3$, $BBr_3$ and $BI_3$. B–N distances obtained from single crystal X-ray diffraction studies as well as gas phase measurements yielded only small variations (X-ray: 158.5–161.0 pm; gas phase: 165.2–167.4 pm),[38] again demonstrating that the thermodynamic stability of group 13/15 adducts not necessarily correlates with their M–E bond length.

In contrast to these trends observed for borane–stibine adducts, the M–Sb bond lengths of alane–, gallane– and indane–stibine adducts are significantly elongated compared to the sum of the covalent radii ($\sum r_{cov}$(AlSb): 266; $\sum r_{cov}$(GaSb): 267; $\sum r_{cov}$(InSb): 285 pm),[48] as is expected for dative bonds.

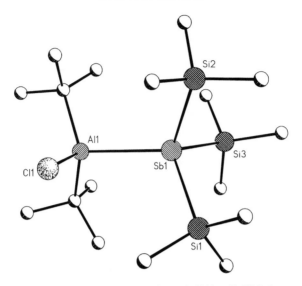

FIG. 7. Solid state structure of $(t\text{-Bu})_2\text{ClAl} \leftarrow \text{Sb}(\text{SiMe}_3)_3$.

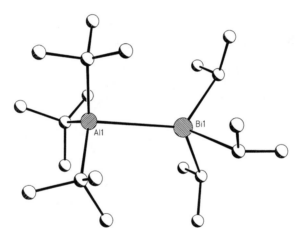

FIG. 8. Solid state structure of $t\text{-Bu}_3\text{Al} \leftarrow \text{Bi}(i\text{-Pr})_3$.

The Al–Sb distances range from 279.8(1) **3** to 292.7(1) pm **7** and the Ga–Sb bond lengths vary between 284.6(5) **11** and 302.7(2) pm **12**, showing significant dependencies on the steric demand of their organic substituents. The longest M–Sb bond distances in both groups were observed for the sterically overcrowded adducts $^t\text{Bu}_3\text{M} \leftarrow \text{SbR}'_3$ (R' = $i$-Pr, SiMe$_3$), whereas the shortest Al–Sb distance was found in the partially chloro-substituted adduct $t\text{-Bu}_2\text{AlCl} \leftarrow \text{Sb}(\text{SiMe}_3)_3$, most likely caused by the

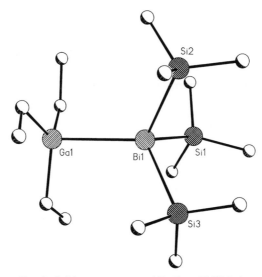

FIG. 9. Solid state structure of Et$_3$Ga ← Bi(SiMe$_3$)$_3$.

electron-withdrawing effect of the Cl-atom. Br$_3$Al ← SbBr$_3$ displays a much shorter Al–Sb bond length in the gas phase (252.2 pm), as was determined by electron diffraction. This bond distance is even shorter than the sum of the covalent radii, which is rather unusual for adducts. Unfortunately, to the best of our knowledge, no other electron diffraction studies of group 13 stibine adducts have been reported to date. Considering the almost equal covalent radii of Al and Ga, the question arises if Lewis acidity has a significant influence on the geometry of the R$_3$M ← SbR$_3'$ adducts, in particular on the M–Sb bond lengths. Comparisons between identically substituted Al–Sb and Ga–Sb adducts may qualitatively reveal such an influence because structural differences between such adducts should be based only on different electronic properties of the Lewis acid, whereas steric effects are negligible according to the almost equal covalent radii of Al and Ga. However, as was observed for the aforementioned amine–borane adducts, our experimental results clearly state that there is *no correlation* between Lewis acidity and M–Sb bond distance. While M–Sb bond distances of sterically crowded adducts such as *t*-Bu$_3$M ← Sb(*i*-Pr)$_3$ differ by almost 4 pm (M = Al: 292.7(1) pm **7**, Ga: 296.2(1) pm **15**), adducts of the type *t*-Bu$_3$M ← SbEt$_3$ (M = Al: 284.5(1) pm **6**; Ga: 284.8(5) pm **14**) and Et$_3$M ← Sb(SiMe$_3$)$_3$ (M = Al: 284.1(1) pm **1**; Ga: 284.6(5) pm **11**) show almost the same M–Sb distances.

(Me$_3$SiCH$_2$)$_3$In ← Sb(SiMe$_3$)$_3$ **13**, to date the only structurally characterized In–Sb adduct, shows an In–Sb bond distance of 300.8(1) pm.[41c]

This bond distance is supposed to be at the lower end of the In–Sb dative bond range since the covalent radius of In ($r_{cov}$: 143 pm) is about 17 pm larger than those of the lighter elements Al and Ga. Consequently, In–Sb dative bonds are expected to vary from 300 to 320 pm. Due to the lack of other structurally characterized In–Sb adducts, no further structural comparisons are possible.

*Bismuthine Adducts*

The central atoms in group 13 bismuthine adducts $R_3M \leftarrow BiR'_3$ (M = Al, Ga) also reside in distorted tetrahedral environments[43] and the Al–Bi (292.1(2) pm **16**; 308.8(1) pm **18**; 294.0(1) pm **20**) and Ga–Bi bond distances (292.1(2) pm **17**; 313.5(1) pm **19**; 296.6(1) pm **21**) are significantly elongated compared to the sum of the covalent radii ($\sum r_{cov}$(AlBi): 275; $\sum r_{cov}$(GaBi): 276 pm).[48] Al–Bi bond lengths tend to be shorter than Ga–Bi distances of identically substituted adducts, which is in contrast to our observations on M–Sb adducts. In accordance with the increased atomic radius of Bi (Sb: 141 pm, Bi: 150 pm), the M–Bi bond lengths are longer than the M–Sb bond lengths of analogously substituted adducts. However, the increase was found to be much more pronounced than was expected. In particular, the elongations of 16 (M = Al) and 17 pm (M = Ga), respectively, detected for the $t$-Bu$_3$M $\leftarrow$ Bi($i$-Pr)$_3$ adducts (M = Al: 308.8(1) pm **18**; M = Ga: 313.5(1) pm **19**) clearly exceed the difference between the covalent radii of 9 pm.

*Structural Trends within Alane and Gallane Adducts*

According to the VSEPR model (see also Section 2.1.3 and Fig. 2), the adduct formation process between a group 13 Lewis acid $MR_3$ and a group 15 Lewis base $ER'_3$ is expected to have a strong influence on the M–C bond lengths and C–M–C bond angles. We investigated identically substituted adduct families of the type $Et_3M \leftarrow E(SiMe_3)_3$ and $t$-Bu$_3$M $\leftarrow E(i$-Pr)$_3$ (M = Al, Ga; E = P, As, Sb, Bi) by single crystal X-ray diffraction in order to study the influence of the central group 13/15 elements on such specific structure parameters, which are summarized in Tables IX and X.

The trends observed for the average M–C bond lengths and the sum of the C–M–C bond angles are displayed in Figs. 10 and 11. In each adduct family, the M–C bond distances and the C–M–C bond angles correlate with the atomic number of the group 15 element except for $t$-Bu$_2$($i$-Bu)Ga $\leftarrow$ P($i$-Pr)$_3$, most likely caused by the replacement of one $t$-Bu group by a sterically less demanding $i$-Bu group. The values of uncomplexed $t$-Bu$_3$Al (200.4(3) pm; 120°) and $t$-Bu$_3$Ga (203.4(2) pm; 120°) have recently been determined by electron diffraction studies.[49]

TABLE IX

SELECTED BONDING PARAMETERS [pm, °] OF $Et_3M \leftarrow E(SiMe_3)_3$ ADDUCTS

| E = | M–E | M–X (av.) | E–X (av.) | $\sum$ X–E–X | $\sum$ X–M–X |
|-----|-----|-----------|-----------|------------|------------|
| | | $Et_3Al \leftarrow E(SiMe_3)_3$ | | | |
| P 22 | 255.5(2) | 198.9 | 227.3 | 319.9 | 340.7 |
| As 23 | 265.4(2) | 198.8 | 236.5 | 316.1 | 342.2 |
| Sb 1 | 284.1(1) | 198.4 | 256.0 | 310.8 | 347.3 |
| Bi 16 | 292.1(2) | 197.8 | 263.2 | 305.7 | 350.8 |
| | | $Et_3Ga \leftarrow E(SiMe_3)_3$ | | | |
| P 24 | 258.2(1) | 200.4 | 227.0 | 319.0 | 343.4 |
| As 25 | 268.4(1) | 200.2 | 236.3 | 315.0 | 345.7 |
| Sb 11 | 285.4(1) | 199.5 | 255.9 | 309.3 | 349.5 |
| Bi 17 | 296.6(1) | 199.0 | 263.5 | 303.5 | 353.9 |

TABLE X

SELECTED BONDING PARAMETERS [pm, °] OF $t\text{-}Bu_3M \leftarrow E(i\text{-}Pr)_3$ ADDUCTS

| E = | M–E | M–X (av.) | E–X (av.) | $\sum$ X–E–X | $\sum$ X–M–X |
|-----|-----|-----------|-----------|------------|------------|
| | | $t\text{-}Bu_3Al \leftarrow E(i\text{-}Pr)_3$ | | | |
| P 26 | 266.7(2) | 206.9 | 186.1 | 322.2 | 337.8 |
| As 27 | 283.9(1) | 204.0 | 199.4 | 307.3 | 342.9 |
| Sb 7 | 292.7(1) | 203.0 | 218.2 | 294.1 | 348.7 |
| Bi 18 | 308.8(1) | 201.8 | 229.5 | 286.5 | 350.4 |
| | | $t\text{-}Bu_3Ga \leftarrow E(i\text{-}Pr)_3$ | | | |
| P[a] 28 | 272.0(2) | 204.4 | 186.2 | 312.9 | 346.3 |
| As 29 | 290.5(1) | 205.0 | 199.5 | 306.3 | 344.8 |
| Sb 15 | 296.2(1) | 204.2 | 218.4 | 300.5 | 347.6 |
| Bi 19 | 313.5(1) | 203.0 | 229.2 | 286.1 | 352.1 |

[a]Data for $t\text{-}Bu_2(i\text{-}Bu)Ga \leftarrow P(i\text{-}Pr)_3$.

These experimentally observed structural trends were confirmed by computational calculations on $H_3Al$ and $Me_3Al$ adduct families. In each adduct family, the amount of increase of the M–X bond lengths and decrease of the X–M–X bond angles (X = H, Me) compared to uncomplexed $MX_3$ diminishes with the atomic number of the group 15 element. The structural trends very well reflect the trends observed for the thermodynamic stability of such adducts, as is illustrated in Tables XI and XII.

It should also be stated that the M–E bond distances within the $Et_3M \leftarrow E(SiMe_3)_3$ and $t\text{-}Bu_3M \leftarrow E(i\text{-}Pr)_3$ adduct families increase by about 40 pm according to the increase of the covalent radii of the group 15 element ($r_{cov}$(P) 110 pm, $r_{cov}$(Bi) 150 pm). Within the sterically overcrowded $t\text{-}Bu_3M \leftarrow E(i\text{-}Pr)_3$ adduct families the arsine and bismuthine adducts show

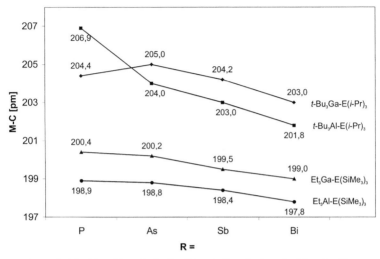

FIG. 10. M–C bond lengths [pm] of identically substituted adduct families.

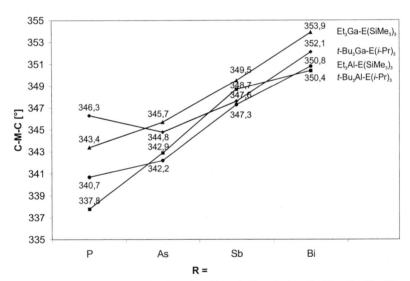

FIG. 11. Sum of the C–M–C bond angles of identically substituted adduct families [°].

significantly elongated distances compared to the phosphine and stibine adducts, whereas the increase between the arsine and stibine adducts is much less pronounced than was expected according to the increase of the covalent radii (20 pm). Figure 12 displays the trends observed for the M–E bond distances.

## TABLE XI

B3LYP/SDD Optimized Structural Parameters [pm, °] and Dissociation Enthalpies $D_e$ [kcal/mol] of $H_3Al \leftarrow ER_3'$ Adducts

| Adduct | Al–E | Al–R | R–Al–R | E–R' | R'–E–R' | $D_e$ |
|---|---|---|---|---|---|---|
| | | | $H_3Al \leftarrow EH_3$ | | | |
| E = P | 265.4 | 159.5 | 118.7 | 142.5 | 98.8 | 13.5 |
| E = As | 276.5 | 159.3 | 119.0 | 151.8 | 96.4 | 9.5 |
| E = Sb | 304.9 | 159.2 | 119.3 | 169.8 | 94.9 | 6.4 |
| E = Bi | 317.4 | 159.0 | 119.7 | 179.9 | 93.4 | 2.9 |
| | | | $H_3Al \leftarrow EEt_3$ | | | |
| E = P | 256.9 | 160.2 | 117.4 | 189.6 | 103.3 | 21.3 |
| E = As | 266.2 | 160.0 | 117.8 | 199.0 | 101.5 | 16.6 |
| E = Sb | 290.5 | 159.8 | 118.3 | 217.5 | 99.3 | 12.6 |
| E = Bi | 300.6 | 159.5 | 118.8 | 227.7 | 97.5 | 9.2 |
| | | | $H_3Al \leftarrow E(i\text{-}Pr)_3$ | | | |
| E = P | 257.5 | 160.3 | 117.1 | 193.2 | 104.3 | 22.3 |
| E = As | 266.4 | 160.1 | 117.5 | 202.1 | 103.2 | 17.7 |
| E = Sb | 289.8 | 159.8 | 118.1 | 220.1 | 100.9 | 13.6 |
| E = Bi | 299.3 | 159.7 | 118.5 | 229.9 | 99.4 | 10.3 |

## TABLE XII

B3LYP/SDD Optimized Structural Parameters [pm, °] and Dissociation Enthalpies $D_e$ [kcal/mol] of $Me_3Al \leftarrow ER_3'$ Adducts

| Adduct | Al–E | Al–R | R–Al–R | E–R' | R'–E–R' | $D_e$ |
|---|---|---|---|---|---|---|
| | | | $Me_3Al \leftarrow EH_3$ | | | |
| E = P | 279.1 | 198.8 | 118.5 | 142.9 | 97.6 | 7.4 |
| E = As | 298.5 | 198.3 | 119.0 | 152.5 | 94.8 | 4.3 |
| E = Sb | 345.3 | 196.1 | 119.4 | 170.7 | 93.0 | 2.0 |
| E = Bi | 382.2 | 197.7 | 120.0 | 180.9 | 91.5 | 0.4 |
| | | | $Me_3Al \leftarrow EEt_3$ | | | |
| E = P | 266.1 | 199.6 | 116.9 | 189.9 | 102.5 | 12.6 |
| E = As | 279.3 | 199.3 | 117.2 | 199.4 | 100.3 | 8.8 |
| E = Sb | 311.6 | 198.9 | 118.3 | 218.4 | 97.6 | 5.7 |
| E = Bi | 328.4 | 198.4 | 119.0 | 228.3 | 95.8 | 3.6 |
| | | | $Me_3Al \leftarrow E(i\text{-}Pr)_3$ | | | |
| E = P | 269.4 | 199.7 | 116.4 | 193.1 | 100.7 | 13.6 |
| E = As | 283.8 | 199.3 | 115.7 | 202.8 | 103.0 | 9.9 |
| E = Sb | 310.0 | 199.0 | 117.7 | 220.7 | 99.4 | 2.3 |
| E = Bi | 324.8 | 198.6 | 118.8 | 230.5 | 98.2 | 4.4 |

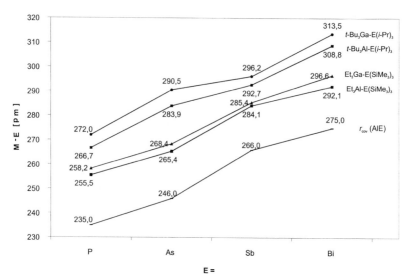

FIG. 12. M–E bond lengths [pm] of identically substituted adduct families.

## C. *Hypercoordinated Adducts of the Type* $R'_3E \rightarrow MR_3 \leftarrow ER'_3$

In addition to 1:1 molar adducts $R_3M \leftarrow ER'_3$ as described in the previous section, adducts containing higher coordinated triele centers, typically referred to as hypercoordinated compounds, have been prepared. In particular, alane $AlH_3$ tends to coordinate a second Lewis base. This tendency is already reflected by the fact that $H_3Al \leftarrow NMe_3$, which is monomeric in the gas phase, adopts a dimeric structure in the solid state (Al–H bridges) in order to increase its coordination number from 4 to 5.[50] Dimerization was found to be typical for amine–alane adducts in the solid state.[51] In contrast, the corresponding gallane–amine adduct $H_3Ga \leftarrow NMe_3$ is monomeric both in the gas phase and in the solid state.[52]

The tendency of group 13 compounds $MR_3$ to react with two Lewis bases $ER'_3$, under formation of fivefold-coordinated adducts of the type $R'_3E \rightarrow MR_3 \leftarrow ER'_3$, was demonstrated for the first time in 1963. It was shown that alane reacts with two equivalents of $NMe_3$ to give $Me_3N \rightarrow H_3Al \leftarrow NMe_3$,[53] whose structure has been determined by X-ray diffraction.[54] Since these initial studies, comparable reactions were observed for alanes, gallanes and indanes $R_3M$ (M = Al, Ga, In; R = H, halogen) with amines and phosphines, yielding bis-amine, bis-phosphine and mixed amine/ phosphine adducts. Figure 13 presents the different coordination modes that have been observed so far.

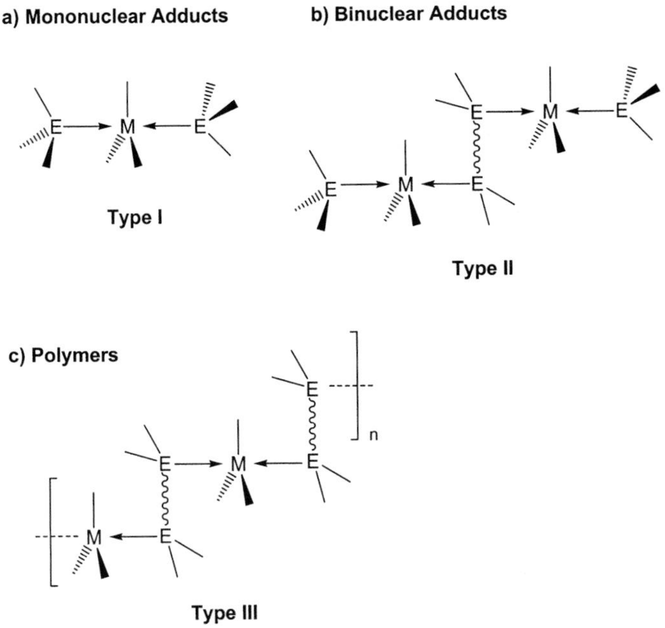

**a) Mononuclear Adducts**   **b) Binuclear Adducts**

**Type I**

**Type II**

**c) Polymers**

**Type III**

Fig. 13. Coordination modes observed for hypercoordinated adducts.

Unlike $MX_3$ compounds ($M = Al$, Ga, In; $X = H$, Hal), trialkylalanes, -gallanes and -indanes $MR_3$ failed to add a second Lewis base. Boranes $BX_3$ ($X = $ Hal, H, alkyl) generally do not add a second Lewis base. For a long time, this was explained by the lack of any d-orbital contribution and by steric interactions due to the small boron atom. However, Goldman *et al.* recently demonstrated by computational calculations that differences between borane $BH_3$ and alane $AlH_3$ are in fact based on the different electronegativities of boron and aluminum: $BH_3$ tends to form a predominantly covalent bond with ammonia, while the bond between $AlH_3$ and $NH_3$ shows substantial electrostatic character, hence increasing the stability of a bisadduct.

In all hypercoordinated bisadducts the central acceptor atom displays a trigonal bipyramidal coordination geometry with the donor molecules in axial position. These structural findings are in accordance to the VSEPR model, predicting the steric requirements of a "donor" electron pair to be greater than those of a "regular" covalent bonding electron pair. The dative M–E bond distances in 2:1 adducts are significantly elongated compared to 1:1 adducts, as can be realized by comparing $H_3Al \leftarrow NMe_3$ (206.3(7) gas phase), $[H_3Al \leftarrow NMe_3]_2$ (209.2 solid state) and $Me_3N \rightarrow H_3Al \leftarrow NMe_3$ (218.1 solid state). This trend, which is also expected from the VSEPR

TABLE XIII

Selected Bond Lengths [pm] and Angles [°] of Selected Bisadducts

| Adduct | M–E (av.) |
|---|---|
| $R'_3N \rightarrow MR_3 \leftarrow NR'_3$ | |
| $AlH_3$; $NMe_3$ | 218.1 |
| $AlCl_3$; $NMe_3$ | 216.6; 215.8 |
| $AlCl_3$; $N(H)Me_2$ | 207.8 |
| $InCl_3$; $NMe_3$ | 231.9; 234.5 |
| $R'_3As \rightarrow MR_3 \leftarrow AsR'_3$ | |
| $InI_3$; $AsPh_3$ | 292.7; 299.1 |
| $R'_3P \rightarrow MR_3 \leftarrow PR'_3$ | |
| $AlI_3$; $PEt_3$ | 252.1; 252.7 |
| $InCl_3$; $PMe_3$ | 257.5; 257.6 |
| $InCl_3$; $PPh_3$ | |
| $InBr_3$; $PMe_2Ph$ | 261.4; 262.2 |
| $InI_3$; $PPh_3$ | 285.5; 298.7 |
| $InH_3$; $P(cy\text{-}Hex)_3$ | 298.7(1) |

model, reflects both the increased steric interactions and the reduced acidity of the group 13 metal after the coordination of the first Lewis base. Table XIII summarizes central bonding parameters of selected alane–, gallane– and indane–amine,[55] –phosphine[56] and –arsine[57] bisadducts of Type I (Fig. 13).

No hypercoordinated adducts containing stibines or bismuthines have been prepared, so far. This is most likely a result of the insufficient electrostatic character of the bonding interaction in such adducts. Consequently, strongly polarizing Lewis bases such as amines and phosphines and highly Lewis acidic $MX_3$ compounds ($X = H$, Hal) are required to allow the addition of a second molecule Lewis base. To date, even no simple $MX_3 \leftarrow ER'_3$ adducts ($E = Sb$, Bi) have been prepared. In our hands, reactions of $MX_3$ with $SbR'_3$ or $BiR'_3$ always induced the decomposition of the stibine and bismuthine without yielding the desired adducts.[58]

## D. Adducts of Low-Valent Organopenteles

The potential of group 15 trialkyls $ER'_3$ to form Lewis acid–base adducts with group 13 compounds has been demonstrated as shown previously. In contrast, that of low-valent organopenteles such as dipenteles $R'_2E-ER'_2$ and cyclopenteles $[ER']_x$ ($E = P$, As, Sb, Bi) has been studied to a far lesser extent.[59] Tetraalkyldipenteles $E_2R_4$ belong to the oldest metalorganic main

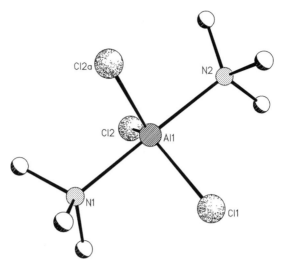

FIG. 14. Solid state structure of $AlCl_3(NMe_3)_2$.

group element compounds, whose first synthesis goes back to Cadet's initial discovery of the *"fuming liquid"* in 1757. The major components of the *fuming liquid* are tetramethyldiarsine $Me_4As_2$ and $[Me_2As]_2O$, which are also known as *cacodyl* and *cacodyl oxide*, as was shown later by Bunsen.[60] Dipenteles $R_2'E–ER_2'$ and cyclopenteles $[ER']_x$ are available by reduction of organoelement halides $REX_2$ with alkali metals or alkaline earth metals (Wurtz-coupling). Each group 15 element in both types of compounds contains an electron *lone pair* which is potentially active for further complexation reactions. Consequently, reactions with transition metal complexes have been known for many years, in particular, those of cyclophosphines and cyclostibines[61] as well as those of diphosphines, diarsines and distibines.[62] Reactions of the latter have been shown to occur either with preservation or cleavage of the E–E bond. Consequently, several monodentate (Types A and B) and bidentate complexes (Types C and D) as well as heterocycles (Types E and F) as shown in Fig. 15 have been synthesized and structurally characterized. In contrast, comparable transition metal complexes of dibismuthines are unknown, to date.[63] This is mainly based on their lability toward disproportionation into elemental E and $R_3E$.[64]

In sharp contrast to the intensely studied reactions of dipenteles with transition metal compounds, reactions with group 13 metal compounds are almost unknown. Only two diphosphine–borane bisadducts of Type C ($[H_3B]_2[Me_4P_2]$, $[H_2(Br)B]_2[Me_4P_2]$) have been synthesized and structurally characterized[65] but no diarsine, distibine or dibismuthine adducts. We, therefore, became interested in the synthesis of such compounds, focusing

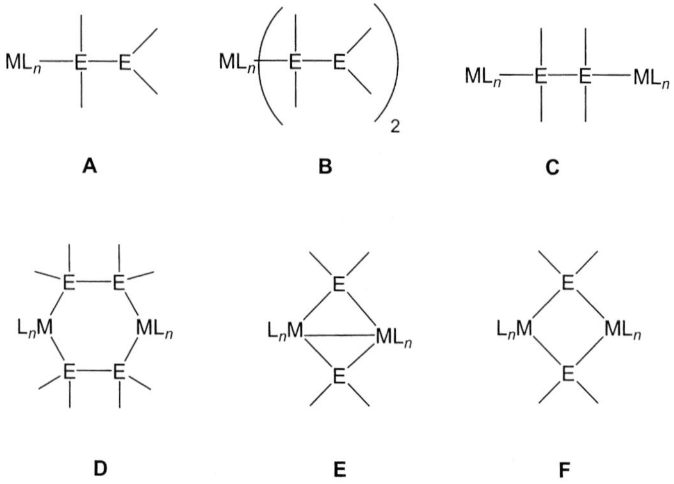

FIG. 15. Possible coordination MODII within dipentele–transition metal complexes.

on reactions of tetraalkyldistibines $Sb_2R'_4$ with trialkylalanes and -gallanes $MR_3$. Stable distibine adducts $[R_3M]_x[E_2R'_4]$ ($x = 1, 2$) of Types A and C were obtained and the solid state structures of one monoadduct and four bisadducts were determined by X-ray diffraction.[66] In addition, for the first time the potential of dibismuthines to take part in complex formation reactions was demonstrated. Reactions of $Bi_2Et_4$ with $t$-$Bu_3M$ (M = Al, Ga) gave the corresponding bisadducts $[t$-$Bu_3M]_2[Bi_2Et_4]$, whose structures in the solid state were confirmed by single crystal X-ray analyses.[67] These adducts are stable in the pure form, but they easily undergo consecutive reactions in solution at ambient temperature. Distibine adducts tend to react under E–E bond cleavage, while dibismuthine adducts decompose under formation of elemental Bi. Figures 16–18 display the solid state structures of three representative distibine and dibismuthine adducts. The most important structural parameters of the adducts and of selected pure distibines and dibismuthines are given in Tables XIV and XV.

## 1. Distibine Adducts

The ligands bound to the central Sb atoms adopt a staggered conformation in relation to one another, with the bulky $M(t$-$Bu)_3$ groups arranged in the *trans*-position. These findings are most likely caused by repulsive interactions between the extremely bulky organic substituents. The central Sb–Sb bond distances of the bisadducts vary between 281.1(1) **30** and 283.9(1) pm **33**, whereas that of the monoadduct is slightly elongated

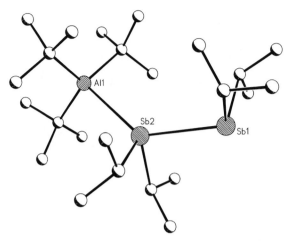

Fig. 16. Solid state structure of $[t\text{-}Bu_3Al][Sb_2(i\text{-}Pr)_4]$.

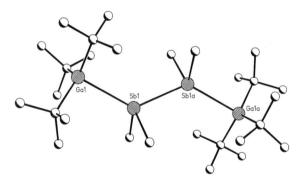

Fig. 17. Solid state structure of $[t\text{-}Bu_3Ga]_2[Sb_2Me_4]$.

(285.5(1) pm **34**). Comparable Sb–Sb bond distances were found in transition metal complexes such as $\{[I_2Cd]_2[Sb_2Et_4]\}_n$ (278.4(2) pm),[62g] $[(CO)_5W]_2[Sb_2Ph_4]$ (286.1(1) pm)[62e] and $[(CO)_5Cr]_2[Sb_2Ph_4]$ (286.6(1) pm)[62f] as well as in pure distibines such as $Sb_2(SiMe_3)_4$ (286.6 pm).[68] Both the Al–Sb and Ga–Sb bond distances of the $Sb_2Et_4$ bisadducts $[t\text{-}Bu_3M]_2[Et_4Sb_2]$ (300.1(1) pm **32**, 302.2(2) pm **33**) were found to be significantly elongated compared to those of the $Sb_2Me_4$ bisadducts $[t\text{-}Bu_3M]_2[Me_4Sb_2]$ (291.9(1) pm **30**, 291.9(1) pm **31**), in accordance with minor steric hinderance between the organic substituents. The M–Sb bond distances found for **34** (300.3(2) pm) and **33** (302.2(1) pm) are the longest M–Sb bond lengths ever observed. They are significantly elongated (about 35 pm) compared to the sum of the covalent radii ($\sum r_{cov}$(AlSb): 266 pm;

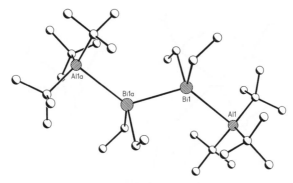

FIG. 18. Solid state structure of [$t$-Bu$_3$Al]$_2$[Bi$_2$Et$_4$].

## TABLE XIV
BOND LENGTHS [pm] AND ANGLES [°] OF DISTIBINE ADDUCTS AND PURE DISTIBINES

| M =; R' = | M–Sb | Sb–Sb | $\sum$ C–M–C | $\sum$ CSbC + CSbSb |
|---|---|---|---|---|
| | [$t$-Bu$_3$M]$_2$[Sb$_2$R$_4'$] | | | |
| Al; Me **30** | 291.9(1) | 281.1(1) | 351.1 | 295.1 |
| Ga; Me **31** | 291.9(1) | 281.4(1) | 352.2 | 292.9 |
| Al; Et **32** | 300.1(1) | 283.8(1) | 350.2 | 292.9 |
| Ga; Et **33** | 302.2(1) | 283.9(1) | 351.1 | 291.1 |
| | [$t$-Bu$_3$M]$_2$[Sb$_2$R$_4'$] | | | |
| Al; $i$-Pr **34** | 300.3(2) | 285.5(1) | 347.4 | 300.9; 288.2 |
| | R$_4$Sb$_2$ | | | |
| R = Me | – | 2.830(1) | – | 283.1 |
| R = SiMe$_3$ | – | 286.6(1) | – | 283.4 |
| R = Ph | – | 284.4(2) | – | 285.2 |

## TABLE XV
BOND LENGTHS [pm] AND ANGLES [°] OF DIBISMUTHINE ADDUCTS AND PURE DIBISMUTHINES

| M = | M–Bi | Bi–Bi | $\sum$ C–M–C | $\sum$ CBiC + CBiBi |
|---|---|---|---|---|
| | [$t$-Bu$_3$M]$_2$[Bi$_2$Et$_4$] | | | |
| Al **35** | 308.4(2) | 298.3(1) | 352.7 | 287.7 |
| Ga **36** | 311.4(2) | 298.4(1) | 353.4 | 285.9 |
| | R$_4$Bi$_2$ | | | |
| Ph$_4$Bi$_2$ | – | 299.0(1) | – | 283.1 |
| (Me$_3$Si)$_4$Bi$_2$ | – | 303.5(1) | – | 287.6 |

$\sum r_{\text{cov}}(\text{GaSb})$: 267 pm). As was found for identically substituted stibine adducts of the type $R_3M \leftarrow ER'_3$, the Ga–Sb and Ga–C distances found for the gallane–distibine bisadducts $[t\text{-Bu}_3\text{Ga}]_2[\text{Sb}_2\text{R}_4]$ are slightly elongated and the sum of the C–Ga–C bond angles is slightly greater than the respective bonding parameters of the alane–distibine bisadducts $[t\text{-Bu}_3\text{Al}]_2[\text{Sb}_2\text{R}_4]$. These observations are in accordance with the reduced Lewis acidity of $t\text{-Bu}_3\text{Ga}$ compared to $t\text{-Bu}_3\text{Al}$.

The $\text{Sb}_2\text{Me}_4$ bisadducts feature slightly shorter Sb–C distances (214.6 pm **30**; 214.4 pm **31**) but significantly increased C–Sb–C and C–Sb–Sb bond angles (295.1° **30**; 292.9° **31**) compared to uncomplexed $\text{Me}_4\text{Sb}_2$ (216.2 pm; 283.1°). In accordance with the slightly bigger Et substituents, the analogous parameters of the $\text{Sb}_2\text{Et}_4$ bisadducts (216.7 pm; 292.9° **32**; 216.8 pm; 291.1° **33**) differ slightly. Both the shortening of the Sb–C bond lengths and the increase of the sum of the bond angles in the $\text{Sb}_2\text{Me}_4$ bisadducts $[t\text{-Bu}_3\text{M}]_2$ $[\text{Me}_4\text{Sb}_2]$ are in accordance with a partial rehybridization of the Sb atoms, as was described previously for simple $R_3M \leftarrow ER'_3$ adducts. The former Sb electron *lone pair*, which had a high *s*-character, gets more *p*-character upon coordination to the Lewis acid. Simultaneously, the *s*-character of the former Sb–C and Sb–Sb bonding electron pairs increases, leading to a widening of the C–Sb–C and C–Sb–Sb bond angles and a shortening of the Sb–C and Sb–Sb bond distances. Unfortunately, the solid state structure of $\text{Sb}_2\text{Et}_4$ has not yet been determined, allowing no structural comparison between the uncomplexed and complexed distibine unit.

## 2. Dibismuthine Adducts

The tetraalkyldibismuthine adducts exhibit similar structural parameters. The Bi–Bi bond distances were found to be almost equal (298.3(1) pm **35** and 298.4(1) pm **36**), while the M–Bi bond distances range from 308.4(2) pm **35** to 311.4(2) pm **36**. They are comparable to those found for $(t\text{-Bu})_3$ $M \leftarrow \text{Bi}(i\text{-Pr})_3$ (M = Al: 308.8(1) pm **18**, Ga: 313.5(1) pm **19**) and significantly elongated compared to the sum of the covalent radii ($\sum r_{\text{cov}}(\text{AlBi})$: 275 pm; $\sum r_{\text{cov}}(\text{GaBi})$: 276 pm). As was observed for the distibine bisadducts, the gallane bisadduct $[t\text{-Bu}_3\text{Ga}]_2[\text{Bi}_2\text{Et}_4]$ features slightly larger C–Ga–C bond angles and Ga–Bi and Ga–C distances than the alane bisadduct. The Bi–C distances (average: 226.7 pm **35**; 227.9, 228.6 pm **36**) as well as the sum of the C–Bi–C and C–Bi–Bi bond angles (287.7° **35**; 285.9, 291.6° **36**) are comparable. However, the sums of the bond angles are slightly larger than those of the identically substituted distibine bisadducts $[t\text{-Bu}_3\text{M}]_2[\text{Sb}_2\text{Et}_4]$ (M = 292.9; 291.1°), pointing to a higher *s*-character of the dative Bi–M bonding electron pairs and a higher *p*-character of the Bi–C and Bi–Bi bonding electron pairs compared to the distibine bisadducts. This is confirmed by the sums of the C–M–C bond angles (352.8° **35**; 354.3, 353.4° **36**) which are also slightly larger compared to those of the distibine derivatives (M = Al, 350.2° **32**; Ga, 351.1° **33**), again indicating tetraalkyldibismuthines to be slightly weaker Lewis bases than tetraalkyldistibines.

## E. Adducts of Low-Valent Organotrieles

Low-valent $MX_2$ compounds of group 13 elements adopt different structures. Tetrachlorodiborane, $B_2Cl_4$, contains a covalent B–B bond, while tetrachlorodigallane $Ga_2Cl_4$ and indane $In_2Cl_4$ are mixed M(I/III) halides in the solid and in the molten state.[69] They can be transformed into "real" digallanes and diindanes $Cl_2M–MCl_2$ by addition of a Lewis base such as dioxane.[70] Schnöckel and coworkers demonstrated that the corresponding dialanes $Al_2X_4$, which have been unknown for a long time, can be obtained from metastable AlX solutions by addition of a Lewis base. This particular reaction makes use of the tendency of Al(I) species to disproportionate into elemental Al and either Al(II) or Al(III) compounds, respectively, depending on the solvent, the temperature and the Lewis base. To date, several Lewis base stabilized dialanes,[71] digallanes[72] and diindanes[73] have been synthesized and structurally characterized, as shown in Table XVI. The dialanes were found to be more sensitive toward disproportionation into elemental aluminum and Al(III)-halides than the digallanes. It was found that the central M–M bond is strongly affected by the Lewis base: the stronger the Lewis base, the longer the M–M bond. This can clearly be seen when comparing $Ga_2I_4(EEt_3)_2$ (E = N 249.8 pm, P 243.6 pm, As 242.8 pm), showing a Ga–Ga bond shortening of 7 pm. At the same time, the X–Ga–X bond angles increase from 101.8 ($NEt_3$) to 109.0 ($AsEt_3$), demonstrating that a weaker donor bond favors a more $sp^2$-like geometry around the metal center. Such findings agree very well with those observed for simple $R_3M \leftarrow ER'_3$ adducts, which show an increase of the R–M–R bond angles toward 120° with decreasing Lewis basicity (Table XVI).

It was also demonstrated that M(I) compounds (M = Al, Ga) are accessible via base stabilization. Schnöckel and coworkers have reported on the structures of three novel heterocyclic compounds of the type $(MX)_x$. $Al_4Br_4(NEt_3)_4$,[74] $Al_4I_4(PEt_3)_4$[75] and $Ga_8I_8(PEt_3)_6$[72e] were obtained from metastable solutions of AlBr, AlI and GaI in toluene in the presence of the corresponding Lewis base. In addition, a unique compound containing a $Ga_3$ chain, $Ga_3I_5(PEt_3)_3$, was obtained from reaction of "GaI" and $PEt_3$ in toluene and its solid state structure determined by single crystal X-ray

TABLE XVI

SELECTED BOND LENGTHS [pm] AND ANGLES [°] OF ADDUCTS OF THE TYPE $M_2X_4(ER'_3)_2$

| Compound | M–E (av.) | M–X (av.) | M–M | X–M–X (av.) |
|---|---|---|---|---|
| $M_2X_4(NR_3)_2$ | | | | |
| $Al_2Cl_4(NMe_2SiMe_3)_2$ | 200.1 | 216.8 | 257.3 | 107.3 |
| $Al_2Br_4(NMe_2SiMe_3)_2$ | 199.9 | 233.2 | 256.4 | 107.4 |
| $Ga_2Cl_4(NMe_3)_2$ | 204.6 | 219.0 | 242.1 | 106.4 |
| $Ga_2Cl_4(py)_2$ | 200.2 | 220.0 | 240.3 | 105.7 |
| $Ga_2Cl_4(p\text{-}Me\text{-}py)_2$ | 200.4 | 219.5 | 241.4 | 106.2 |
| $Ga_2Br_4(py)_2$ | 202.5 | 235.0 | 242.0 | 105.9 |
| $Ga_2I_4(NEt_3)_2$ | 209.4 | 259.7 | 249.8 | 101.9 |
| $M_2X_4(PR_3)_2$ | | | | |
| $Al_2I_4(PEt_3)_2$ | 244.0 | 256.2 | 254.6 | 108.7 |
| $Ga_2I_4(PEt_3)_2$ | 241.3 | 258.8 | 243.6 | 107.6 |
| $Ga_2I_4(PPh_3)_2$ | 244.5 | 256.8 | 244.4 | 107.2 |
| $In_2I_4(PPr_3)_2$ | 258.5 | 274.8 | 274.5 | 107.8 |
| $M_2X_4(AsR_3)_2$ | | | | |
| $Ga_2I_4(AsEt_3)_2$ | 248.4 | 257.1 | 242.8 | 109.0 |

analysis.[72f] This compound features a central Ga(I) unit, to which two Ga(II) units are connected (Figs. 19–21).

As already was observed for hypercoordinated adducts $MX_3(ER'_3)_2$, no stibine and bismuthine adducts of low-valent alanes, gallanes or indanes have been prepared, to date. According to the lability of low-valent group 13 compounds toward disproportionation into M(III) and elemental M, stibines and bismuthines are expected to be too weak as Lewis bases, preventing them from the stabilization of such compounds.

# III

## COMPOUNDS OF THE TYPE $[R_2MER'_2]_x$

### A. *Four- and Six-Membered Heterocycles*

Syntheses, structures and properties of heterocyclic group 13/15 compounds of the type $[R_2MER'_2]_x$ have been the subject of intense studies for many years. In particular, aminoalanes and -gallanes $[R_2MNR'_2]_x$ (M = Al, Ga) have received considerable attention. The interest in these particular compounds has increased within the last two decades due to their potential application as *single source precursors* for the deposition of AlN and GaN

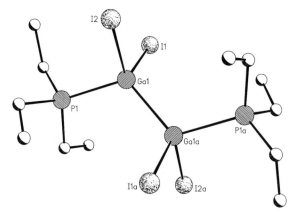

FIG. 19. Solid state structure of $Ga_2I_4(PEt_3)_2$.

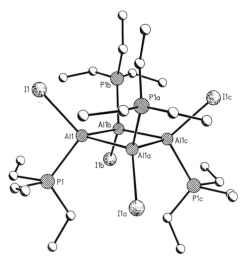

FIG. 20. Solid state structure of $Al_4I_4(PEt_3)_4$.

films by MOCVD process. Therefore, it is not surprising that they represent by far the most investigated group 13/15 heterocyclic compounds. In addition, phosphinogallanes and -indanes as well as arsinogallanes and -indanes $[R_2MER'_2]_x$ (M = Ga, In; E = P, As) have been studied intensely. Due to their large number, these heterocycles will not be reviewed in detail here. Interested readers are referred to several reviews dealing with the synthesis and structures of such compounds.[76] Instead, general pathways for the synthesis of long time unknown group 13 stibides and bismuthides will

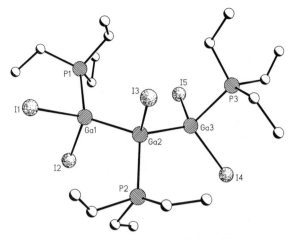

FIG. 21. Solid state structure of $Ga_3I_5(PEt_3)_3$.

$$R_2MH + HER'_2 \xrightarrow[-H_2]{} [R_2MER'_2]_x \qquad (7)$$

M = Al, Ga, In; E = N, P, (As)
R, R' = H, alkyl, aryl

$$R_3M + HER'_2 \xrightarrow[-RH]{} [R_2MER'_2]_x \qquad (8)$$

M = Al, Ga, In; E = N, P, (As)
R = Me, Et, R' = H, alkyl, aryl

$$R_2MX + M'ER'_2 \xrightarrow[-M'X]{} [R_2MER'_2]_x \qquad (9)$$

M = Al, Ga, In; E = N, P, As
M' = Li, (Na); R, R' = alkyl, aryl

SCHEME 4. Synthesis of ME-heterocycles by standard hydrogen, alkane and salt elimination reactions.

be presented and their solid state structures compared with those of analogous amides, phosphides and arsenides. In addition, the general potential of heterocycles $[R_2MER'_2]_x$ (M = Al, Ga; E = P, As, Sb, Bi) to serve as precursors for the synthesis of Lewis base-stabilized monomeric compounds of the type base $\rightarrow M(R_2)ER'_2$ will be shown. The group 15 elements of such monomers contain an electron *lone pair* that is active for complexation reactions both with transition metal complexes and group 13 trialkyls. Most of the compounds presented in the following chapter have been synthesized within the last 5 years in our group and by Wells *et al.*

## 1. General Synthetic Pathways

Several general pathways for the synthesis of group 13/15-heterocycles of the type $[R_2MER'_2]_x$ have been developed over the years. The most important are alkane, hydrogen and salt elimination reactions, as summarized in Scheme 4.

While these specific reaction types were found to be generally useful for the synthesis of heterocycles containing the lighter elements of group 15, N and P, they failed almost completely for the synthesis of the corresponding stibides and bismuthides. Hydrogen and alkane elimination reactions are inappropriate according to the non-acidic E–H group. In contrast, hydrogen substituents of the corresponding amines and phosphines $REH_2$ and $R_2EH$ (E = N, P), respectively, are positively polarized. Salt elimination reactions between chloroalanes, -gallanes or -indanes and Li salts of stibides or bismuthides $LiER'_2$ often occured under disproportionation and formation of elemental Sb and Bi. Consequently, only a very few reports on the synthesis and solid state structures of heterocyclic group 13 stibides have been known for many years and group 13 bismuthides remained completely unknown. Cowley *et al.* reported on the synthesis of four heterocyclic stibinogallanes and -indanes. $[Me_2GaSb(t-Bu)_2]_3$ **45**, $[Cl_2GaSb(t-Bu)_2]_3$ **44**, $[Me_2InSb(t-Bu)_2]_3$ **53**, $[t-Bu_2Sb(Cl)In-\mu-Sb(t-Bu)_2]_2$ **57** were obtained in rather low yield by salt elimination reaction between $Me_2MCl$ (M = Ga, In) and $t-Bu_2SbLi$ and dehalosilylation reactions, respectively.[77] In addition, Roesky and coworkers reported on the unique reaction of the low-valent alane ($[Cp^*Al]_4$) and stibine ($[t-BuSb]_4$), yielding $(Cp^*Al)_3Sb_2$[78] (Scheme 5).

Therefore, alternative synthetic pathways had to be developed to obtain a deeper knowledge about this class of compounds. To date, three different reaction types have been found to yield the desired heterocycles, which in the following will be presented in detail.

*Dehalosilylation Reaction*

Wells *et al.* introduced the *dehalosilylation reaction* for the synthesis of group 13/15 compounds in 1986. Reactions between silylarsines of the type $R'_2AsSiMe_3$ with chlorogallanes occurred with elimination of $Me_3SiCl$, leading to the formation of arsinogallanes of the types $[Cl_2GaAsR'_2]_x$[79] (Scheme 6).

The most striking advantages of this powerful reaction pathway that has been established as the most versatile method for the synthesis of arsinogallanes, are the easy workup of the reaction products and the possibility to obtain also bisarsinogallanes as well as trisarsinogallanes of the types $RGa(AsR_2)_2$ and $Ga(AsR_2)_3$ via multiple $Me_3SiCl$ elimination.[76c]

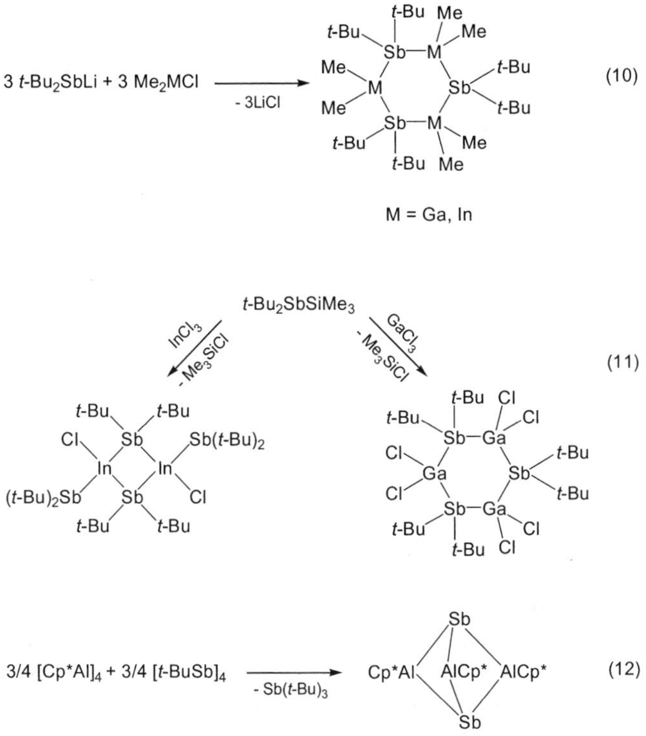

$$3\ t\text{-Bu}_2\text{SbLi} + 3\ \text{Me}_2\text{MCl} \xrightarrow[-\ 3\text{LiCl}]{} \qquad (10)$$

M = Ga, In

$$ \qquad (11)$$

$t\text{-Bu}_2\text{SbSiMe}_3$

InCl₃ / - Me₃SiCl          GaCl₃ / - Me₃SiCl

$$3/4\ [\text{Cp*Al}]_4 + 3/4\ [t\text{-BuSb}]_4 \xrightarrow[-\ \text{Sb}(t\text{-Bu})_3]{} \qquad (12)$$

**SCHEME 5.** Synthesis of compounds containing regular M–Sb bonds.

$$\text{R}_2\text{AsSiMe}_3 + \text{GaCl}_3 \xrightarrow[-\ \text{Me}_3\text{SiCl}]{} 1/x\ [\text{Cl}_2\text{GaAsR}_2]_x \qquad (13)$$

R = Me₃SiCH₂, Mes

$$2\ \text{R}_2\text{AsSiMe}_3 + \text{GaCl}_3 \xrightarrow[-\ 2\text{Me}_3\text{SiCl}]{} 1/x\ [\text{ClGa}(\text{AsR}_2)_2]_x \qquad (14)$$

R = Me₃SiCH₂, Mes

$$3\ \text{Mes}_2\text{AsSiMe}_3 + \text{GaCl}_3 \xrightarrow[-\ 3\text{Me}_3\text{SiCl}]{} \text{Ga}(\text{AsR}_2)_3 \qquad (15)$$

**SCHEME 6.** Wells initial studies on the synthesis of arsinogallanes by dehalosilylation reaction.

In 1988, Cowley *et al.* demonstrated this specific reaction to be also useful for the preparation of heterocyclic stibinogallanes and -indanes.[77a,77b] [Cl₂GaSb(*t*-Bu)₂]₃ **44** and [*t*-Bu₂Sb(Cl)In-μ-Sb(*t*-Bu)₂]₂ **57** were obtained by reaction of MCl₃ and *t*-Bu₂SbSiMe₃ (see also Scheme 5) and structurally characterized by single crystal X-ray diffraction. In 1996, we and Wells

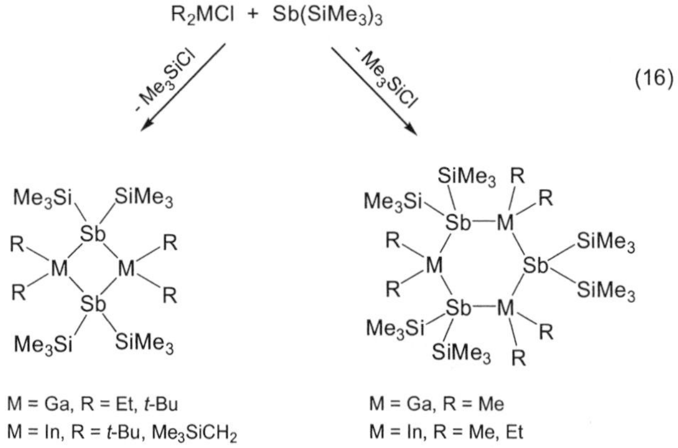

$$R_2MCl + Sb(SiMe_3)_3$$

(16)

M = Ga, R = Et, t-Bu
M = In, R = t-Bu, Me₃SiCH₂

M = Ga, R = Me
M = In, R = Me, Et

SCHEME 7. Synthesis of heterocyclic stibinogallanes and -indanes by dehalosilylation reaction.

started to investigate the potential of this particular reaction type in more detail, leading to the synthesis of several four-membered Ga–Sb and In–Sb heterocycles ([Et₂GaSb(SiMe₃)₂]₂ **47**, [t-Bu₂GaSb(SiMe₃)₂]₂ **48**, [(t-Bu₂Ga)₂(Cl)Sb(SiMe₃)₂] **49**, [t-Bu₂InSb(SiMe₃)₂]₂ **58** and [Me₃SiCH₂)₂ InSb(SiMe₃)₂]₂ **59**) as well as six-membered stibinogallanes and -indanes ([Me₂GaSb(SiMe₃)₂]₃ **46**, [Me₂InSb(SiMe₃)₂]₃ **55**, [Et₂InSb(SiMe₃)₂]₃ **56**)[80] (Scheme 7).

In contrast, several attempts to synthesize Al–Sb heterocycles by dehalosilylation reactions failed. Only the reaction of Me₂AlCl with Sb(SiMe₃)₃ gave the six-membered stibinoalane [Me(Cl)AlSb(SiMe₃)₂]₃ **38**.[42b] Obviously, **38** was not formed by dehalosilylation but by a tetramethylsilane elimination reaction. Its formation most likely results from the increased dissociation energy of the Al–Cl bond ($D_{298}^{\circ}$ [kJ/mol]: Al–Cl $511 \pm 1$; Ga–Cl $481 \pm 13$; In–Cl $439 \pm 8$)[81] and the higher Lewis acidity of the aluminum center compared to analogous chlorogallanes and -indanes. Such thermodynamic and electronic differences favor the formation of simple Lewis acid–base adducts, as was observed for the reaction of R₂AlCl with Sb(SiMe₃)₃ (R = Et, t-Bu; see also Section 2).[42b] Comparable trends were observed in reactions of R₂MCl (M = Al, Ga, In; R = Et, i-Bu) with P(SiMe₃)₃ and As(SiMe₃)₃, yielding the adducts in the case of M = Al[82] and the heterocycles only in the case of M = Ga, In.[83] In contrast to the stibine–alane adducts, the phosphine– and arsine–alane adducts could be converted into the corresponding heterocycles by thermal activation.[84]

*Dehydrosilylation Reaction*

Al–Sb heterocycles were obtained for the first time in our group from reactions of dialkylalanes R₂AlH and tris(trimethylsilyl)stibine Sb(SiMe₃)₃.

$$3\ Cl_2AlH + 3\ Et_2PSiMe_3 \xrightarrow[- 3Me_3SiH]{} [Cl_2AlPEt_2]_3 \qquad (17)$$

$$3\ ClAlH_2 + 3\ Et_2PSiMe_3 \xrightarrow[- 3Me_3SiH]{} [Cl(H)AlPEt_2]_3 \qquad (18)$$

$$2Me_2AlH + 2\ HP(SiMe_3)_2 \xrightarrow[- Me_3SiH]{} [Me_2AlP(H)SiMe_3]_2 \qquad (19)$$

$$2\ i\text{-}Bu_2AlH + 2\ HP(SiMe_3)_2 \xrightarrow[- 2H_2]{} [i\text{-}Bu_2AlP(SiMe_3)_2]_2 \qquad (20)$$

SCHEME 8. Initial dehydrosilylation reactions.

Dialkylalanes $R_2AlH$ are less Lewis acidic and the hydrogen atom is less strongly bound than the chlorine atom of dialkylchloroalanes $R_2AlCl$, hence depressing the tendency to form adducts but to react under $Me_3SiH$ elimination. This reaction type, usually called *dehydrosilylation reaction*, was first reported by Nöth et al. Reaction of diborane $B_2H_6$ and $R_2PSiMe_3$ yielded the six-membered phosphinoborane $[H_2BPR_2]_3$.[85] Detailed studies of Fritz et al. and Krannich et al. demonstrated the dehydrosilylation reaction to be the favored reaction pathway in reactions of $Et_2PSiMe_3$ with $HAlCl_2$ and $H_2AlCl$, and of $Me_2AlH$ with $HP(SiMe_3)_2$.[86] In contrast, the reaction of $i\text{-}Bu_2AlH$ with $HP(SiMe_3)_2$ was shown to occur with $H_2$ elimination[87] (Scheme 8).

Reactions of $R_2AlH$ with $Sb(SiMe_3)_3$ and dialkylsilylstibines $R_2SbSiMe_3$ gave the desired four- and six-membered stibinoalanes of the type $[R_2AlSbR'_2]_x$ ($R' = SiMe_3$, R = Me **37**, Et **39**, i-Bu **40**; $R' = t\text{-}Bu$, R = Me **42**, Et) in excellent yields.[42b,88] The reactions of $Me_2AlH$ with $P(SiMe_3)_3$ and $As(SiMe_3)_3$ also resulted in the formation of the corresponding heterocycles $[Me_2AlP(SiMe_3)_2]_2$ and $[Me_2AlAs(SiMe_3)_2]_2$.[89] Moreover, six-membered bismuthinoalane and -gallane $[Me_2MBi(SiMe_3)_2]_3$ (M = Al **43**, Ga **52**) were obtained for the first time by reactions of $Bi(SiM_3I_3)$ with $Me_2AlH$[90] and $Me_2GaH$,[91] respectively, demonstrating the dehydrosilylation reaction to be generally applicable for the synthesis of group 13/15 heterocycles (Scheme 9). The high yields and the mild reaction conditions, in particular, the low reaction temperatures, are the major advantages of this reaction type, which was found to be essential for the synthesis of the bismuthinogallane $[Me_2GaBi(SiMe_3)_2]_3$ **52**. **52** is thermolabile and decomposes in solution at temperatures above $-20\,°C$.

*Distibine Cleavage Reaction*

It is well known in organoantimony chemistry that tetraalkyldistibines $Sb_2R'_4$ tend to react with electrophilic reagents under Sb–Sb bond cleavage.[64] This tendency was already observed in reactions with transition metal

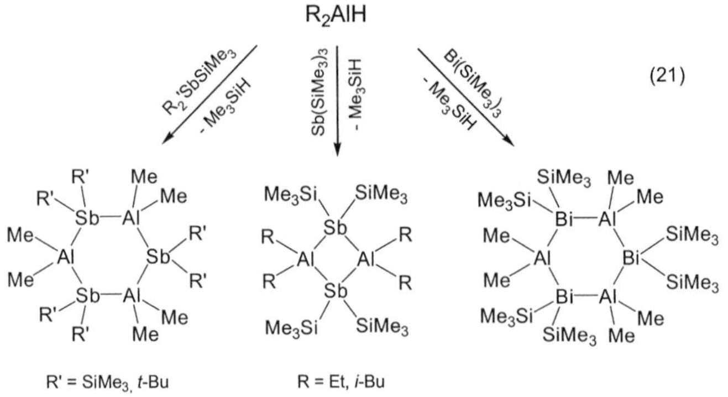

$$R_2AlH$$

(21)

Scheme 9. Formation of heterocyclic stibino- and bismuthinoalanes by dehydrosilylation reaction.

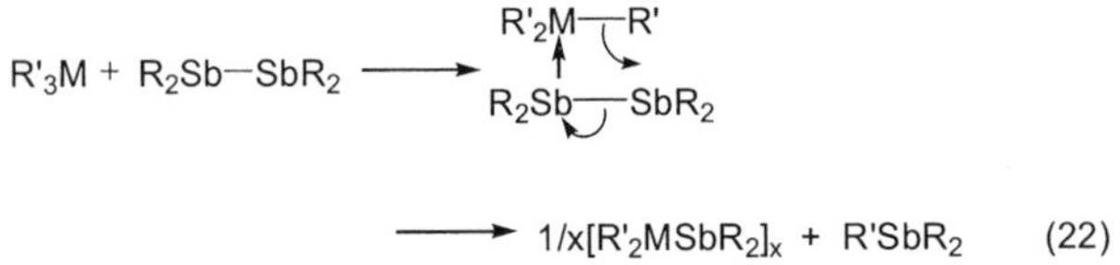

$$R'_3M + R_2Sb\!-\!SbR_2 \longrightarrow$$

$$\longrightarrow 1/x[R'_2MSbR_2]_x + R'SbR_2 \quad (22)$$

Scheme 10. Proposed reaction mechanism for the distibine cleavage reaction.

complexes, leading to the formation of heterocyclic compounds.[92] In contrast, reactions of distibines and group 13 compounds were unknown until Breunig et al. described the reaction of $Me_4Sb_2$ with $(Me_3SiCH_2)_3In$, leading to the formation of the six-membered stibinoindane $[(Me_3SiCH_2)_2InSbMe_2]_3$ **54**.[93] In contrast to our observations on the reactions of $Sb_2Me_4$ and $Sb_2Et_4$ with trialkylalanes and -gallanes $MR_3$ (M = Al, Ga), the formation of a distibine adduct was not observed. However, its (intermediary) formation is most likely since we recently demonstrated that the distibine bisadducts $['Bu_3Ga]_2[Sb_2R'_4]$ (R = Me, Et), which are stable in the pure form below $0\,°C$ under inert conditions, rearrange in solution at ambient temperature to give the corresponding heterocycles $[t\text{-}Bu_2GaSbMe_2]_3$ **51** and $[t\text{-}Bu_2GaSbEt_2]_2$ **50** in excellent yields.[66] Comparable reaction patterns were observed for $[R_3Ga]_2[Sb_2Me_4]$, $[R_3Ga]_2[Sb_2Et_4]$ (R = Me, Et) and for $['Bu_3Al]_2[Sb_2R'_4]$ (R' = Me, Et) bisadducts. However, the cleavage reactions of the alane bisadducts require higher temperatures. The most probable reaction mechanism includes the formation of either 1:1 or 1:2 adducts, followed by simultaneous M–C and Sb–Sb bond cleavage as shown in Scheme 10. According to this reaction pathway, "mixed-substituted" stibines of the type $R_2SbR'$ are formed as

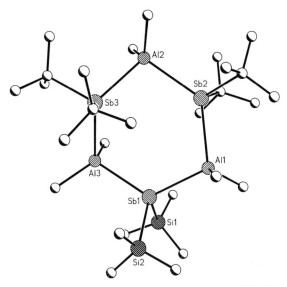

FIG. 22. Solid state structure of $(Me_2Al)_3(t\text{-}Bu_2Sb)_2Sb(SiMe_3)_2$.

byproducts. This was experimentally verified by multinuclear NMR studies ($^1H$, $^{13}C$).

The most striking feature of the distibine cleavage reaction is the simple formation of completely alkyl-substituted M–Sb heterocycles, which have potential applications to serve as *single source precursors* in MOCVD processes (*metal organic chemical vapor deposition*) for the preparation of thin films of the corresponding semiconducting MSb-materials (see also Section 5). Such heterocycles were previously only available by dehalo-silylation (Ga–Sb) and dehydrosilylation reactions (Al–Sb), respectively. Unfortunately, the use of silylsubstituted dialkylstibines $R_2SbSiMe_3$ sometimes yielded partially silyl-substituted heterocycles, as was demonstrated for instance in the reaction of $t\text{-}Bu_2SbSiMe_3$ and $Me_2AlH$. Under non-specific reaction conditions, the silyl/alkyl mixed heterocycle $(Me_2Al)_3$ $(t\text{-}Bu_2Sb)_2Sb(SiMe_3)_2$ **41** was obtained and its solid state structure determined by single crystal X-ray analysis (Fig. 22).[42]

### 2. Single Crystal X-ray Structures

M–E heterocycles of the type $[R_2MER'_2]_x$ generally adopt either four- or six-membered ring geometries ($x = 2, 3$), as was shown already for a large number of M–N heterocycles. This is illustrated in Table XVII, summarizing the number of structurally characterized four- and six-membered heterocycles $[R_2MER'_2]_x$ (E = N, P, As).[94]

TABLE XVII
Number of Structurally Characterized Four- and Six-membered Group 13/15
Heterocycles of the Type $[R_2MER'_2]_x$

| Ring size | Al | | Ga | | In | |
|---|---|---|---|---|---|---|
| | 4 | 6 | 4 | 6 | 4 | 6 |
| N | 82 | 9 | 35 | 5 | 12 | 0 |
| P | 5 | 3 | 19 | 6 | 19 | 7 |
| As | 5 | 2 | 14 | 2 | 7 | 1 |

TABLE XVIII
Selected Bond Lengths [pm] and Angles [°] of Heterocyclic Stibino- and
Bismuthinoalanes of the Type $[R_2AlER'_2]_x$ (E = Sb, Bi)

| Heterocycle | M–E | Al–E–Al | E–Al–E |
|---|---|---|---|
| [Me$_2$AlSb(SiMe$_3$)$_2$]$_3$ **37** | 270.3(1)–273.6(1) | 118.5(1)–128.2(1) | 103.5(1)–106.5(1) |
| [Me(Cl)AlSb(SiMe$_3$)$_2$]$_3$ **38** | | | |
| [Et$_2$AlSb(SiMe$_3$)$_2$]$_2$ **39** | 272.3(1), 272.9(1) | 91.7(1) | 88.3(1) |
| [$i$-Bu$_2$AlSb(SiMe$_3$)$_2$]$_2$ **40** | 274.3(1), 274.6(1) | 93.7(1) | 86.3(1) |
| (Me$_2$Al)$_3$(Sb$t$-Bu$_2$)$_2$Sb(SiMe$_3$)$_2$ **41** | 271.9(2)–278.0(2) | 115.4(1)–128.4(1) | 103.1(1)–106.9(1) |
| [Me$_2$AlSb($t$-Bu)$_2$]$_3$ **42** | 271.9(1)–278.4(1) | 115.3(1)–128.9(1) | 102.8(1)–108.2(1) |
| [Me$_2$AlBi(SiMe$_3$)$_2$]$_3$ **43** | 275.5(3)–279.3(3) | 121.7(1)–130.5(1) | 101.0(1)–104.1(1) |

As the solid state structures of heterocyclic group 13 amides, phosphides and arsenides have been reviewed in the past,[76] they will not be discussed here. Instead, the solid state structures of group 13 stibides and bismuthides will be described in detail.

### Al–E Heterocycles

Heterocyclic stibino- and bismuthinoalanes, -gallanes and -indanes adopt structures in the solid state analogous to those observed for the corresponding amides, phosphides and arsenides. Their central structural parameters are summarized in Table XVIII.

[Me$_2$AlSb(SiMe$_3$)$_2$]$_3$ **37** and [Me$_2$AlSb($t$-Bu)$_2$]$_3$ **42** are non-planar heterocycles and adopt distorted twist-boat conformations with the Al and Sb atoms arranged in tetrahedral environments. The Al–Sb bond lengths range from 270 to 278 pm (270.3(1)–273.8(1) pm **37**, 271.9(1)–278.4(1) pm **42**). As was expected, they are elongated compared to the sum of the covalent radii ($\sum r_{cov}$(AlSb) = 266 pm), but significantly shortened in respect to the Al–Sb bond lengths of alane–stibine adducts $R_3Al \leftarrow SbR'_3$ and [$t$-Bu$_3$Al]$_2$[Sb$_2$R$_4$] (Al–Sb: 280–300 pm). The smaller variation of the

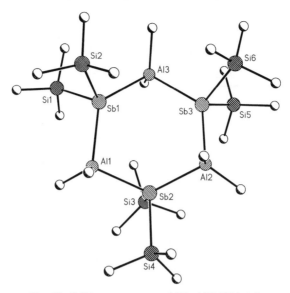

FIG. 23. Solid state structure of [Me$_2$AlSb(SiMe$_3$)$_2$]$_3$.

endocyclic bond angles and the smaller exocyclic bond angles (Si–Sb–Si: 100.7(1)–102.3(1)° **37**, C–Sb–C: 104.9(2)–106.0(2)° **42**) indicate that the six-membered ring of 37 is less strained than that of **42** due to the replacement of a tertiary C atom by a larger Si atom (CMe$_3$ vs. SiMe$_3$) despite the larger effective sterical parameter of a Me$_3$Si group compared to a $t$-Bu group (Me$_3$Si: 1.40; $t$-Bu: 1.24[95]). The endocyclic Sb–Al–Sb bond angles are smaller (103.5(1)–106.5(1)° **37**; 102.8(1)–108.2(1)° **42**) and the Al–Sb–Al bond angles larger (118.5(1)–128.2(1)° **37**; 115.3(1)–128.9(1)° **42**) than tetrahedral. Comparable structural parameters were reported for other six-membered heterocycles such as [Me$_2$AlAs(CH$_2$SiMe$_3$)Ph]$_3$ (Al–As–Al: 118.2(2)–122.2(2)°, As–Al–As: 102.6(2)–104.8(2)°), [Me$_2$AlAsPh$_2$]$_3$ · (C$_7$H$_8$)$_2$ (Al–As–Al: 118.1(1)–122.7(1)°, As–Al–As: 99.1(1)–101.1(1)°), [Me$_2$AlN(CH$_2$)$_2$]$_3$ (Al–N–Al: 119.9(5)°, N–Al–N: 102.0(5)°).[96]

The increased steric demand of $i$-Bu groups compared to Et groups is reflected by the central bonding parameters of the planar, four-membered Al–Sb heterocycles [Et$_2$AlSb(SiMe$_3$)$_2$]$_2$ **39** and [$i$-Bu$_2$AlSb(SiMe$_3$)$_2$]$_2$ **40**. The latter shows elongated Al–Sb bond distances (272.3(1)–272.9(1) pm **39**, 274.3(1)–274.6(1) pm **40**) and larger endocyclic Al–Sb–Al bond angles (91.7(1)° **39**; 93.7(1)° **40**), whereas the Sb–Al–Sb angles are smaller (88.3(1)° **39**; 86.3(1)° **40**). Comparable structural trends were previously found for heterocyclic phosphino- and arsinoalanes: the more demanding the

TABLE XIX

SELECTED BOND LENGTHS [pm] AND ANGLES [°] OF FOUR-MEMBERED PHOSPHINO-, ARSINO- and STIBINOALANES [R$_2$AlE(SiMe$_3$)$_2$]$_2$

| Heterocycle | Al–E | Al–E–Al | E–Al–E | R–Al–R | Si–E–Si |
|---|---|---|---|---|---|
| [Me$_2$AlP(SiMe$_3$)$_2$]$_2$ | 245.7 | 90.6 | 89.4 | 113.4 | 108.4 |
| [Et$_2$AlP(SiMe$_3$)$_2$]$_2$ | 245.7 | 90.2 | 89.8 | 114.2 | 108.0 |
| [i-Bu$_2$AlP(SiMe$_3$)$_2$]$_2$ | 247.6 | 91.0 | 89.0 | 117.1 | 106.3 |
| [Me$_2$AlAs(SiMe$_3$)$_2$]$_2$ | 253.6 | 91.7 | 88.3 | 115.0 | 108.1 |
| [Et$_2$AlAs(SiMe$_3$)$_2$]$_2$ | 253.5 | 91.0 | 89.0 | 115.0 | 107.6 |
| [i-Bu$_2$AlAs(SiMe$_3$)$_2$]$_2$ | 255.0 | 92.2 | 87.8 | 118.8 | 105.6 |
| [Et$_2$AlSb(SiMe$_3$)$_2$]$_2$ | 272.6 | 91.7 | 88.3 | 117.3 | 107.3 |
| [i-Bu$_2$AlSb(SiMe$_3$)$_2$]$_2$ | 274.4 | 93.7 | 86.3 | 121.2 | 102.7 |

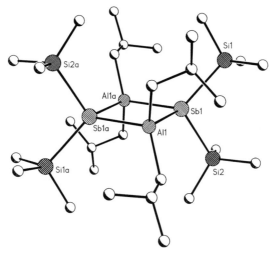

FIG. 24. Solid state structure of [i-Bu$_2$AlSb(SiMe$_3$)$_2$]$_2$.

substituents bound to Al, the wider the C–Al–C and Al–E–Al bond angles, whereas the E–Al–E bond angles decrease as shown in Table XIX (Fig. 24).

The Al–Bi heterocycle [Me$_2$AlBi(SiMe$_3$)$_2$]$_3$ **43** exhibits almost the same bonding parameters as were found in other six-membered heterocycles of the type [R$_2$AlE(SiMe$_3$)$_2$]$_3$ (E = P, As, Sb). The only differences are the increased endocyclic Al–Bi–Al bond angles (121.7(1)–130.5(1)°) and the decreased Bi–Al–Bi bond angles (101.0(1)–104.1(1)°). The Al–Bi bond distances range from 275.5(3) to 279.3(3) pm, which is slightly longer than the sum of the covalent radii of 275 pm, but significantly shorter than the distances found in the Al–Bi Lewis acid–base adducts R$_3$Al ← BiR$'_3$ and [t-Bu$_3$Al]$_2$[Bi$_2$Et$_4$] (292–308 pm **35**) (Fig. 25).

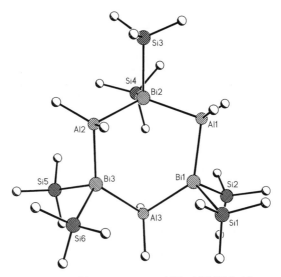

FIG. 25. Solid state structure of [Me₂AlBi(SiMe₃)₂]₃.

TABLE XX

SELECTED BOND LENGTHS [pm] AND ANGLES [°] OF HETEROCYCLIC STIBINO- AND
BISMUTHINOGALLANES OF THE TYPE [R₂GaER′₂]ₓ (E = Sb, Bi)

| Heterocycle | Ga–E | Ga–E–Ga | E–Ga–E |
|---|---|---|---|
| [Cl₂GaSb(t-Bu)₂]₃ **44** | 265.9(2)–266.2(2) | 109.9 (av.) | 108.4 (av.) |
| [Me₂GaSb(t-Bu)₂]₃ **45** | No data available | | |
| [Me₂GaSb(SiMe₃)₂]₃ **46** | 267.7(1)–271.4(1) | 118.3(1)–127.6(1) | 103.6(1)–107.3(1) |
| [Et₂GaSb(SiMe₃)₂]₂ **47** | 271.8(1), 272.9(1) | 92.7(1) | 87.3(1) |
| [t-Bu₂GaSb(SiMe₃)₂]₂ **48** | 276.5(1), 276.8(1) | 94.4(1), 94.5(1) | 85.5(1) |
| (t-Bu₂Ga)₂ClSb(SiMe₃)₂ **49** | 273.4(2) | 85.7(1) | – |
| [t-Bu₂GaSbEt₂]₂ **50** | 273.1(1), 273.5(1) | 96.6(1) | 83.4(1) |
| [t-Bu₂GaSbMe₂]₃ **51** | 271.3(1)–275.1(1) | 127.3(1)–133.3(1) | 96.7(1)–102.8(1) |
| [Me₂GaBi(SiMe₃)₂]₃ **52** | 274.4(1)–278.3(1) | 121.6(1)–130.7(1) | 100.7(1)–103.8(1) |

*Ga–E Heterocycles*

Central bonding parameters of structurally characterized four- and six-membered Ga–E heterocycles are summarized in Table XX.

As was observed for comparable six-membered Al–E heterocycles [Me₂AlER₂]₃, the Ga–E heterocycles [Me₂GaSb(t-Bu)₂]₃ **45**, [Cl₂GaSb(t-Bu)₂]₃ **44**, [Me₂GaSb(SiMe₃)₂]₃ **46**, [t-Bu₂GaSbMe₂]₃ **51** and [Me₂Ga Bi(SiMe₃)₂]₃ **52** also adopt distorted twist-boat type conformations. [Cl₂GaSb(t-Bu)₂]₃ features the shortest Ga–Sb bond distances (average

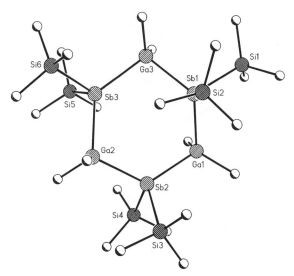

FIG. 26. Solid state structure of [Me$_2$GaSb(SiMe$_3$)$_2$]$_3$.

Ga–Sb: 266.1 pm), obviously resulting from the strong Lewis-acidic character of the Ga atoms due to the electron-withdrawing effect of the Cl substituents. The distortion of the six-membered rings is reflected both by the wide range of the Ga–Sb bond lengths and endocyclic Ga–Sb–Ga and Sb–Ga–Sb bond angles. As was observed for Al–E heterocycles, the bond angles at the Ga centers are larger and those at the Sb atoms smaller than tetrahedral (Fig. 26).

[Me$_2$GaBi(SiMe$_3$)$_2$]$_3$, the only structurally characterized bismuthinogallane, is isostructural to its alane-analogue [Me$_2$AlBi(SiMe$_3$)$_2$]$_3$. The Ga–Bi bond distances range from 274.4(1) to 278.3(1) pm, corresponding well to those of the bismuthinoalane 43 as well as to the sum of the covalent radii for Ga and Bi (276 pm). Lewis acid–base adducts R$_3$Ga ← BiR$_3'$ and [$t$-Bu$_3$Ga]$_2$[Bi$_2$Et$_4$] show significantly longer Ga–Bi distances (297–314 pm) according to their dative bonding character. The endocyclic Ga–Bi–Ga bond angles are significantly larger than the Bi–Ga–Bi bond angles, ranging from 121.6(1) to 130.7(1)° and 100.7(1) to 103.8(1)°, respectively, while the exocyclic C–Ga–C (115.4(3)–123.4(3)°) and Si–Bi–Si bond angles (99.2(1)–101.1(1)°) consequently display the opposite trend (Fig. 27).

The central structural parameters of the four-membered stibinogallanes [Et$_2$GaSb(SiMe$_3$)$_2$]$_2$ 47, [$t$-Bu$_2$GaSb(SiMe$_3$)$_2$]$_2$ 48, [($t$-Bu$_2$Ga)$_2$(Cl)Sb (SiMe$_3$)$_2$] 49 and [$t$-Bu$_2$GaSbEt$_2$]$_2$ 50 are similar to those found in four-membered stibinoalanes (Fig. 28).

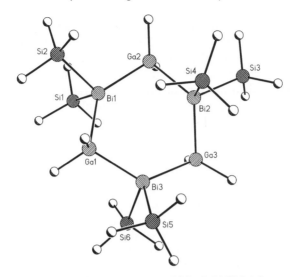

FIG. 27. Solid state structure of [Me$_2$GaBi(SiMe$_3$)$_2$]$_3$.

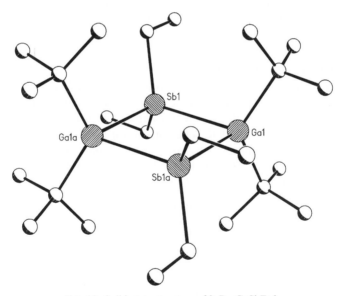

FIG. 28. Solid state structure of [$t$-Bu$_2$GaSbEt$_2$]$_2$.

The Ga–Sb bond lengths range from 272.3 to 276.7 pm. Again, steric interactions between the substituents have a significant influence on the Ga–Sb distances. Consequently, the longest Ga–Sb bond length was observed in **48** (276.7 pm). The asymmetric heterocycle [($t$-Bu$_2$Ga)$_2$

Sb(SiMe$_3$)$_2$Cl] **49**, which contains both a Cl- and a Sb(SiMe$_3$)$_2$-bridge, show slightly shorter Ga–Sb bond lengths (273.4 pm) compared to **48**, most likely caused by the electron withdrawing influence of the Cl-atom. The average Ga–Sb distances observed for the four-membered heterocycles are both elongated compared to the sum of the covalent radii (267 pm) and to the average Ga–Sb bond distances of the six-membered rings, most likely caused by the increased ring strain.

*In–Sb Heterocycles*

Stibinoindanes [R$_2$InSbR$_2'$]$_x$ display structural parameters comparable to those of the corresponding stibinoalanes and -gallanes. The In–Sb bond distances of four-membered rings are also slightly elongated compared to those of the six-membered rings, but somewhat longer than the sum of the covalent radii (283 pm). The longest In–Sb bond distances were observed in the sterically overcrowded heterocycle [*t*-Bu$_2$InSb(SiMe$_3$)$_2$]$_2$ **58**. The In–Sb bond distances of [*t*-Bu$_2$Sb(Cl)In-µ-Sb(*t*-Bu)$_2$]$_2$ **57** differ significantly. The In–Sb bond distance of the terminal *t*-Bu$_2$Sb fragment (279.7 pm) is about 7 pm shorter than those of the bridging *t*-Bu$_2$Sb moiety (286.5 pm). This is caused by the lower coordination number (3 (terminal moiety) vs. 4 (bridging moiety)) (Table XXI, Figs. 29 and 30).

*Factors Affecting the Ring Size*

The ring size (degree of oligomerization *x*) and conformation of M–E heterocycles strongly depends on steric effects of the substituents (repulsive interactions), ring strain effects and on entropy factors. This was shown for instance by Beachley and Racette for several heterocyclic aminoalanes [R$_2$AlNR$_2'$]$_x$[97] and confirmed by our own results. However, predictions whether four- or six-membered heterocycles will be formed are

TABLE XXI

SELECTED BOND LENGTHS [pm] AND ANGLES [°] OF HETEROCYCLIC STIBINOINDANES OF THE TYPE [R$_2$InSbR$_2'$]$_x$

| Heterocycle | In–Sb | In–Sb–In | Sb–In–Sb |
|---|---|---|---|
| [Me$_2$InSb(*t*-Bu)$_2$]$_3$ **53** | 282.2(1)–288.9(1) | 115.8(1)–127.8(1) | 103.7(1)–109.4(1) |
| [(Me$_3$SiCH$_2$)$_2$InSbMe$_2$]$_3$ **54** | 285.2(1)–286.9(1) | 129.2(1)–137.7(1) | 92.4(1)–98.6(1) |
| [Me$_2$InSb(SiMe$_3$)$_2$]$_3$ **55** | 284.4(1)–287.0(1) | 119.8(1)–127.0(1) | 102.7(1)–106.8(1) |
| [Et$_2$InSb(SiMe$_3$)$_2$]$_3$ **56** | 282.4(2)–291.1(2) | 119.1(1)–129.7(1) | 102.4(1)–106.2(1) |
| ([*t*-Bu$_2$Sb(Cl)In-µ-Sb(*t*-Bu)$_2$]$_2$ **57** | 286.5(1), 287.0(1) | 94.9(1) | 85.1(1) |
| [*t*-Bu$_2$InSb(SiMe$_3$)$_2$]$_2$ **58** | 292.7(1), 293.4(1) | 94.8(1), 95.1(1) | 85.0(1) |
| [(Me$_3$SiCH$_2$)$_2$InSb(SiMe$_3$)$_2$]$_2$ **59**[*] | 288 | 95.2 | 84.8 |

[*]The poor quality of the crystals did not facilitate a complete data set collection.

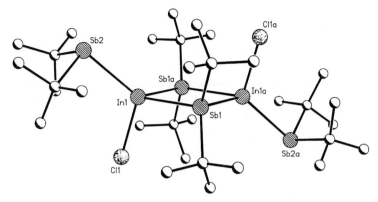

FIG. 29. Solid state structure of $[t\text{-Bu}_2\text{Sb(Cl)In-}\mu\text{-Sb}(t\text{-Bu})_2]_2$.

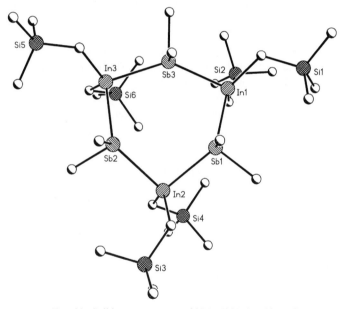

FIG. 30. Solid state structure of $[(\text{Me}_3\text{SiCH}_2)_2\text{InSbMe}_2]_2$.

not simple since ring strain effects, which generally favor the less strained six-membered heterocycles, and entropy effects, which generally favor the formation of the less-aggregated four-membered rings, are counteractive. If both energies have comparable values, sterical interactions between the substituents may control the degree of cyclization. Substituents in four-membered heterocycles generally have more space available due to the (idealized) 90° endocyclic bond angle of the central group 13/15 ring atoms, while six-membered rings display endocyclic bond angles of 109.5 (ideal chair conformation) or 120° (ideal planar ring), respectively. Additional repulsive interactions between the substituents in 1,3-position may also affect the ring geometry/conformation. Consequently, Et-substituted heterocycles of the type $[Et_2ME(SiMe_3)_2]_x$ tend to adopt dimeric structures, while Me-substituted heterocycles prefer trimeric structures. However, steric pressure not only depends on the size of the substituent but also on the size of the central atoms. Investigations on the ring size of a complete family of analogously substituted $[Me_2ME(SiMe_3)_2]_x$ and $[Et_2ME(SiMe_3)_2]_x$ heterocycles (M = Al, Ga, In; E = N, P, As, Sb, Bi) in the solid state clearly revealed the influence of the atomic radii of the central group 13 and 15 elements[89] (Table XXII).

In the case of the Me-substituted heterocycles $[Me_2ME(SiMe_3)_2]_x$, N, P and As yield four membered ring structures $[Me_2ME(SiMe_3)_2]_2$ in the solid state, whereas the somewhat larger Sb and Bi atoms favor the formation of six-membered rings $[Me_2ME(SiMe_3)_2]_3$. Obviously, steric interactions become less important with increasing atomic radii of the central ring atoms. The influence of the group 13 element can clearly be seen when comparing Et-substituted heterocycles $[Et_2ME(SiMe_3)_2]_x$. Only $[Et_2InSb(SiMe_3)_2]_3$, containing the largest group 13 element, adopts a six-membered ring structure in the solid state. Evidently, the larger In atoms within the ring sufficiently diminish repulsive interactions between the organic substituents,

TABLE XXII

RING SIZES OF $[Me_2ME(SiMe_3)_2]_x$ AND $[Et_2ME(SiMe_3)_2]_x$ HETEROCYCLES

| M | E | x | M–E | E–M–E | M–E–M | C–M–C | Si–E–Si |
|---|---|---|-----|-------|-------|-------|---------|
| Al | N* | 2 | 199.6 | 90.1 | 89.9 | 109.7 | 117.5 |
|    | P | 2 | 245.7 | 89.4 | 90.6 | 113.4 | 108.3 |
|    | As | 2 | 253.6 | 88.3 | 91.7 | 115.0 | 108.1 |
|    | Sb | 3 | 271.8 | 104.9 | 124.0 | 117.9 | 101.7 |
|    | Bi | 3 | 277.4 | 102.3 | 126.8 | 119.2 | 100.5 |
| Ga | N | 2 | 208.2 | 89.8 | 90.3 | 113.8 | 118.7 |
|    | P | 2 | 245.0 | 88.2 | 91.8 | 114.4 | 108.0 |
|    | As | 2 | 253.0 | 87.0 | 93.0 | 116.8 | 107.7 |
|    | Sb | 3 | 269.1 | 105.2 | 123.6 | 118.1 | 101.6 |
| In | N | 2 | 230.4 | 89.7 | 90.3 | 108.8 | 111.4 |
|    | P | 2 | 263.0 | 86.7 | 93.3 | 116.9 | 109.8 |
|    | As | 2 | 270.1 | 85.5 | 94.5 | 118.8 | 109.4 |
|    | Sb | 3 | 285.3 | 104.1 | 124.3 | 120.5 | 103.0 |

*$[Me_2AlN(SiHMe_2)_2]_2$.

resulting in the formation of the less-strained six-membered ring. In this context it should be interesting to know, if the analogous Bi-containing heterocycles $[Et_2MBi(SiMe_3)_2]_x$ (M = Al, Ga, In), which are currently unknown, either form four- or six-membered rings.

## B. *Lewis Base-Stabilized Monomeric Compounds*

In contrast to amino-, phosphino- and arsinoboranes $R_2B-ER'_2$ (E = N, P, As), which exhibit strong $\pi$-bonding interactions due to an overlap of the electron *lone pair* of the group 15 element and an empty p-orbital of boron leading to a formally double bonded species,[98] $\pi$-bonding in compounds containing heavier elements of group 13 is less favored.[99] This is indicated by significantly smaller rotational barriers around the M–E bond. Consequently, such compounds tend to form *head-to-tail adducts*, yielding four- and six-membered rings. Only the use of sterically demanding organic substituents such as Mes, Dipp ($C_6H_3(i\text{-}Pr)_2$), Trip ($C_6H_2(i\text{-}Pr)_3$) or Mes* ($C_6H_2(t\text{-}Bu)_3$) inhibits the formation of heterocyclic structures of the types $[R_2MER'_2]_x$ ($x \geq 2$) and allows the isolation of *kinetically stabilized*, monomeric compounds $R_2MER'_2$. This has been demonstrated, in particular, by the group of P. P. Power, who prepared and structurally characterized several monomeric group 13–amides, –phosphides and – arsenides. Interested readers are referred to an excellent review dealing with this particular class of compounds.[100]

In contrast, the synthesis of monomeric stibides and bismuthides $R_2MER'_2$ failed until we recently introduced an alternative, generally applicable pathway for the synthesis of monomeric group 13/15 compounds. We found that heterocycles may generally serve as starting compounds for the generation of their monomeric forms. In accordance with their formulation as *"head-to-tail adducts"*, strong Lewis bases are able to cleave the heterocycles by coordinating to the group 13 element, leading to the formation of *electronically stabilized* monomers of the type base $\rightarrow$ $M(R_2)ER'_2$. Prior to our studies, a very few compounds of this type were obtained from reactions of $H_3Al \leftarrow NMe_3$ with sterically demanding secondary amines $R_2NH$ and $As(SiMe_3)_3$, respectively.[101] In addition, a salt elimination reaction between $H_2AlCl \leftarrow NMe_3$ and $LiPMes_2$ yielded $Me_3N \rightarrow Al(H_2)PMes_2$ (Scheme 11).[102] The group 13 element is coordinatively saturated by interaction with the Lewis base $NMe_3$, while the electron *lone pair* of the group 15 atom is potentially active for further coordination chemistry. Unfortunately, no general pathway for the synthesis of this particular class of compounds was known. The reactions were limited to $Me_3N$-stabilized $AlH_3$ and $H_2AlCl$ as well as amines, phosphines and

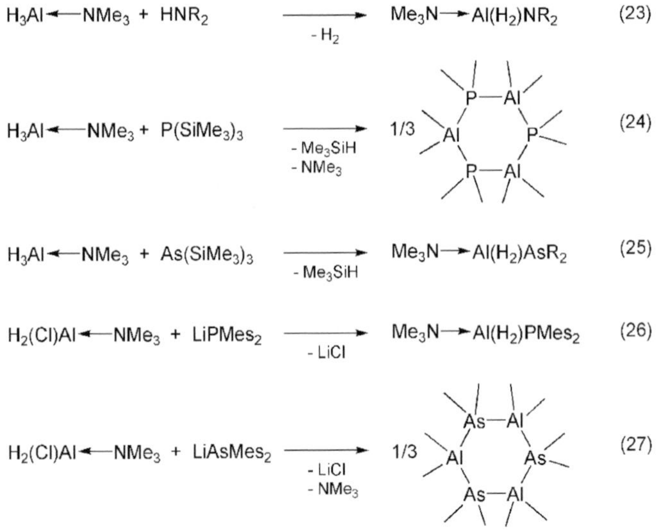

$$H_3Al \leftarrow NMe_3 + HNR_2 \xrightarrow[-H_2]{} Me_3N \rightarrow Al(H_2)NR_2 \quad (23)$$

$$H_3Al \leftarrow NMe_3 + P(SiMe_3)_3 \xrightarrow[\substack{-Me_3SiH \\ -NMe_3}]{} 1/3 \; \text{[ring]} \quad (24)$$

$$H_3Al \leftarrow NMe_3 + As(SiMe_3)_3 \xrightarrow[-Me_3SiH]{} Me_3N \rightarrow Al(H_2)AsR_2 \quad (25)$$

$$H_2(Cl)Al \leftarrow NMe_3 + LiPMes_2 \xrightarrow[-LiCl]{} Me_3N \rightarrow Al(H_2)PMes_2 \quad (26)$$

$$H_2(Cl)Al \leftarrow NMe_3 + LiAsMes_2 \xrightarrow[\substack{-LiCl \\ -NMe_3}]{} 1/3 \; \text{[ring]} \quad (27)$$

SCHEME 11. Reactions of $H_3Al \leftarrow NMe_3$ and $H_2(Cl)Al \leftarrow NMe_3$ with $E(SiMe_3)_3$ and $LiEMes_2$ (E = P, As).

arsines as starting compounds. In addition, no prediction as to whether monomeric or heterocyclic structures would form was possible. This was shown for instance for the reactions of $H_3Al \leftarrow NMe_3$ and $H_3Ga \leftarrow NMe_3$[103] with $E(SiMe_3)_3$ (E = P, As), either leading to heterocycles or monomers.

We introduced two straightforward synthetic pathways to such intermolecularly stabilized compounds: they are generally available either by ring cleavage reaction of the M–E heterocycle (M = Al, Ga) with 4-dimethylaminopyridine (dmap) or by reaction of dialkylalanes $R_2AlH$ with $E(SiMe_3)_3$ in solution *in the presence* of dmap.[43a,104] The Al-containing monomers dmap $\rightarrow$ Al(Me$_2$)E(SiMe$_3$)$_2$ (E = P **60**, As **61**, Sb **63**, Bi **65**) and dmap $\rightarrow$ Al(Et$_2$)E(SiMe$_3$)$_2$ (E = P, As **62**, Sb **64**, Bi **66**), which were obtained in almost quantitative yield, are very stable compounds both in solution and in the pure form under an inert gas atmosphere at ambient temperature. In contrast, the Ga-containing monomers dmap $\rightarrow$ Ga(R$_2$)E(SiMe$_3$)$_2$ (R = Me, E = P **67**, As **68**, Sb; R = Et, E = Sb **69**) are temperature labile, decomposing significantly at room temperature both in solution and in the solid state. Several attempts to synthesize In-containing compounds have failed to date (Scheme 12).

The compounds are very sensitive toward air and moisture, in particular in solution. $^1H$ and $^{13}C$ NMR spectra prove the formation of 1 : 1 adducts of the type dmap $\rightarrow$ M(R$_2$)E(SiMe$_3$)$_2$. The proton-resonances of the dmap molecule are shifted to higher field, as was observed for similar borane

$$1/x\ [R_2ME(SiMe_3)_2]_x + dmap \longrightarrow dmap{\longrightarrow}M(R_2)E(SiMe_3)_2 \qquad (28)$$

$$M = Al,\ Ga;\ E = P,\ As,\ Sb,\ (Bi)$$

$$R_2AlH + E(SiMe_3)_3 + dmap \xrightarrow[-\ Me_3SiH]{} dmap{\longrightarrow}Al(R_2)E(SiMe_3)_2 \qquad (29)$$

$$E = P,\ As,\ Sb,\ Bi$$

SCHEME 12. Synthesis of dmap-stabilized monomers.

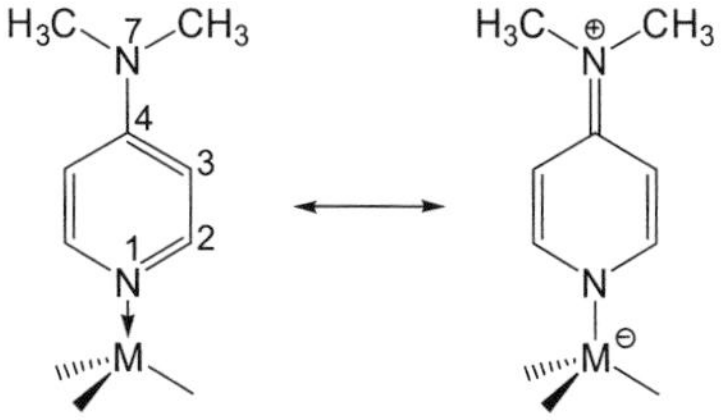

$$(30)$$

M = Al, Ga

SCHEME 13. Charge distribution within the dmap molecule.

adducts,[105] pointing to a partial rearrangement of the charge distribution within the aromatic ring as shown in Scheme 13.

## 1. Solid State Structures—General Trends

Single crystal X-ray analyses clearly reveal the formation of monomeric, dmap-stabilized compounds. The most important bonding parameters of the alane and gallane derivatives are given in Table XXIII.

### Al–E Monomers

The substituents bound to Al and the pentele atom E generally adopt a staggered conformation relative to one another. The Al–N bond distances are relatively short. They range from 197.2(4) to 198.8(3) pm, showing no significant electronic influence of the specific group 15 element on the acceptor property of the Al-fragment. As a result of the increased coordination number ($3 \rightarrow 4$) of the nitrogen center, $Me_3N$-stabilized monomers exhibit significantly longer Al–N distances as can clearly be seen in $Me_3N \rightarrow Al(H_2)N(tmp)_2$ (205.8(2) pm), $Me_3N \rightarrow Al(H)(Cl)N(SiMe_3)_2$ (201.6(5) pm), $Me_3N \rightarrow Al(H_2)As(SiMe_3)_2$ (199.8(7) pm) or $Me_3N \rightarrow Al(H_2)PMes_2$ (200.9(8) pm)[102] (Figs. 31 and 32).

TABLE XXIII

SELECTED BONDING PARAMETERS [pm, °] OF dmap-STABILIZED MONOMERIC GROUP 15 ALANES
AND GALLANES

| Monomer | M–E | M–N | M–R (av.) | $\sum$X–E–X |
|---|---|---|---|---|
| dmap → Al(Me₂)P(SiMe₃)₂ **60** | 237.9(1) | 198.4(2) | 197.5 | 309.1 |
| dmap → Al(Me₂)As(SiMe₃)₂ **61** | 247.2(2) | 197.5(4) | 196.8 | 304.1 |
| dmap → Al(Et₂)As(SiMe₃)₂ **62** | 247.3(1) | 198.8(3) | 197.7 | 306.6 |
| dmap → Al(Me₂)Sb(SiMe₃)₂ **63** | 269.1(1) | 197.8(2) | 197.0 | 302.4 |
| dmap → Al(Et₂)Sb(SiMe₃)₂ **64** | 268.0(1) | 198.0(2) | 198.0 | 298.9 |
| dmap → Al(Me₂)Bi(SiMe₃)₂ **65** | 275.5(2) | 197.2(4) | 197.2 | 296.8 |
| dmap → Al(Et₂)Bi(SiMe₃)₂ **66** | 275.0(2) | 197.8(5) | 198.8 | 293.4 |
| dmap → Ga(Me₂)P(SiMe₃)₂ **67** | 237.2(1) | 208.0(2) | 198.5 | 305.3 |
| dmap → Ga(Me₂)As(SiMe₃)₂ **68** | 245.5(1) | 208.2(2) | 198.2 | 300.2 |
| dmap → Ga(Et₂)Sb(SiMe₃)₂ **69** | 264.8(1) | 206.6(2) | 199.4 | 298.0 |

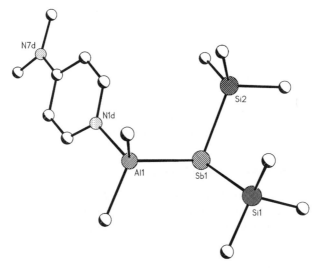

FIG. 31. Solid state structure of dmap → Al(Me₂)Sb(SiMe₃)₂.

The Al–pentele bond lengths found for the dmap-stabilized monomers are generally short and comparable to the sum of the Al–E covalent
radii (235 (Al–P); 246 (Al–As), 266 (Al–Sb), 275 pm (Al–Bi)). dmap → Al
(Me₂)P(SiMe₃)₂ **60** shows an Al–P bond distance of 237.9(1) pm.
Comparable Al–P bond distances were previously only reported for the
monomeric phosphinoalanes Trip₂AlP(Ada)(SiPh₃) (234.2(2) pm)[106] and
Tmp₂AlPPh₂ (237.7(1) pm),[107] containing both threefold-coordinated Al
and P atoms. In contrast, the NMe₃-stabilized phosphinoalane Me₃N →
Al(H₂)PMes₂ shows a slightly elongated Al–P bond distance of 240.9(3) pm.

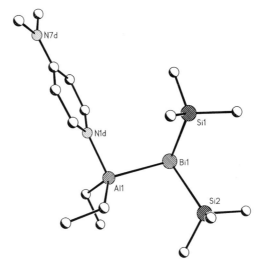

FIG. 32. Solid state structure of dmap → Al(Et$_2$)Bi(SiMe$_3$)$_2$.

The Al–As bond distances of dmap → Al(R$_2$)As(SiMe$_3$)$_2$ (Me: 247.2(2) **61**; Et: 247.3(1) pm **62**) are comparable to those observed for Tmp$_2$AlAsPh$_2$ (248.5(2) pm). Both Me$_3$N → Al(H$_2$)As(SiMe$_3$)$_2$ (243.8 (2) pm), which is sterically less hindered and contains a more Lewis-acidic alane fragment caused by the weaker +I−effect of a H-compared to a Me-substituent, and the base-stabilized heterocycle [Me$_3$N → (H)AlAsSi($i$Pr)$_3$]$_2$[108] show slightly elongated Al–As distances. The Al–Sb (Me: 269.1 (1) pm **63**; Et: 268.0(1) pm **64**) and Al–Bi bond lengths (Me: 275.5(2) pm **65**; Et: 275.0(2) pm **66**) found in dmap → Al(R$_2$)E(SiMe$_3$)$_2$ are the shortest Al–E bond distances reported to date. Those observed for heterocycles of the type [R$_2$AlER$_2'$]$_x$ range from 273 to 278 pm (Al–Sb) and 275 to 279 pm (Al–Bi), respectively.

Interestingly, the degree of the Al–E bond length shortening between the Lewis base-stabilized monomers and their corresponding heterocycles strongly depends on the atomic radii of the group 15 element. Al–P and Al–As bond distances of the base-stabilized monomers are shortened by 8 (Al–P) and 6 pm (Al–As), while this effect diminishes for the monomeric stibides (3–5 pm) and bismuthide (2 pm), respectively. This trend most likely results from reduced intramolecular repulsive interactions between the organic substituents, which become less important in heterocycles with increasing atomic radius of the group 15 elements.

The most striking structural difference between analogously substituted, base-stabilized monomeric alanes is reflected by the degree of pyramidalization of the group 15 element. The sum of the bond angles of the pentele center (sum of the Si–E–Si and Si–E–Al bond angles) of the Me-substituted

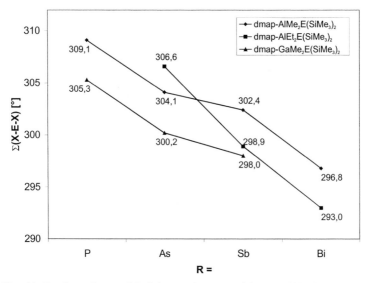

FIG. 33. Bond angular sum [°] of the pentele center of dmap-stabilized monomers.

derivatives dmap → Al(Me$_2$)E(SiMe$_3$)$_2$ (E = P, As, Sb and Bi) steadily decreases from the phosphine (309.1° **60**) toward the higher homologues (304.1° **61**, 302.4° **63**, 296.8° **65**). Comparable trends are found for the Et-substituted Lewis base-stabilized monomers dmap → Al(Et$_2$)E(SiMe$_3$)$_2$ (306.6° **62**, 298.9° **64**, 293.4° **66**), as is illustrated in Fig. 33.

These findings agree very well with generally observed structural trends within analogously substituted group 15 triorganyls such as EH$_3$, EPh$_3$ and EMe$_3$, also showing steadily decreasing bond angular sums with increasing atomic number. The observed trend most likely results from decreasing steric interactions between the SiMe$_3$ groups and the alane fragment with increasing atomic radii of the group 15 element as well as from the increased s-character of the pentele electron *lone pair* and the increased p-character of the bonding electron pairs due to relativistic effects and the lanthanoid-contraction (*inert-pair effect*).[20]

*Ga–E Monomers*

The structural parameters of monomeric gallium derivatives dmap → Ga(R$_2$)E(SiMe$_3$)$_2$ (E = P, As, Sb), are comparable to those of the corresponding alanes except for the (dative) Ga–N bond. Despite the almost identical covalent radii of Al and Ga, the Ga–N bond distances (206.6(2)–208.2(2) pm) are significantly elongated compared to the Al–N bond lengths, most likely a result of the reduced Lewis acidity of the R$_2$Ga- compared to the R$_2$Al-fragment. Almost the same Ga–N distances were found for 2-(methylamino)pyridine → (Me$_2$)GaCl (206.6(3) pm),[109] Me$_3$N → GaH$_3$ (208.1(4) pm),[110] 4-(methyl)pyridine → Ga(Mes)$_2$SeMes (209.5(3) pm)[111] and for quinuclidine-stabilized amido- and azidogallanes (206–210 pm).[112] The Ga–E bond distances (237.2(1) pm **67**, 245.5(1) pm **68**, 264.8(1) pm **69**) are shorter than those of the corresponding Ga–E heterocycles and those of the analogously substituted dmap-stabilized Al–E monomers. They are among the shortest Ga–E bond lengths ever observed.

## 2. Reactivity of dmap-Stabilized Monomeric Compounds

*Reactions with Transition Metal Carbonyls*

Lewis base-stabilized compounds base → M(R$_2$)ER$_2'$ are promising candidates for coordination reactions due to the presence of an uncomplexed, coordinatively active electron *lone pair* on the group 15 element. In particular, reactions of base-stabilized alane-penteles with transition metal complexes were found to yield the corresponding "bimetallic" species. This is very interesting since despite the steadily growing number of intermetallic complexes containing a *direct bond* between a transition metal and a group 13 metal,[113] those containing a group 13 metal fragment (R$_x$M) and a transition metal fragment (M'L$_n$) *bridged* by a pentele atom are almost unknown.[114] The first structurally characterized compound of this specific type, Me$_3$N → Al(CH$_2$SiMe$_3$)$_2$PPh$_2$–Cr(CO)$_5$ **73**, was obtained in a ring cleavage reaction between [(Me$_3$SiH$_2$C)$_2$AlPPh$_2$]$_2$ and (Me$_3$N)Cr(CO)$_5$.[115] Unfortunately, several attempts to expand this specific reaction pathway to the synthesis of other complexes failed. Beachley *et al.* suggested that the reactivity of the group 13/15 heterocycle toward (Me$_3$N)Cr(CO)$_5$ depends on both the degree of association and the Lewis acidity of the heterocycle in solution.[116] Very recently, Scheer and coworkers demonstrated that the coordination of both a Lewis base and a transition metal complex may stabilize highly unstable compounds such as monomeric phosphanylalane and -gallane H$_2$MPH$_2$ (M = Al, Ga).[117] Me$_3$N → M(H$_2$)PH$_2$–W(CO)$_5$ (M = Al **74**, Ga **75**) were obtained in good yields from reactions of W(CO)$_5$PH$_3$ and Me$_3$N → MH$_3$ with elimination of dihydrogen (Scheme 14). Theoretical calculations proved that the coordination of the Lewis base NMe$_3$ (108 kJ/mol) and the Lewis acid W(CO)$_5$ (154 kJ/mol) to H$_2$AlPH$_2$

$$H_3Al \leftarrow NMe_3 \; + \; H_3P\text{-W(CO)}_5 \quad \xrightarrow{\; -H_2 \;} \quad Me_3N \rightarrow Al(H_2)PH_2\text{-W(CO)}_5 \quad (31)$$

$$H_3Ga \leftarrow NMe_3 \; + \; H_3P\text{-W(CO)}_5 \quad \xrightarrow{\; -H_2 \;} \quad Me_3N \rightarrow Ga(H_2)PH_2\text{-W(CO)}_5 \quad (32)$$

SCHEME 14. Synthesis of monomeric phosphanylalane and -gallane.

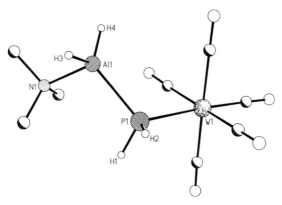

FIG. 34. Solid state structure of $Me_3N \rightarrow AlH_2PH_2-W(CO)_5$.

stabilizes the monomeric unit by 262 kJ/mol, while the dimerization of phosphanylalane $H_2AlPH_2$, yielding the heterocycle $[H_2AlPH_2]_2$, is exothermic by only 74 kJ/mol. Interestingly, the $Me_3N$-stabilized phosphanylalane $Me_3N \rightarrow M(H_2)PH_2$, which seems to be fairly stable according to these calculations, has not been prepared, to date.

The two compounds are not isostructural. $Me_3N \rightarrow Ga(H_2)PH_2-W(CO)_5$ **75** is monomeric in the solid state, whereas $Me_3N \rightarrow Al(H_2)PH_2-W(CO)_5$ **74** adopts a H-bridged dimeric structure as shown in Fig. 34. The Al–P (236.7(1) pm) and Ga–P (234.9(1) pm) bond distances are relatively short.

Scheer's interesting synthetic pathway for the synthesis of the desired compounds requires the synthesis of transition metal–phosphine complexes $L_nM-PH_3$. Unfortunately, analogous reactions of arsine ($AsH_3$) and in particular of stibine ($SbH_3$) transition metal complexes $L_nM-EH_3$ (E = As, Sb) are expected not to give the corresponding arsanyl or stibanylalane complexes due to the significantly less acidic H-substituent. Therefore, Scheer's particular reaction pathway seems only to be useful for the synthesis of phosphine complexes. In sharp contrast, Lewis base-stabilized compounds dmap $\rightarrow M(R_2)ER'_2$, which already contain the group 13/15 element backbone as well as a coordinatively active electron *lone pair* at the pentele atom, are generally useful starting compounds for the synthesis of the desired class of compounds. This was demonstrated in several reactions with different transition metal carbonyls such as $Ni(CO)_4$, $Fe_2(CO)_9$ and $(Me_3N)Cr(CO)_5$. Complexes of the type dmap $\rightarrow M(Me_2)E(SiMe_3)_2-M'(CO)_n$ of both the alane (M = Al, E = P, $M'(CO)_n = Ni(CO)_3$, $Fe(CO)_4$, $Cr(CO)_5$; E = As, $M'(CO)_n = Ni(CO)_3$, $Cr(CO)_5$, E = Sb, $M'(CO)_n = Ni(CO)_3$) and gallane derivatives (M = Ga, $M'(CO)_n = Ni(CO)_3$, E = P, As, Sb) were synthesized and their solid state structures investigated by single crystal X-ray diffraction[118] (Scheme 15).

dmap⟶M(Me$_2$)E(SiMe$_3$)$_2$ + Ni(CO)$_4$ $\xrightarrow[-CO]{}$ dmap⟶M(Me$_2$)E(SiMe$_3$)$_2$-Ni(CO)$_3$ $\qquad$ (33)

M = Al, E = P, As, Sb; Ga, E = As, Sb

dmap⟶Al(Me$_2$)P(SiMe$_3$)$_2$ + Fe$_2$(CO)$_9$ $\xrightarrow[-Fe(CO)_5]{}$ dmap⟶Al(Me$_2$)E(SiMe$_3$)$_2$-Fe(CO)$_4$ $\qquad$ (34)

dmap⟶Al(Me$_2$)E(SiMe$_3$)$_2$ + Me$_3$N-Cr(CO)$_5$ $\xrightarrow[-Me_3N]{}$ dmap⟶Al(Me$_2$)E(SiMe$_3$)$_2$-Cr(CO)$_5$ $\qquad$ (35)

E = P, As

SCHEME 15. Reaction of base-stabilized monomers with transition metal complexes.

TABLE XXIV

$^1$H NMR RESONANCES, IR VIBRATION BANDS AND SELECTED BOND DISTANCES (AV. VALUES [pm]) OF TRANSITION METAL COMPLEXES

| | Ni–C | C–O | $^{13}$C | IR |
|---|---|---|---|---|
| Ni(CO)$_4$ | 181.6 | 112.7 | 191.9 | 2057 |
| dmap → Al(Me$_2$)P(SiMe$_3$)$_2$–Ni(CO)$_3$ **70** | 179.6 | 113.9 | 199.5 | 2048 |
| dmap → Al(Me$_2$)As(SiMe$_3$)$_2$–Ni(CO)$_3$ **71** | 179.1 | 113.9 | 199.8 | 2046 |
| dmap → Al(Me$_2$)Sb(SiMe$_3$)$_2$–Ni(CO)$_3$ **72** | 177.9 | 114.6 | 200.9 | 2042 |

Carbonyl resonances in the $^{13}$C NMR spectra and in the infrared spectra clearly prove the formation of the transition metal complexes. The influence of the specific pentele atom on the complex properties can be seen by comparing the Ni-complexes dmap → Al(Me$_2$)E(SiMe$_3$)$_2$–M′(CO)$_n$ (E = P **70**, As **71**, Sb **72**) and dmap → Ga(Me$_2$)E(SiMe$_3$)$_2$–M′(CO)$_n$ (E = P, As **79**, Sb **80**). The steadily increasing downfield shift of the carbonyl resonances and the steadily decreasing wave numbers of the A$_1$ vibration bands with increasing atomic number of the pentele atom as shown in Table XXIV point to a decrease of the C–O and an increase of the Ni–C bond order. dmap → Al(Me$_2$)P(SiMe$_3$)$_2$–Ni(CO)$_3$ **70** for instance was found to have a negative Tolman parameter value of $\chi = -8$.[119] Thus the π-acceptor ability of this particular ligand seems to be extremely weak and the phosphorus–transition metal interaction has almost exclusively P → M′ σ-dative character. These findings are most likely caused by the influence of the highly electropositive group 13 metals in combination with the two SiMe$_3$ groups, which strongly increase the electron density on the phosphorus atom and rendering additional π back donation from the transition metal to the phosphorus center improbable. Comparable findings were observed for the corresponding arsine and stibine complexes dmap → M(R$_2$)E(SiMe$_3$)$_2$–Ni(CO)$_3$ (M = Al, Ga; E = As, Sb). According to the synergic σ-donor/π-acceptor bonding concept, the findings are consistent with a slight increase

## TABLE XXV
### TRANSITION METAL COMPLEXES OF BASE-STABILIZED Al–E AND Ga–E MONOMERS

| | M–E | E–M' | M–N | $\sum$ X–E–Y |
|---|---|---|---|---|
| Me$_3$N → Al(CH$_2$SiMe$_3$)$_2$PPh$_2$–Cr(CO)$_5$ **73** | 248.5(1) | 248.2(1) | 204.9(3) | 308.3 |
| Me$_3$N → Al(H$_2$)PH$_2$–W(CO)$_5$ **74** | 236.7(1) | 254.9(1) | 203.6(3) | – |
| Me$_3$N → Ga(H$_2$)PH$_2$–W(CO)$_5$ **75** | 234.9(2) | 253.7(2) | 203.9(7) | – |
| dmap → Al(Me$_2$)P(SiMe$_3$)$_2$–Ni(CO)$_3$ **70** | 240.0(2) | 231.9(2) | 196.1(5) | 326.0 |
| dmap → Al(Me$_2$)P(SiMe$_3$)$_2$–Fe(CO)$_4$ **76** | 243.2(1) | 237.7(1) | 196.1(2) | 318.9 |
| dmap → Al(Me$_2$)P(SiMe$_3$)$_2$–Cr(CO)$_5$ **77** | 242.8(1) | 252.8(1) | 196.3(2) | 313.5 |
| dmap → Al(Me$_2$)As(SiMe$_3$)$_2$–Ni(CO)$_3$ **71** | 247.9(1) | 241.9(1) | 196.6(2) | 317.7 |
| dmap → Al(Me$_2$)As(SiMe$_3$)$_2$–Cr(CO)$_5$ **78** | 251.2(1) | 260.0(1) | 195.5(2) | 313.0 |
| dmap → Al(Me$_2$)Sb(SiMe$_3$)$_2$–Ni(CO)$_3$ **72** | 268.0(2) | 255.6(1) | 196.5(4) | 314.3 |
| dmap → Ga(Me$_2$)As(SiMe$_3$)$_2$–Ni(CO)$_3$ **79** | 246.5(1) | 241.9(1) | 204.5(2) | 316.3 |
| dmap → Ga(Me$_2$)Sb(SiMe$_3$)$_2$–Ni(CO)$_3$ **80** | 264.7(1) | 255.4(1) | 204.6(2) | 312.8 |

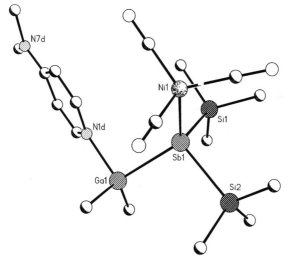

FIG. 35. Solid state structure of dmap → Ga(Me$_2$)Sb(SiMe$_3$)$_2$–Ni(CO)$_3$.

in σ-donor/π-acceptor ratio with increasing atomic number of the group 15 element. Comparable trends were observed by Bodner *et al.* for more than 100 transition metal complexes (R$_3$E)M'L$_n$ (E = group 15 element).[120] These trends were confirmed by single crystal X-ray diffraction studies, showing an increase of the Ni–C bond order and a decrease of the C–O bond order compared to uncomplexed Ni(CO)$_4$.[121]

The most important structural parameters of the transition metal complexes are summarized in Table XXV and Figs. 35–37 display the solid state structures of three selected complexes.

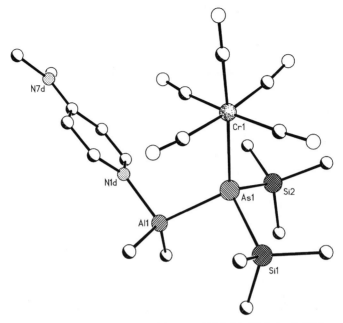

FIG. 36.  Solid state structure of dmap → Al(Me$_2$)As(SiMe$_3$)$_2$–Cr(CO)$_5$.

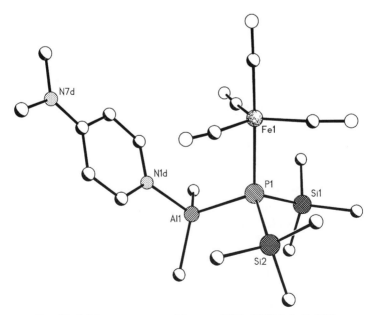

FIG. 37.  Solid state structure of dmap → Al(Me$_2$)P(SiMe$_3$)$_2$–Fe(CO)$_4$.

In each complex, the ligands bound to the group 13 and group 15 elements adopt a staggered conformation in relation to one another. Compared to the starting monomeric compounds dmap $\rightarrow$ M(Me$_2$)E(SiMe$_3$)$_2$, the M–C, M–N and E–Si bond lengths display no significant variations. The Al–E and Ga–E bond distances are either slightly elongated (Al–P) or almost equal (Al–As, Ga–As, Ga–Sb) compared to the uncomplexed monomers. Only dmap $\rightarrow$ Al(Me$_2$)Sb(SiMe$_3$)$_2$–Ni(CO)$_3$ features a slightly shorter Al–Sb bond distance than the corresponding uncomplexed monomer dmap $\rightarrow$ Al(Me$_2$)Sb(SiMe$_3$)$_2$. Interestingly, the Al–P distance found in dmap $\rightarrow$ Al(Me$_2$)P(SiMe$_3$)$_2$–Cr(CO)$_5$ (242.8(1) pm) is almost 6 pm shorter than that of Beachley's complex Me$_3$N $\rightarrow$ Al(CH$_2$SiMe$_3$)$_2$PPh$_2$–Cr(CO)$_5$ (248.5(1) pm) but 6 pm elongated compared to Scheer's phosphanylalane (236.7(1) pm). These differences are most likely based on different repulsive interactions between the organic substituents.

In contrast, complexation of the Lewis base-stabilized monomers indeed has a strong effect on the degree of pyramidalization of the pentele atom as can clearly be seen for the tricarbonylnickel complexes. The sum of the Si–E–Si and M–E–Si bond angles of such complexes are widened by 12–17°, respectively. These findings are in agreement with the larger steric demand of an electron *lone pair* in dmap $\rightarrow$ M(Me$_2$)E(SiMe$_3$)$_2$ (M = Al, Ga) compared to that of an E–M' bonding electron pair as was expected from the VSEPR concept. In addition, it indicates a partial rehybridization of the E–Si and M–E bonding electron pairs, whose *s*-character increases, and the former electron *lone pair*, developing more *p*-character. Comparable trends were observed for simple Lewis acid–base adducts as shown in Section 2. However, the degree of pyramidalization of the pentele atom is strongly influenced by the steric demand of the transition metal carbonyl complex, as was shown for Ni(CO)$_3$ **70**, Fe(CO)$_4$ **76** and Cr(CO)$_5$ complexes **77** of dmap $\rightarrow$ Al(Me)$_2$P(SiMe$_3$)$_2$. The steric demand of the transition metal carbonyl fragment increases with increasing number of CO ligands. Consequently, the sum of the Al–E–Si and Si–E–Si bond angles decrease from the Ni(CO)$_3$ complex (tetrahedral environment, 326.0° **70**) to the Fe(CO)$_4$ complex (trigonal bipyramidal, 318.9° **76**) and finally the Cr(CO)$_5$ complex (octahedral environment, 313.5° **77**). The comparable arsine complexes dmap $\rightarrow$ Al(Me$_2$)As(SiMe$_3$)$_2$–Ni(CO)$_3$ **71** and dmap $\rightarrow$ AlMe$_2$As(SiMe$_3$)$_2$–Cr(CO)$_5$ **78** show a much smaller decrease (317.7° **71** to 313.0° **78**). This observation coincides to the increased atomic radius of As compared to P, leading to reduced steric interactions between the Cr(CO)$_5$ and the R$_3$As fragment. Further experiments are needed to give a more detailed insight into the bonding situation of such complexes.

The pentele–transition metal bond distances of all complexes are exceptionally long. In accordance with the observations in solution by NMR and IR spectroscopy, the phosphorus atom in dmap → Al(Me$_2$)P(SiMe$_3$)$_2$–Fe(CO)$_4$ **76** occupies an axial position, which is preferred by ligands with strong σ-donor and weak π-acceptor abilities.[122] The P–Fe(CO)$_4$ bond distance (237.7(1) pm) is about 4 pm elongated compared to that of (Me$_3$Si)$_3$P–Fe(CO)$_4$ (233.8(4) pm) and is one of the longest reported so far (observed range for pentacoordinated iron: 216–237 pm[123]). A CSD database search confirmed only one compound containing a phosphorus atom bound to a terminal iron complex featuring a longer P–Fe bond.[124] The P–Ni bond length found in dmap → Al(Me$_2$)P(SiMe$_3$)$_2$–Ni(CO)$_3$ (231.5 pm **70**) is even 2 pm longer than that of $t$-Bu$_3$P–Ni(CO)$_3$ containing the sterically demanding and very weak π-accepting $t$-Bu$_3$P ligand.[125] The pentele–Cr bond distances (P–Cr 252.8(1) pm **77**, As–Cr 260.0(1) pm **78**) are also among the longest E–Cr bonds observed for E–Cr(CO)$_5$ fragments so far (typical range for E = P: 224–258 pm, E = As: 237–263 pm; cp. 242.2 pm for Ph$_3$P–Cr(CO)$_5$ and 249.7 pm for Ph$_3$As–Cr(CO)$_5$).[126] Only [(Me$_2$N)$_2$C=As{Fe(CO)$_2$Cp}{Cr(CO)$_5$}] features a longer As–Cr(CO)$_5$ bond (262.8(1) pm).[127] In combination with the rather short Cr–CO(*trans*) bond distances (184.4(2), 183.7(2) pm) these results confirm the comparatively weak Cr → E π back bonding as was found by IR spectroscopy.[128]

*Reactions with Group 13 Trialkyls*

Lewis base-stabilized monomers dmap → M(R$_2$)E(SiMe$_3$)$_2$ are also interesting candidates for reactions with group 13-trialkyls, yielding group 13/15 compounds containing both M–E σ- and M–E dative bonds within a single molecule (Scheme 16).

Only N-bridged compounds of this particular type, most of them containing an heterocyclic iminoborane moiety, have been structurally characterized to date.[129] These reactions may be important for material sciences, since they offer the possibility to open a general pathway for the synthesis of metalorganic group 13/15 compounds containing two *different* group 13 elements bridged by one group 15 element. Compounds of this

dmap→Al(Me$_2$)Sb(SiMe$_3$)$_2$ + $t$-Bu$_3$M

$$\longrightarrow \text{dmap→Al(Me}_2\text{)Sb(SiMe}_3\text{)}_2\text{→M($t$-Bu)}_3 \quad (36)$$

M = Al, Ga

dmap→Al(Me$_2$)P(SiMe$_3$)$_2$ + Me$_3$Ga

$$\longrightarrow \text{dmap→Al(Me}_2\text{)P(SiMe}_3\text{)}_2\text{→GaMe}_3 \quad (37)$$

SCHEME 16. Reaction of Lewis base-stabilized monomers with group 13 trialkyls.

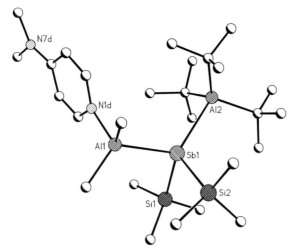

FIG. 38. Solid state structure of dmap → Al(Me₂)Sb(SiMe₃)₂ → Al(t-Bu)₃.

type may serve as *single source precursors* for the synthesis of ternary group 13/15 materials.

Initial studies on the reaction of dmap → Al(Me₂)Sb(SiMe₃)₂ **63** with t-Bu₃Al and t-Bu₃Ga resulted in the formation of the expected adducts dmap → AlMe₂Sb(SiMe₃)₂ → M(t-Bu)₃ (M = Al **82**, Ga **83**). Their solid state structures were determined by single crystal X-ray diffraction[130] (Fig. 38).

As was observed for the transition metal complexes, the Al–N and Al–C bond distances as well as the C–Al–C and N–Al–C bond angles of both compounds are almost unchanged compared to the uncomplexed monomer dmap → AlMe₂Sb(SiMe₃)₂. The Sb–Si and Al–Sb bond distances (272.5(1) pm **82**; 272.6(3) pm **83**) are slightly elongated, most likely caused by the increase in the coordination number from 3 to 4 (Sb) and the steric interactions between the SiMe₃ groups and the bulky M(t-Bu)₃ fragment. The dative M–Sb bond lengths (286.9(1) pm **82**; 288.9(1) pm **83**) are unexpectedly short with regard to the very bulky M(t-Bu)₃ group. For instance, the Ga–Sb bond distance found in the very similar adduct t-Bu₃Ga ← Sb(SiMe₃)₃ (302.7(2) pm **12**) is about 14 pm longer than that in dmap → AlMe₂Sb(SiMe₃)₂ → Ga(t-Bu)₃. These findings point to a high Lewis basicity of the Sb-atom in both compounds, most likely caused by the presence of the electropositive Al-fragment, leading to strong attractive acid–base interactions. The M(t-Bu)₃ fragments show no surprising structural parameters: the increased M–C bond lengths and the decreased C–M–C bond angles in respect to free Al(t-Bu)₃ and Ga(t-Bu)₃ are in accordance with observations made for simple Lewis acid–base adducts.

The most surprising structural features of **82** and **83** are the sums of the Si–Sb–Si and Si–Sb–Al bond angles (298.3° **82**, 298.2° **83**). The degree of pyramidalization of the Sb atom decreases upon coordination of the Lewis acid M(*t*-Bu)$_3$ to dmap → Al(Me$_2$)Sb(SiMe$_3$)$_2$ (302.4°). This is in sharp contrast to findings generally observed for group 13/15 Lewis acid–base adducts as well as for the transition metal complexes. According to the VSEPR model, assuming an electron *lone pair* to be sterically more demanding than a donor electron pair, and a partial rehybridization of the Sb *lone pair* upon adduct formation (*p*-character increases), the sums of the bond angles in **82** and **83** were expected to increase. Evidently, the steric interactions between M(*t*-Bu)$_3$ and the organic ligands, in particular the SiMe$_3$ groups which are severely intensified by the short dative M–Sb bond are responsible for these observations.

Unfortunately, reactions of dmap → Al(Me$_2$)Sb(SiMe$_3$)$_2$ with sterically less demanding trialkylalanes or -gallanes such as AlMe$_3$, GaMe$_3$ and InMe$_3$ did not yield the respective adducts but dmap → AlMe$_3$ and the corresponding heterocycle [Me$_2$MSb(SiMe$_3$)$_2$]$_3$ (M = Al, Ga, In). Only the reaction with AlMe$_3$ could be explained by a simple Lewis acid–base exchange reaction (dmap coordinates to stronger acidic AlMe$_3$). However, the reaction pathway in case of the reactions with GaMe$_3$ and InMe$_3$ is more complex. In a first step, the formation of the expected adduct is proposed. This adduct reacts even under low temperature conditions with M–Me bond cleavage and Me-group transfer to the AlMe$_2$-fragment as illustrated in Scheme 17. Obviously, the adduct dmap → AlMe$_2$Sb(SiMe$_3$)$_2$ → MMe$_3$ is thermodynamically less stable than dmap → AlMe$_3$ and the corresponding heterocycle [Me$_2$MSb(SiMe$_3$)$_2$]$_3$ (M = Ga, In).

This reaction pathway was recently confirmed by the reaction of dmap → AlMe$_2$P(SiMe$_3$)$_2$ with GaMe$_3$, yielding the corresponding adduct dmap → AlMe$_2$P(SiMe$_3$)$_2$ → GaMe$_3$ **81**.[130] Complex **81** is stable below

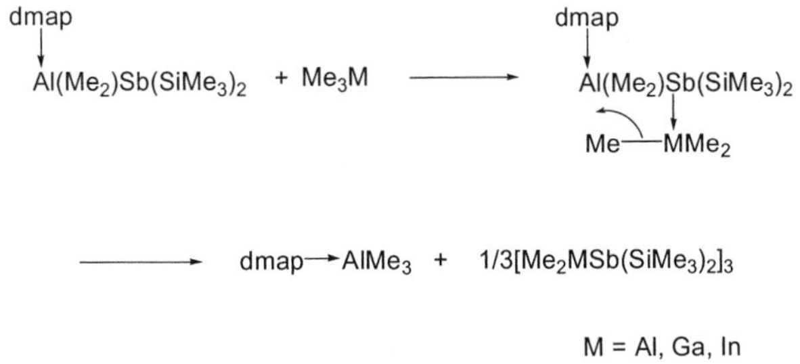

$$\text{(38)}$$

M = Al, Ga, In

SCHEME 17. Proposed reaction mechanism for the reaction of dmap → AlMe$_2$Sb(SiMe$_3$)$_2$ and MMe$_3$ (M = Ga, In).

$-10\,°C$ as was demonstrated by temperature-dependent $^{1}H$, $^{13}C$ and $^{31}P$ NMR spectroscopy. At ambient temperature, resonances both due to the adduct and due to dmap $\rightarrow$ AlMe$_3$ and [Me$_2$GaP(SiMe$_3$)$_2$]$_2$ were detected. A single crystal X-ray analysis confirmed the formation of **81**. The investigated crystal was found to contain both **81** and the heterocycle [Me$_2$GaP (SiMe$_3$)$_2$]$_2$ co-crystallized in the elemental cell. The Al–N (197.4(2) pm) and Al–P (241.6(1) pm) bond lengths show almost no changes compared to the uncomplexed monomer dmap $\rightarrow$ AlMe$_2$P(SiMe$_3$)$_2$ **60** (198.4(2), 237.9(1) pm). In contrast, the sum of the bond angles at the P center (318.9°) of **81** is significantly wider compared to that found for **60** (309.1°). These findings, which are in sharp contrast to those observed for the $t$-Bu$_3$M adducts, strongly underline the influence of steric interactions between the organic substituents on the sum of the bond angles of the pentele center. As was found for simple Lewis acid base adducts, the Ga–C bond distances (av. 200.0 pm) are elongated and the C–Ga–C bond angular sum (339.2°) is

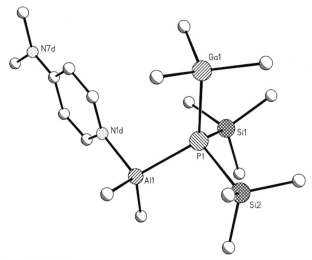

Fɪɢ. 39. Solid state structure of dmap $\rightarrow$ Al(Me)$_2$P(SiMe$_3$)$_2$ $\rightarrow$ GaMe$_3$.

TABLE XXVI

Sᴇʟᴇᴄᴛᴇᴅ Bᴏɴᴅ Lᴇɴɢᴛʜs [pm] ᴀɴᴅ Aɴɢʟᴇs [°] ᴏғ dmap $\rightarrow$ Al(Me)$_2$E(SiMe$_3$)$_2$ $\rightarrow$ MR$_3$

| | M–E | E–M′ | M–N | $\sum$ Si–E–X |
|---|---|---|---|---|
| dmap $\rightarrow$ Al(Me$_2$)P(SiMe$_3$)$_2$–GaMe$_3$ **81** | 242.8(1) | 252.8(1) | 196.3(2) | 313.5 |
| dmap $\rightarrow$ Al(Me$_2$)Sb(SiMe$_3$)$_2$–Al($t$-Bu)$_3$ **82** | 272.5(1) | 286.9(1) | 196.8(3) | 298.3 |
| dmap $\rightarrow$ Al(Me$_2$)Sb(SiMe$_3$)$_2$–Ga($t$-Bu)$_3$ **83** | 272.6(3) | 288.9(1) | 196.1(7) | 298.2 |

decreased compared to pure GaMe$_3$ (av. 195.7 pm; 359.9°).[131] Obviously, adducts containing sterically demanding trialkylalanes and -gallanes such as $t$-Bu$_3$M force the compound to follow Gutman's rule whereas those of sterically less demanding group 13 trialkyls (GaMe$_3$) show structural trends which agree to Haaland's rule (see also Section 2.1.3). However, further studies are neccessary in the near future to obtain a more detailed insight into the reactivity, stability and solid state structure of such compounds (Fig. 39, Table XXVI).

## IV

## MONOMERS, HETEROCYCLES AND CAGES OF THE TYPE [RMER']$_x$

As was already mentioned for the reaction of AlH$_3$ and NH$_3$ (Scheme 1), elimination of two equivalents of dihydrogen yields iminoalanes of the type [HAlNH]$_x$. Comparable results were obtained for the analogous Ga system. Ammonothermal conversation of cyclotrigallazane [H$_2$GaNH$_2$]$_3$ into GaN yields [HGaNH]$_x$ as an intermediate.[132] Unfortunately, the solid state structures of these specific compounds have not been determined to date, most likely caused by their insolubility in organic solvents. In order to investigate such compounds in more detail, reactions of alanes, gallanes and indanes such as Me$_3$N → MH$_3$ (M = Al, Ga), RAlH$_2$ and MR$_3$ (M = Al, Ga, In) with primary amines H$_2$NR' have been studied in detail. Typically the formation of dimeric[133] and trimeric heterocyclic[134] compounds showing threefold-coordinated group 13/15 elements and tetrameric[135] cubane-type

$$MMe_3 + H_2ER' \xrightarrow[-2CH_4]{} 1/x[MeMER']_x \quad (39)$$

$$M = Al, Ga, In; E = N, P, As; x = 3, 4$$

$$AlCp_3 + H_2ER' \xrightarrow[-2CH_4]{} 1/2[CpAlNR']_2 \quad (40)$$

$$Me_3N\text{-}AlH_3 + H_2ER' \xrightarrow[-2H_2, -NMe_3]{} 1/6[HAlER']_6 \quad (41)$$

$$E = N, P, As$$

$$Me_3N\text{-}AlH_3 + H_2NR' \xrightarrow[-2H_2, -NMe_3]{} 1/x[HAlNR']_x \quad (42)$$

$$x = 4, 5, 6, 7, 8, 15, 16$$

$$RGaCl_2 + Li_2PR' \xrightarrow[-2LiCl]{} 1/2[RGaPR']_2 \quad (43)$$

SCHEME 18. Synthetic pathways for the preparation of group 13/15 oligomers.

compounds of the type $[RMNR']_x$ $(x = 2\text{--}4)$, containing fourfold-coordinated central atoms was observed.[136] Comparable results have been reported for analogous reactions with primary phosphines $H_2PR$ and arsines $H_2AsR$[133b,137] as well as for reactions of organochlorogallanes $RGaCl_2$ with dilithium phosphides $Li_2PR$ as shown in Scheme 18.[138]

The degree of oligomerization $(x)$ of the resulting compounds $[RMER']_x$ was found to strongly depend on the steric demand of the organic substituents R and R': small ligands yield highly aggregated oligomers $(x > 4)$ whereas large ligands generally stabilize low-aggregated compounds $(x = 2, 3, 4)$. For instance, reactions of $Me_3N \rightarrow AlH_3$ with sterically demanding primary amines, phosphines and arsines yields dimeric, trimeric or tetrameric compounds $(x = 2, 3, 4)$, whereas reactions with $EtNH_2$ or $n\text{-}PrNH_2$ or $i\text{-}PrNH_2$ yield larger clusters (up to $x = 16$[139]) as was demonstrated, in particular, by Cesari and Cuccinella.[140] Compounds with $x = 5$,[141] $6$,[135b,142]

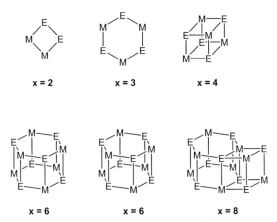

FIG. 40. Structurally characterized group 13/15 polyeder of the type $[RMER']_x$.

$$2GaCl_3 + 4i\text{-}Pr_3SiAsLi_2 \xrightarrow[-6LiCl]{} [\{Li(thf)_3\}_2Ga_2\{As(Si i\text{-}Pr_3)\}_4] \quad (44)$$

$$[\{HC(CMeDippN\}_2M] + N_3\text{-}2,6\text{-}Trip_2C_6H_3]$$

$$\xrightarrow[-N_2]{} [\{HC(CMeDippN\}_2M]N\text{-}2,6\text{-}Trip_2C_6H_3] \quad (45)$$

$$M = Al, Ga$$

$$[Cp^*Al]_4 + 4N_3R \xrightarrow[-4N_2]{} [Cp^*AlNR]_2 \quad (46)$$

$$R = Si(i\text{-}Pr)_3, Si(t\text{-}Bu)_3, SiPh_3$$

SCHEME 19. Synthesis of double bonded group 13/15 compounds and reactions of $[Cp*Al]_4$ with organoazides.

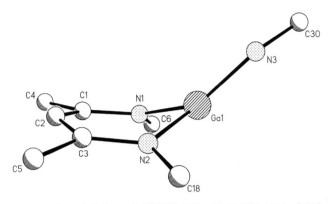

FIG. 41. Central skeleton of [{HC(CMeDippN)$_2$Ga]N-2,6-Trip$_2$C$_6$H$_3$].

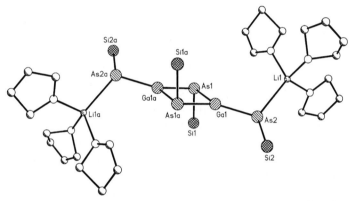

FIG. 42. Solid state structure of [{Li(thf)$_3$}$_2$Ga$_2${As(Si$i$-Pr$_3$)}$_4$]. $i$-Pr groups bound to As are omitted for clarity.

7[143] and 8[144] have been structurally characterized by single crystal X-ray analysis. Their central cage-type structures are shown in Fig. 40.

Monomeric compounds ($x = 1$) containing M=E double bonds remained unknown until Power and Roesky *et al.* reported in 2001 on the synthesis of the first monomeric iminoalane and -gallane by oxidative addition of an organoazide to a sterically encumbered low-valent Al(I) compound.[4] This reaction type has been shown previously to yield heterocyclic compounds by use of sterically less hindered [Cp*Al]$_4$.[145] Very recently, von Hänisch and Hampe synthesized the first GaAs compound featuring a Ga=As double bond[5] (Scheme 19, Figs. 41 and 42).

In sharp contrast to compounds containing the lighter elements of group 15, N, P and As, which will not be discussed in detail here, analogous compounds of the heavier elements Sb and Bi are almost completely unknown.

$$2(\text{Ph}(n\text{-Pr})_2\text{P})\text{GaCl}_3 \; + \; 2(i\text{-Pr})_3\text{SiSb}(\text{SiMe}_3)_2$$

$$\xrightarrow[-\,4\text{Me}_3\text{SiCl}]{} \; [(\text{Ph}(n\text{-Pr})_2\text{P})\text{Ga}(\text{Cl})\text{SbSi}(i\text{-Pr})_3]_2 \qquad (47)$$

$$4(\text{Ph}(n\text{-Pr})_2\text{P})\text{GaCl}_3 \; + \; 4(i\text{-Pr})_3\text{SiSb}(\text{SiMe}_3)_2$$

$$\xrightarrow[\substack{-\,8\text{Me}_3\text{SiCl} \\ -\,2\text{P}(n\text{-Pr})_2\text{Ph}}]{} \; [(\text{Ph}(n\text{-Pr})_2\text{P})_2(\text{GaCl})_4(\text{SbSi}(i\text{-Pr})_3)_4] \quad (48)$$

SCHEME 20. Synthesis of GaSb heterocycles containing low coordinated Sb centers.

Very recently, von Hänisch *et al.* reported on the syntheses and crystal structures of cyclic and polycyclic GaSb compounds featuring low coordinated Sb centers.[146] Reaction of Ph($n$-Pr)$_2$P → GaCl$_3$ with $i$-Pr$_3$SiSb(SiMe$_3$)$_2$ yielded [Ph($n$-Pr)$_2$P → Ga(Cl)SbSi($i$-Pr)$_3$]$_2$ **84** and [(GaCl)$_4$(P($n$-Pr)$_2$Ph)$_2$(SbSi($i$-Pr)$_3$)$_4$] **85**, respectively[147] (Scheme 20).

The solid state structures of both compounds were determined by single crystal X-ray analysis. **84** adopts a four-membered non-planar Ga$_2$Sb$_2$ ring caused by the threefold-coordinated Sb centers. In contrast, the Ga centers are fourfold-coordinated. Both the Si($i$-Pr)$_3$ groups and the coordinated phosphines adopt a *cis*-orientation related to one another as was previously observed in comparable GaP and GaAs compounds [GaCl(P($t$-Bu)$_2$Me)ESiMe$_3$]$_2$ (E = P, As).[148] Complex **85** contains three four-membered rings which are arranged in a chair-like structure. The most striking structural findings are the short Ga–Sb bond distances. The average value of 264.1 pm found for the four-membered ring **(84)** is the shortest Ga–Sb bond length reported to date. A slightly longer Ga–Sb bond length (264.8(1) pm) was found in dmap → Ga(R$_2$)E(SiMe$_3$)$_2$. In contrast, the Ga–Sb bond distances observed for **85** range from 261.8(1) to 271.7(1) pm with the central four-membered ring displaying significantly elongated distances (269.1(1) and 271.7(1) pm) caused by the increased coordination number of the Sb centers (4 vs. 3). In addition, the average Sb–Ga–Sb (98.7°) and Ga–Sb–Ga bond angles (75.2°) observed for **84** differ significantly from those reported for four-membered stibinogallanes of the type [R$_2$GaSbR$_2'$]$_2$, typically showing Ga–Sb–Ga bond angles of 93–97° and Sb–Ga–Sb bond angles of 83–87° (see also Table XX). These differences are most likely caused by the lower coordination number of the Sb centers in **84**. In contrast, the endocyclic Ga–Sb–Ga (80.4(1)–86.3(1)°) and Sb–Ga–Sb bond angles (93.7(1)–99.0(1)°) observed within the four-membered rings of **85** are in between the values reported for **84** and four-membered stibino-gallanes, whereas those found between the rings are significantly widened

TABLE XXVII

Sᴇʟᴇᴄᴛᴇᴅ Bᴏɴᴅɪɴɢ Pᴀʀᴀᴍᴇᴛᴇʀs [pm, °] ᴏꜰ [Ph($n$-Pr)$_2$P → Ga(Cl)SbSi($i$-Pr)$_3$]$_2$ **84** ᴀɴᴅ

[(GaCl)$_4$(P($n$-Pr)$_2$Ph)$_2$(SbSi($i$-Pr)$_3$)$_4$] **85**

| Heterocycle | Ga–Sb | Ga–P | Ga–Sb–Ga | Sb–Ga–Sb |
|---|---|---|---|---|
| **84** | 264.1(1)–265.1(1) | 243.6(2) | 75.1(1)–75.3(1) | 98.4(1)–99.0(1) |
| **85** | 261.8(1)–271.7(1) | 242.9(2) | 80.4(1)–86.3(1); 117.7(1) | 93.3(1)–99.0(1); 110.9(1) |

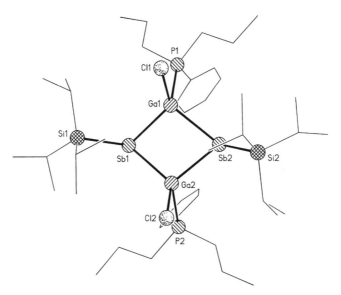

Fɪɢ. 43. Solid state structure of [Ph($n$-Pr)$_2$P → Ga(Cl)SbSi($i$-Pr)$_3$]$_2$.

(Ga(1′)–Sb(2)–Ga(2): 117.7(1); Sb(1′)–Ga(2)–Sb(2): 110.9(1)°). Selected bonding parameters of **84** and **85** are summarized in Table XXVII (Figs. 43 and 44).

The coordination of the phosphine P($n$-Pr)$_2$Ph to the Lewis acidic Ga center is essential for the synthesis of both compounds. In the absence of any Lewis base, the most likely reaction product would be the heterocubane [ClGaSbSi($i$-Pr)$_3$]$_4$. However, in analogy to the results observed for reactions of heterocycles [R$_2$MER$_2'$]$_x$ with Lewis bases, leading to base-stabilized monomeric compounds, both the formation of **84** and **85** can be explained by reaction of such a heterocubane intermediate with the phosphine base. According to the description of heterocycles as *head-to-tail adducts*, heterocubanes may be described as Lewis acid–base adducts between two four-membered rings as shown in Fig. 45.

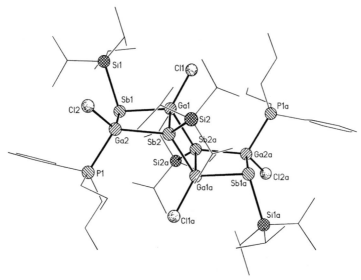

FIG. 44. Solid state structure of [(GaCl)$_4$(P($n$-Pr)$_2$Ph)$_2$(SbSi($i$-Pr)$_3$)$_4$].

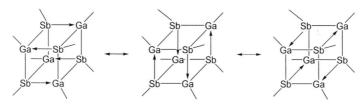

FIG. 45. Possible donor–acceptor interactions in a Ga$_4$Sb$_4$ heterocubane.

The formation of polyclic **85** can be rationalized by coordination of two phosphine molecules to the Lewis acidic Ga centers in the heterocubane [ClGaSbSi($i$-Pr)$_3$]$_4$, displacing the weaker stibine bases and leading to the chair-like structure of **85**. An analogous structure motif was recently observed in [(AlCl)$_4$(PR)$_4$(OEt$_2$)$_2$] (R = Si($i$-Pr)$_3$, Si($i$-Pr)$_2$Me).[149] The formation of cyclic **84** consequently needs the coordination of two more phosphine molecules to the remaining Ga centers, yielding the base-stabilized four-membered heterocycle. The successful synthesis of **84** and **85** clearly prove the concept of electronic stabilization to be very useful not only for the stabilization of monomeric compounds of the type base → MR$_2$ER$_2'$ (see Section 3.2) but also for the stabilization of cyclic compounds containing low coordinated pentele centers such as **84** and **85** or for base-stabilized iminoalanes such as [(Me$_3$N)AlHNDipp]$_2$.[150] Further studies are needed to reveal the full potential of this concept.

# V

# POTENTIAL APPLICATIONS OF GROUP 13/15 ORGANOELEMENT COMPOUNDS IN MATERIAL SCIENCES

Binary group 13/15 materials, usually referred to as III–V materials, are intensely investigated compound semiconductors. Their narrow direct bandgaps between 0.17 (InSb) and 3.44 (GaN) render them very interesting candidates for potential applications in micro- and optoelectronic devices. In addition, the possibility of "tuning" the band gaps by combination of three or four different group 13/15 elements leading to ternary and quarternary materials such as AlGaP, AlGaSb or AlGaAsSb is one of the major advantages of the III–V material system. The most important physical properties of binary III–V materials are summarized in Table XXVIII.[151]

The particular applications very often require the use of thin films of the desired III–V materials. The most important industrial process for the fabrication of such films is the MOCVD process (_metal organic chemical vapor deposition_). This process, introduced by Manasevit in 1968,[152] typically makes use of group 13 trialkyls such as MMe$_3$ or MEt$_3$ (M = Al, Ga, In) and group 15 hydrides EH$_3$ (E = N, P, As). Both components are simultaneously pyrolyzed in specific, reactor systems and the desired material film is deposited on a specific substrate. Numerous investigations on the deposition of group 13–nitrides, –phosphides and –arsenides have been reported, rendering it impossible to present them in detail within this

TABLE XXVIII

SELECTED PHYSICAL PROPERTIES OF BINARY GROUP 13/15 MATERIALS

| Material | Energy gap [eV] | Electron mobility [cm$^2$/V/s] | Hole mobility [cm$^2$/V/s] | Emission λ [nm] |
|---|---|---|---|---|
| AlN | 6.20 | – | 14 | 200 |
| GaN | 3.44 | 440 | – | 380 |
| InN | 2.05 | 250 | – | 620 |
| AlP | 3.62 | 60 | 450 | 500 |
| GaP | 2.27 | 160 | 135 | 550 |
| InP | 1.34 | 5370–5900 | 150 | 950 |
| AlAs | 2.15 | 75–294 | – | 570 |
| GaAs | 1.42 | 8000–9200 | 400 | 860 |
| InAs | 0.35 | 33,000 | 100–450 | 3400 |
| AlSb | 1.61 | 200 | 400 | 770 |
| GaSb | 0.75 | 3750 | 680 | 1800 |
| InSb | 0.17 | 77,000 | 850 | 7700 |

review. Instead, the following chapters concentrate on the deposition of binary group 13–antimonides by MOCVD process.

### A. *Binary Group 13–Antimonide Material Films—Introduction*

Binary Group 13–Sb materials are narrow direct band gap semiconductors. They have small band gaps ranging from 0.16 (InSb) to 1.60 eV (AlSb) and high electron mobilities. Consequently, they are of interest for potential applications in optoelectronic devices. GaSb for instance is used for the production of light-emitting and -detecting devices operating in the 2 μm wavelength range, in field effect transistors, infrared detectors and hot electron transistors.[153] Traditionally, these materials have been prepared by MBE (*molecular beam epitaxy*) and LPE (*liquid phase epitaxy*) processes, whereas the CVD-process is less important due to the following difficulties:[154]

– Stibine $SbH_3$, formally the ideal Sb-source, is too unstable for a successful use in MOCVD. It starts decomposing at temperatures above $-60\,°C$, severely limiting its possible use for industrial processes. Similar problems exist for primary and secondary stibines $RSbH_2$ and $R_2SbH$, in particular, those with small substituents (R = Me, Et). Consequently, trialkylstibines $SbR_3$ are generally used.

– The low vapor pressure of elemental antimony is the most important difference compared to the lighter group 15 elements, N, P and As. Consequently, the deposition of binary group 13–antimonide films requires different process parameters such as a molar ratio of the group 13 precursor and the stibine near unity at the growing surface. The presence of an excess of the trialkylstibine $SbR_3$ has to be strictly avoided due to the possible formation of elemental Sb, that cannot be removed from the substrate under typical CVD film growth conditions. In sharp contrast, group 15 hydrides such as $NH_3$, $PH_3$ or $AsH_3$ are typically used in large excess. Under MOCVD conditions, they generate atomic hydrogen that is important for the saturation of organic radicals that arise from the decomposition of the metalorganic precursor. Consequently, carbon impurities in the resulting material films are significantly reduced.

– Under typical MOCVD growth conditions ($T < 550\,°C$), the growth rate of binary antimonides is in the *kinetically controlled regime* according to the incomplete pyrolysis of the group 13 metalorganic precursor. In sharp contrast, the growth rate of binary nitrides, phosphides and arsenides, that are deposited at higher temperatures

( > 600 °C) under *mass transport controlled* conditions, only depends on the concentration of the group 13 metalorganic precursor.

To date, only GaSb and InSb films of reasonable quality have been obtained by CVD-process.[155] Typically, standard group 13 sources such as GaMe$_3$, GaEt$_3$ and InMe$_3$, respectively, as well as trialkylstibines such as SbMe$_3$, SbEt$_3$ and Sb($i$-Pr)$_3$ are used.[156] However, both the incapacity of trialkylstibines to produce atomic hydrogen during the pyrolysis as well as the very stable metal–carbon bonds favor the incorporation of considerable amounts of carbon into the resulting material. The most promising Sb-sources to date are $i$-Pr$_2$SbH and Sb(NMe$_2$)$_3$.[157] Carbon contaminations of the resulting material films are significantly reduced compared to those obtained with common trialkylstibines due to the presence of potentially active hydrogen ($i$-Pr$_2$SbH) or thermodynamically weak Sb–N bonds. In sharp contrast, the deposition of AlSb films of good quality still is an unsolved problem.[158] Standard trialkylalanes such as AlMe$_3$ and AlEt$_3$ and trialkylstibines such as Me$_3$Sb and Et$_3$Sb tend to generate materials containing unacceptably large concentrations of carbon,[159] mainly resulting from the chemical nature of the alane precursors. AlMe$_3$ and AlEt$_3$ for instance are not completely pyrolyzed at temperatures typically used for the deposition of binary antimonides ( < 550 °C; lower than for nitrides, phosphides and arsenides), leading to high C-contents in the resulting materials. More suitable Al-sources could be alane–amine adducts such as are H$_3$Al ← NMe$_3$, H$_3$Al ← NMe$_2$Et as well as $t$-Bu$_3$Al according to the low Al–C bond energy.[160] However, there are still several problems based on the precursor chemistry ( *pre-reactions* in the reactor, *pre-deposition* at the walls) that have to be overcome, requiring alternative precursors and/or reactor geometries. In addition, the purity of the alanes needs to be improved since their deposition yielded materials containing large amounts of oxygen. Additional drawbacks resulting from the use of two precursors are the formation of Al-droplets on the surface and low AlSb growth rates that can be attributed to the different vapor pressure of the alane and stibine precursors.

The *single source precursor concept* may solve some of these problems. Single source precursors are molecular (metalorganic) compounds containing the specific elements of the desired material connected by a stable chemical bond *pre*-formed in a single molecule. Typical candidates are Lewis acid–base adducts R$_3$M ← EH$_3$ and heterocycles of the type [R$_2$MER$_2'$]$_x$. The most striking advantage of this concept is based to the typically weaker M–C and E–C bonds of single source precursors compared to pure group 13 and group 15 trialkyls, allowing film growth at lower temperatures. It was also demonstrated that new materials or new phases of known materials could be formed by use of single source precursors. According to the lower

$$AlMe_3 + NH_3 \longrightarrow Me_3Al \leftarrow NH_3 \xrightarrow[- CH_4]{} [Me_2AlNH_2]_3$$

$$\xrightarrow[- CH_4]{} [MeAlNH]_x \xrightarrow[- CH_4]{} AlN$$

(49)

$$InMe_3 + PH_3 \longrightarrow Me_3In \leftarrow PH_3 \xrightarrow[- CH_4]{} [Me_2InPH_2]_3$$

$$\xrightarrow[- CH_4]{} [MeInPH]_x \xrightarrow[- CH_4]{} InP$$

(50)

SCHEME 21. Proposed reaction intermediates for the deposition of AlN and InP.

film deposition temperatures, the film growth process may occur under kinetic control rather than under thermodynamic control as was demonstrated for instance for the deposition of cubic GaN,[161] cubic GaS[162] and $In_xSe_y$,[163] respectively. In addition, single source precursors often are less air- and moisture-sensitive and less toxic. However, their usual low vapor pressures are a major drawback, severely limiting their use in MOCVD processes to date.

The capability of Lewis acid–base adducts $R_3M \leftarrow EH_3$ and heterocycles $[R_2MER'_2]_x$ to serve as precursors for the deposition of the corresponding materials by use of the MOCVD process has been demonstrated in the past. This does not surprise remembering the well known tendency of group 13 trialkyls to react under typical MOCVD conditions with group 15 trihydrides under formation of the corresponding adducts $R_3M \leftarrow EH_3$. As-formed adducts on heating consequently eliminate alkane RH to give the corresponding heterocycles $[R_2MER'_2]_x$. For instance, the deposition of InP from $InMe_3$ and $PH_3$[164] or AlN from $AlMe_3$ and $NH_3$[165] most likely proceed by the mechanism as described in Scheme 21. This mechanism is strongly supported by the fact that both the adduct $Me_3Al \leftarrow NH_3$ and the heterocycle $[Me_2AlNH_2]_3$ successfully have been used for the deposition of AlN in the absence of additional $NH_3$.[166]

In the last decade, numerous compounds of these types have been the subject of detailed CVD studies, demonstrating their potential for the deposition of the corresponding binary materials. Most of the work has concentrated on binary nitrides and phosphides, while the deposition of binary MSb films has been studied to a far lesser extent. The lack of potential precursors has been the major problem for the deposition of group 13–antimonide films for many years. Only a very few group 13–Sb compounds have been known until we and Wells established general synthetic pathways as was shown in Sections 2 and 3. Consequently, detailed investigations concerning their potential to serve for the deposition of the desired materials are very limited to date. In the following, the results of initial MOCVD studies using Lewis acid–base adducts $R_3M \leftarrow ER'_3$ and heterocyclic stibinoalanes and -gallanes $[R_2MSbR'_2]_x$ (M = Al, Ga) will be presented.

## B. Deposition of AlSb Films by Use of Single Source Precursors

AlSb films were deposited on Si(100) and polycrystalline $Al_2O_3$ substrates in a cold wall CVD reactor under high vacuum conditions (HV-CVD) by use of heterocyclic stibinoalanes $[Et_2AlSb(SiMe_3)_2]_2$ **39** and $[i\text{-}Bu_2AlSb$ $(SiMe_3)_2]_2$ **40**.[167] AlSb film deposition was studied in the temperature range of 325–500 °C (**39**) and 375–550 °C (**40**), respectively, and a pressure less than $5 \times 10^{-5}$ mbar. The deposited material films showed a strong dependency concerning their composition and crystallinity on the substrate temperature.

### 1. Films Obtained from $[Et_2AlSb(SiMe_3)_2]_2$ **39**

The crystallinity of as-deposited AlSb films steadily increases with increasing deposition temperature. Below 375 °C no crystalline material was present according to X-ray diffraction studies (Fig. 46).

The optimum deposition temperatures for AlSb films were found to range from 375 to 425 °C. Films obtained within this temperature range showed Al : Sb ratios close to 1.0, whereas at higher temperatures Al:Sb ratios up to 1.28 (500 °C) were detected. In addition, at temperatures above 425 °C large amounts of Si were incorporated into the film material (up to

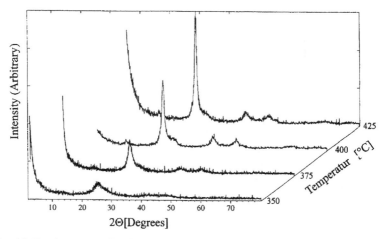

FIG. 46. Powder X-ray diffraction study of AlSb films deposited at different temperatures.

FIG. 47. SEM study of AlSb films obtained at 400 °C.

FIG. 48. SEM study of AlSb films obtained at 450 °C.

15% at 500 °C), most likely due to fragmentation reactions of the $SiMe_3$ groups. The carbon content of the material films was below the detection limit of WDS (*wavelength-dispersive X-ray spectroscopy*), even at these low temperatures. The amount of oxygen stays constant at 1% according to surface oxidation processes.

SEM studies of as-deposited AlSb films revealed the formation of smooth films. No island growth was detected. In addition, the presence of both single and agglomerated crystallites with particle sizes ranging from 300 to 700 nm as is illustrated in Fig. 47 was confirmed. The size distribution was found to be almost independent of the deposition temperature.

### 2. Films Obtained from [*i*-Bu$_2$AlSb(SiMe$_3$)$_2$]$_2$ **40**

The deposition of AlSb film by use of [*i*-Bu$_2$AlSb(SiMe$_3$)$_2$]$_2$ was investigated between 375 and 550 °C and a pressure less than $5 \times 10^{-5}$ mbar. Crystalline AlSb was formed at substrate temperatures above 425 °C, 50 °C higher than for film growth using [Et$_2$AlSb(SiMe$_3$)$_2$]$_2$. As was found for precursor **39**, the composition of the resulting material films strongly depends on the substrate temperature. In the optimum deposition range (425–475 °C), the Al : Sb molar ratio was near unity, whereas at higher temperatures Al : Sb ratios up to 1.7 were detected (525 °C). Simultaneously, the Si content of the film material increased from 1 up to 27%. Again, the carbon content was below the detection limit of WDX and oxygen stayed constant at 1%.

The film growth rates as determined by profilometry ranged from 5 to 9 μm/h. They were found to increase between 400 and 500 °C, whereas they dropped down above 500 °C. Moreover, the precursor flux rate had no influence on the growth rate, indicating that the film growth is not mass transport limited but kinetically controlled. SEM studies proved the formation of smooth AlSb films containing agglomerations of single particles. The particle size varied from 400 to 900 nm. The film exhibits nearly the same surface morphology compared with the film grown using **39** as can be seen in Fig. 48.

These results clearly demonstrate the high potential of the single source precursor concept. In particular, the low deposition temperatures necessary for the stibinoalanes and the low carbon contents of the resulting material films are very promising results. However, the low vapor pressure of the heterocycles, which require sublimation temperatures about 130 °C, are a major drawback.

## C. Deposition of GaSb Films by Use of Single Source Precursors

Reports on the preparation of binary GaSb and InSb films by use of single source precursors are also very rare. The first study was reported by Cowley *et al.* in 1990, investigating the deposition of InSb films on Si(100) wafers using [Me$_2$InSb($t$-Bu)$_2$]$_3$ **53** in a horizontal hot wall reactor at a working pressure of $10^{-3}$ mbar.[77c] The optimum deposition temperature was determined to be 450 °C and the growth rates were as high as 1.0 μm/h. The carbon content of the crystalline InSb material was below the detection limit of XPS (*X-ray photoelectron spectroscopy*). In addition, the corresponding stibinogallane [Me$_2$GaSb($t$-Bu)$_2$]$_3$ **45** was shown to be a promising precursor for the deposition of crystalline GaSb thin films under similar growth conditions. However, no further deposition studies using **53** and **45** have been reported to date.

Wells *et al.* demonstrated in recent years the potential of several four-membered stibinogallanes and indanes to produce GaSb and InSb materials under simple pyrolysis conditions. Crystalline GaSb was produced from [$t$-Bu$_2$GaSb(SiMe$_3$)$_2$]$_2$ **48** between 175 and 400 °C both under *dynamic* vacuum and *static* vacuum conditions.[80b] The precursor decomposition occurs through a β-hydride elimination mechanism and subsequently formed decomposition products such as CH$_4$, H$_2$, Me$_3$SiH and isobutylene were identified by IR spectroscopy. The average particle size of thus-formed GaSb crystallites was estimated to 9 nm. Pyrolysis of [Et$_2$GaSb(SiMe$_3$)$_2$]$_2$ **47**

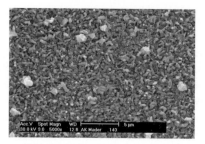

FIG. 49. SEM study of GaSb films obtained at 400 °C.

at 400 °C yielded nanocrystalline GaSb with an average particle size of 10 nm in rather low yield (20%). The material was contamined with carbon and hydrogen (about 2%).[80e] Pyrolysis of $[t\text{-}Bu_2InSb(SiMe_3)_2]_2$ **58** at 400 °C under static vacuum yielded In-rich InSb again through a β-hydride elimination pathway. The material obtained showed no carbon and hydrogen contamination.[80d] In addition, nanocrystalline GaSb and InSb with an average particle size of 11 nm were obtained by thermal decomposition of $t\text{-}Bu_3Ga \leftarrow Sb(SiMe_3)_3$ **12** and $t\text{-}Bu_3In \leftarrow Sb(SiMe_3)_3$ under static vacuum conditions in simple glass tubes at 350 and 400 °C, respectively.

Unfortunately, no statements concerning the Si-contents of the resulting material films are given by Wells *et al.* According to our observations on the AlSb material deposition, silyl-substituted precursors may lead to Si contaminations of the resulting material. Therefore, our interest focused on the use of completely alkyl-substituted precursors to eliminate the possibility of any Si-contamination of the resulting material. Lewis acid–base adducts $t\text{-}Bu_3Ga \leftarrow Sb(i\text{-}Pr)_3$ **15** and $t\text{-}Bu_3Ga \leftarrow Sb(t\text{-}Bu)_3$ containing sterically demanding $i$-Pr and/or $t$-Bu substituents were investigated[168] due to their low metal–C bond energies. The bond energy of metal–C bonds generally decreases with increasing chain length and/or steric branching according to both the increase of the radical stability with increased steric branching and the facile β-H elimination of alkene from higher alkyl compounds. Consequently, the Sb–C bond energy for triorganostibines decrease in the following order: $Me_3Sb$ (233.9 kJ/mol) > $(vinyl)_3Sb$ (205 kJ/mol) > $i\text{-}Pr_3Sb$ (126.8 kJ/mol) > $(allyl)_3Sb$ (90.4 kJ/mol).[169]

1. Decomposition of $t\text{-}Bu_3Ga \leftarrow Sb(t\text{-}Bu)_3$

Pyrolysis of $t\text{-}Bu_3Ga \leftarrow Sb(t\text{-}Bu)_3$ in a sealed glass tube at temperatures between 275 and 450 °C yielded smooth GaSb films with a uniform film morphology as was determined by SEM (Fig. 49).

The film, film morphology and the size of single crystallites (500–1000 nm) were found to be independent from the pyrolysis temperature.

FIG. 50. SEM study of GaSb films obtained at 425 °C.

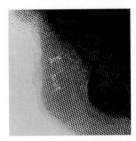

FIG. 51. HR-TEM study of a single GaSb whisker.

TEM-studies confirmed the formation of crystallites with 200–500 nm in length. EDX-studies proved the formation of GaSb particles containing an almost ideal 1 : 1 composition of Ga and Sb. The element-distribution within the crystallites is uniform. Their carbon concentration was below the detection limit of EEL-spectroscopy (*electron energy loss spectroscopy*), whereas oxygen was present as a surface layer according to the sensitivity of GaSb towards surface oxidation. SAED (*selected-area electron diffraction*) showed well-defined, sharp diffraction patterns clearly proving the crystallinity of the particles. They show Debye–Scherrer rings with d-spacings comparing well to literature values of sphalerite-type, cubic GaSb.

## 2. Decomposition of $t$-Bu$_3$Ga $\leftarrow$ Sb($i$-Pr)$_3$

The results from the pyrolysis experiments of $t$-Bu$_3$Ga $\leftarrow$ Sb($i$-Pr)$_3$ are comparable to those from $t$-Bu$_3$Ga $\leftarrow$ Sb($t$-Bu)$_3$. Crystalline GaSb particles of 500–1000 nm in length were obtained as was demonstrated by SEM, EDX and SAED studies. However, films obtained at higher pyrolysis temperatures ( > 400 °C) showed the formation of additional GaSb whiskers (up to 5 μm) at the surface with large length/diameter ratios as can be seen from Fig. 50.

Each GaSb whisker is grown from a ball-shaped center and is capped by a spherule containing amorphous Ga as was shown by SAED (no diffraction patterns) and EDX. In contrast, the Ga:Sb atomic ratio within each whisker was determined by EDX and EELS to be almost equal to 1. The distribution of Sb within the crystalline needle is homogeneous whereas the amorphous tip contains almost no Sb. Carbon could not be detected by EELS within the GaSb whiskers, whereas O is present as a small surface layer. Electron diffraction studies reveal the presence of crystalline cubic-type GaSb. The lattice parameter was determined to 608(2) pm, which agrees very well with the literature value of 609.5 pm.[170] HR-TEM lattice images (Fig. 51) were taken perpendicular to the long axis of the whisker demonstrating its defect-free growth. Only the edge of the whisker was found to be sufficiently electron transparent, while it grows thicker stepwise towards the middle. The measured distances of the lattice planes agree well to the d-spacings of cubic GaSb. The small amorphous border containing oxygen can clearly be observed.

The GaSb whiskers are formed according to the *so-called* VLS-process (*vapor–liquid–solid*) that was introduced by Ellis and Wagner.[27] At the interface between liquid elemental Ga, formed from the pyrolysis of *t*-Bu$_3$Ga, which likely is present in the gas phase due to some dissociation of the starting adduct, and the gas phase thermally induced decomposition reactions of the precursors occur. The Ga droplets serve both as a catalyst for the decomposition of *t*-Bu$_3$Ga and *i*-Pr$_3$Sb, reducing their decomposition temperature significantly, and as a solvent for the decomposition products (elemental Ga and Sb). After Sb has become supersaturated in the Ga droplet, the growth of a single GaSb crystal starts at the interface between the liquid and the solid phase (substrate). The Ga droplet typically remains at the tip of the whisker.[171]

The results obtained with both gallane–stibine adducts are very promising concerning their use as single source precursors in MOCVD studies. In particular, the low deposition temperatures and the absence of carbon in the resulting materials are strong arguments for further detailed investigations on this particular class of compounds.

# VI

## Conclusions and Outlook

This review has described the synthesis, structure and reactivity of important classes of group 13/15 compounds such as Lewis acid base adducts and heterocycles. In addition, their potential to serve as single source precursors for the deposition of the corresponding binary materials by MOCVD process has been demonstrated. Because of the large number of compounds containing the lighter elements of group 15, N, P and As, these could not be discussed in detail. Instead, those containing the heavier group 15 elements, Sb and Bi, which have been almost unknown 6 years ago, have been in the focus of this review. Their synthesis required the development of alternative synthetic pathway such as dehydrosilylation or distibine cleavage reactions. However, in particular Bi-containing compounds are still very rare and their synthesis still is a challenge for preparative chemists working in this field of main group element chemistry. In addition, both monomeric compounds containing a M=E double bond (E = Sb, Bi) as well as cluster-like compounds of the type $[RMER']_4$ ($x > 4$) are unknown and have to be synthesized and structurally characterized for the first time. Alternative concepts such as the *"base-stabilization concept"* may help to realize these goals in the future.

ACKNOWLEDGEMENTS

I am very grateful to my co-workers A. Kuczkowski, F. Thomas and T. Bauer, who spend tremendous time in the preparation of lots of the compounds reported herein. I would also like to thank Dr. M. Nieger, Universität Bonn, for competent structure analyses, Professor P. Schreiner, Universität Gießen, for computational calculations, Professor H. W. Roesky, Universität Göttingen, for the fruitful co-operation on the AlSb thin film deposition as well as Dr. W. Assenmacher, Universität Bonn, for numerous SEM and TEM studies on AlSb and GaSb films. Generous financial support by the DFG, the Fonds der Chemischen Industrie and the Bundesministerium für Bildung, Wissenschaft, Forschung und Technologie (BMBF) as well as by Professor E. Niecke, Universität Bonn, is also gratefully acknowledged.

REFERENCES

(1) Gay-Lussac, J. L.; Thenard, J. L. *Mem. Phys. Chim. Soc. d'Arcueil* **1809**, *2*, 210 as cited in Jonas, G.; Frenking, G. *J. Chem. Soc. Chem. Commun.* **1994**, *116*, 1989.

(2) (a) Stock, A.; Poland, E. *Chem. Ber.* **1926**, *59*, 2215. (b) Bähr, G., (Klemm, W. Ed.) FIAT Review of WWII German Science, 1939–1946, Inorganic Chemistry, Part II, Dieterichsche Verlagsbuchhandlung, Wiesbaden, 1948. pp. 115.

(3) Cesari, M.; Cuccinella, S., (Haiduc, I., Sowerby, D. B. Eds.), *The Chemistry of Inorganic, Homo- and Heterocycles*, Vol. I, Academic Press, London, 1987. pp. 167.

(4) Hardman, N. J.; Cui, C.; Roesky, H. W.; Fink, W. H.; Power, P. P. *Angew. Chem. Int. Ed. Engl.* **2001**, *40*, 2172.

(5) von Hänisch, C.; Hampe, O. *Angew. Chem. Int. Ed. Engl.* **2002**, *41*, 2095.

(6) AlN is better described as an isolator due to its large direct bandgap of 6.28 eV. Its high melting point (3000 °C) and high thermal conductivity (2.6 W/cm/K) renders AlN an useful material for high power applications.

(7) Manasevit, H. M. *Appl. Phys. Lett.* **1968**, *12*, 156. Prior to this study, Didchenko *et al.* as well as Harrison and Tomkins described the synthesis of III–V material powders by reaction of group 13 trialkyl and group 15 hydrides. (a) Didchenko, R.; Alix, J. D.; Toeniskoettler, R. H. *J. Inorg. Chem.* **1960**, *4*, 35. (b) Harrison, B.; Tomkins, E. H. *Inorg. Chem.* **1962**, *1*, 951.

(8) See for example: (a) Jones, A. C.; O'Brien, P. CVD of Compound Semiconductors: Precursor Synthesis, Development and Applications, VCH (Weinheim), 1997. (b) Rees, W. S. CVD of Nonmetals, VCH (Weinheim), 1996. (c) Buhro, W. Adv. Mater. Opt. Electron. 1996, 6, 175.

(9) See for example: Cowley, A. H.; Jones, R. A. Angew. Chem. Int. Ed. Engl. 1989, 28, 1208. (b) Janik, C. F.; Wells, R. L.; Young, V. G. Jr.; Rheingold, A. L.; Guzei, I. A. J. J. Am. Chem. Soc. 1998, 120, 532. (c) Stuczynski, S. M.; Opila, R. L.; Marsh, P.; Brennan, J. G.; Steigerwald, M. L. Chem. Mater. 1991, 3, 379.

(10) For example, see the following and references therein: (a) Beachley, O. T. Jr.; Maloney, J. D. Organometallics 1997, 16, 4016. (b) Park, J. E.; Bae, B. J.; Kim, Y.; Park, J. T.; Suh, I. H. Organometallics 1999, 18, 1059. (c) Krossing, I.; Nöth, H.; Schwenk-Kirchner, H.; Seifert, T.; Tacke, C. Eur. J. Inorg. Chem. 1998, 1925. (d) Müller, J.; Ruschewitz, U.; Indris, O.; Hartwig, H.; Stahl, W. J. Am. Chem. Soc. 1999, 121, 4647. (e) Brain, P. T.; Brown, H. E.; Downs, A. J.; Greene, T. M.; Johnsen, E.; Parsons, S.; Rankin, D. W. H.; Smart, B. A.; Tang, C. Y. J. Chem. Soc. Dalton Trans. 1998, 3685. (f) Wells, R. L.; Baldwin, R. A.; White, P. S. Organometallics 1995, 14, 2123. (g) Wells, R. L.; McPhail, A. T.; Jones, L. J. J. III; Self, M. F.; Butcher, R. J. Organometallics 1992, 11, 2694. (h) Baker, L.-J.; Kloo, L. A.; Rickard, C. E. F.; Taylor, M. J. J. Organomet. Chem. 1997, 545, 249. (i) Beagley, B.; Godfrey, S. M.; Kelly, K. J.; Kungwankunakorn, S.; McAuliffe, C. A.; Pritchard, R. G. J. Chem. Soc. Chem. Commun. 1996, 2179. (j) Wells, R. L.; Foos, E. E.; Rheingold, A. L.; Yap, G. P. A.; Liable-Sands, L. M.; White, P. S. Organometallics 1998, 17, 2869.

(11) Umeyama, H.; Morokuma, K. J. Am. Chem. Soc. 1976, 98, 7208.

(12) The strongest dative bond was reported for $Cl_3Al \leftarrow NMe_3$, showing a bond dissociation energy of 48.3 kcal/mol. Andersen, G. A.; Forgaard, F. R.; Haaland, A. Acta Chem. Scand. 1972, 26, 1947.

(13) Haaland, A. Angew. Chem. Int. Ed. Engl. 1989, 28, 992.

(14) (Weast, R. C. Ed.) CRC Handbook of Chemistry and Physics, 67th ed., CRC Press, Boca Raton, 1987.

(15) (a) Kuchitsu, K. J. Chem. Phys. 1968, 49, 4456. (b) Thorne, L. R.; Suenram, R. D.; Lovas, F. J. J. Chem. Phys. 1983, 78, 167.

(16) See the following and referenced cited therein: (a) Timoshkin, A. Y.; Suvorov, A. V.; Bettinger, H. F.; Schaefer, H. F. III. J. Am. Chem. Soc. 1999, 121, 5687. (c) Anane, H.; Jarid, A.; Boutalib, A. J. Phys. Chem. A 1999, 103, 9847. (d) Jarid, A.; Boutalib, A. J. Phys. Chem. A 2000, 104, 9220.

(17) Guryanova, E. N.; Goldstein, P.; Romm, I. P. Donor–Acceptor Complexes, Wiley, New York, 1975.

(18) Loschen, C.; Voigt, K.; Frunzke, J.; Diefenbach, A.; Diedenhofen, M.; Frenking, G. Z. Anorg. Allg. Chem. 2002, 628, 1294.

(19) For recent reviews on alane and gallane chemistry see for example: (a) Jones, C.; Koutsantonis, G. A.; Raston, C. L. Polyhedron 1993, 12, 1829. (b) Gardiner, M. G.; Raston, C. L. Coord. Chem. Rev. 1997, 166, 1. (c) Downs, A. J. Coord. Chem. Rev. 1999, 189, 59. (d) Aldridge, S.; Downs, A. J. Chem. Rev. 2001, 101, 3305.

(20) Kutzelnigg, W. Angew. Chem. Int. Ed. Engl. 1984, 23, 272.

(21) (a) Holleman, A. F.; Wiberg, E. Lehrbuch der Anorganischen Chemie, 101 ed., Walter de Gruyter, Berlin, 1995. pp. 361. (b) Jache, A. W.; Blevins, G. S.; Gordy, W. Phys. Rev. 1955, 97, 680. (c) Jerzembek, W.; Bürger, H.; Constantin, L.; Margulès, L.; Demaison, J.; Breidung, J.; Thiel, W. Angew. Chem. Int. Ed. Engl. 2002, 41, 2659.

(22) (a) Beagley, B.; Hewitt, T. G. Trans. Faraday Soc. 1968, 64, 2561. (b) Bruckmann, J.; Krüger, C. Acta Crystallogr. Sect. C 1995, 51, 1155. (c) Blom, R.; Haaland, A. J.; Seip, R.

*Acta Chem. Scand. Ser. A* **1983**, *37*, 595. (d) Beagley, B.; Medwid, A. R. *J. Mol. Struct.* **1977**, *38*, 229. (e) Beagley, B.; McAloon, K. T. *J. Mol. Struct.* **1973**, *17*, 429.

(23) (a) Sobolev, A. N.; Belsky, V. K.; Romm, I. P.; Chernikova, N. Y.; Guryanova, E. N. *Acta Crystallogr. Sect. C* **1985**, *41*, 967. (b) Dunne, B. J.; Orpen, A. G. *Acta Crystallogr. Sect. C* **1991**, *47*, 345. (c) Sobolev, A. N.; Belsky, V. K.; Chernikova, N. Y.; Akhmadulina, F. Y. *J. Organomet. Chem.* **1983**, *244*, 129. (d) Adams, E. A.; Kolis, J. W.; Pennington, W. T. *Acta Crystallogr. Sect. C* **1990**, *46*, 917. (e) Effendy, D.; Grigsby, W. J.; Hart, R. D.; Raston, C. L.; Skelton, B. W.; White, A. H. *Aust. J. Chem.* **1997**, *50*, 675. (f) Holmes, N. J. *J. Organomet. Chem.* **1997**, *545*, 111. (g) Aroney, M. J. *J. Chem. Soc. Dalton Trans.* **1994**, 2827.

(24) Haaland, A., (Robinson, G. H. Ed.) Coordination Chemistry of Aluminum, VCH Publishers, Inc, New York, 1993. pp. 2–3.

(25) For an explanation see the following and references cited therein: Rowsell, B. D.; Gillespie, R. J.; Heard, G. L. *Inorg. Chem.* **1999**, *38*, 4659.

(26) (a) Anane, H.; Jarid, A.; Boutalib, A.; Nebot-Gil, I.; Tomàs, F. *Theochem. J. Mol. Struct.* **1998**, *455*, 51. (b) Anane, H.; Jarid, A.; Boutalib, A.; Nebot-Gil, I.; Tomàs, F. *Chem. Phys. Lett.* **1998**, *296*, 277. (c) Jarid, A.; Boutalib, A.; Nebot-Gil, I.; Tomàs, F. *Theochem. J. Mol. Struct.* **2001**, *572*, 161.

(27) Malkova, A. S.; Suvorov, A. V. *Russ. J. Inorg. Chem.* **1969**, *14*, 1049.

(28) Coates, G. E. *J. Chem. Soc.* **1951**, 2003.

(29) (a) Mente, D. C.; Mills, J. L.; Mitchell, R. E. *Inorg. Chem.* **1975**, *14*, 123. (b) Mente, D. C.; Mills, J. L. *Inorg. Chem.* **1975**, *14*, 1862.

(30) Gutmann, V. The Donor–Acceptor Approach to Molecular Interactions, Plenum, New York, 1978.

(31) Gelbrich, T.; Sieler, J.; Dümichen, U. *Z. Kristallogr.* **2000**, *215*, 127.

(32) Clegg, W.; Klingebiel, U.; Niemann, J.; Sheldrick, G. M. *J. Organomet. Chem.* **1983**, *249*, 47.

(33) (a) Almenningen, A.; Gundersen, G.; Haugen, T.; Haaland, A. *Acta Chem. Scand.* **1972**, *26*, 3928. (b) Müller, J.; Ruschewitz, U.; Indris, O.; Hartwig, H.; Stahl, W. *J. Am. Chem. Soc.* **1999**, *121*, 4647.

(34) (a) Thorne, L. R.; Suenron, R. D.; Lovas, F. J. *J. Chem. Phys.* **1983**, *78*, 167. (b) Bühl, M.; Steinke, T.; v. Ragué Schleyer, P.; Boese, R. *Angew. Chem. Int. Ed. Engl.* **1991**, *30*, 1160.

(35) (a) Dvorak, M. A.; Ford, R. S.; Suenram, R. D.; Lovas, F. J.; Leopold, K. R. *J. Am. Chem. Soc.* **1992**, *114*, 108. (b) Swanson, B.; Shriver, D. F.; Ivers, J. A. *Inorg. Chem.* **1969**, *8*, 2182. (c) Hoard, J.-L.; Owen, T. B.; Buzzell, A.; Salmon, O. N. *Acta Crystallogr.* **1950**, *3*, 130.

(36) Burns, W. A.; Leopold, K. R. *J. Am. Chem. Soc.* **1993**, *115*, 11, 622.

(37) Atwood, J. L.; Bennett, F. R.; Elms, F. M.; Jones, C.; Raston, C. L.; Robinson, K. D. *J. Am. Chem. Soc.* **1991**, *113*, 8183.

(38) Jonas, V.; Frenking, G.; Reetz, M. T. *J. Am. Chem. Soc.* **1994**, *116*, 8741.

(39) (a) Coates, G. E. *J. Chem. Soc.* **1951**, 2003. (b) Nemirowskii, L. M.; Kozyrkin, B. I.; Lanstov, A. F.; Gribov, B. G.; Skvortsov, I. M.; Sredinskaya, I. A. *Dokl. Akad. Nauk. SSSR* **1974**, *214*, 590. (c) Hewitt, F.; Holliday, A. K. *J. Chem. Soc.* **1952**, 530. (d) Denniston, M. L.; Martin, D. R. *J. Inorg. Nucl. Chem.* **1974**, *36*, 2175. (e) Spiridonov A.; Malkova A. S. *Zh. Strukt. Khim.* **1969**, *10*, 33; *J. Struct. Chem. USSR* **1969**, *10*, 303. (f) Coleman, A. P.; Nieuwenhuyzen, M.; Rutt, H. N.; Seddon, K. R. *J. Chem. Soc. Chem. Commun.* **1995**, 2369.

(40) Lube, M. S.; Wells, R. L.; White, P. S. *J. Chem. Soc. Dalton. Trans.* **1997**, 285.

(41) (a) Baldwin, R. A.; Foos, E. E.; Wells, R. L.; White, P. S.; Rheingold, A. L.; Yap, G. P. A. *Organometallics* **1996**, *15*, 5035. (b) Wells, R. L.; Foos, E. E.; White, P. S.; Rheingold,

A. L.; Liable-Sands, L. M. *Organometallics* **1997**, *16*, 4771. (c) Foos, E. E.; Wells, R. L.; Rheingold, A. L. *J. Cluster Sci.* **1999**, *10*, 121.

(42) (a) Schulz, S.; Nieger, M. *J. Organomet. Chem.* **1998**, *570*, 275. (b) Schulz, S.; Nieger, M. *Organometallics* **1999**, *18*, 315. (c) Schulz, S.; Kuczkowski, A.; Nieger, M. *J. Organomet. Chem.* **2000**, *604*, 202. (d) Schulz, S.; Nieger, M. *J. Chem. Soc. Dalton. Trans.* **2000**, 639. (e) Schulz S.; Nieger M. Private communication to the Cambridge Crystallographic Data Centre, CCDC No. 138649.

(43) (a) Kuczkowski, A.; Thomas, F.; Schulz, S.; Nieger, M. *Organometallics* **2000**, *19*, 5758. (b) Kuczkowski, A.; Schulz, S.; Nieger, M. *Eur. J. Inorg. Chem.* **2001**, 2605.

(44) (a) Barron, A. R. *J. Chem. Soc. Dalton. Trans.* **1988**, 3047. (b) Leib, A.; Emerson, M. T.; Oliver, J. P. *Inorg. Chem.* **1965**, *4*, 1825. (c) Bradley, D. C.; Dawes, H.; Frigo, D. M.; Hursthouse, M. B.; Hussian, B. *J. Organomet. Chem.* **1987**, *325*, 55.

(45) Their $^1$H NMR spectra feature one resonance due to the Al–R groups at ambient temperature, whereas at $-70\,^\circ$C two resonances of the Al–R groups in a relative intensity of 1:2 appear, obviously caused by the presence of $(R_3Al)_2$ dimers (terminal and bridging substituents) in solution. Both resonances coalesce between $-30$ and $-50\,^\circ$C (Al–Me) and $-25$ and $-40\,^\circ$C (Al–Et), respectively, as was previously described for $(Me_3Al)_2$ and $(Et_3Al)_2$.

(46) The B–N distances in as-described borane–amine adducts were found to be shorter in the solid state (156.4; 161.6 pm) than in the gas phase (167.2; 163.7 pm). Moreover, it was found that the B–N distance of such adducts in solution is strongly affected by the polarity of the solvent: the more polar the solvent the shorter the B–N bond. Both findings indicate the strong influence of a dipolar field, either within a crystal or in polar solvents, on dative B–N bond distances.

(47) Kuczkowski, A.; Schulz, S.; Nieger, M.; Schreiner, P. S. *Organometallics* **2002**, *21*, 1408.

(48) Holleman, A. F.; Wiberg, E. *Lehrbuch der Anorganischen Chemie*, 101, Walter de Gruyter, Berlin, 1995. pp. 1838 ff.

(49) Rankin, D. W. H. Personal communication.

(50) Atwood, J. L.; Bennett, F. R.; Elms, F. M.; Jones, C.; Raston, C. L.; Robinson, K. D. *J. Am. Chem. Soc.* **1991**, *113*, 8183.

(51) Atwood, J. L.; Bennett, F. R.; Jones, C.; Koutsantonis, G. A.; Raston, C. L.; Robinson, K. D. *J. Chem. Soc. Chem. Commun.* **1992**, 541.

(52) Brain, P. T.; Brown, H. E.; Downs, A. J.; Greene, T. M.; Johnsen, E.; Parsons, S.; Rankin, D. W. H.; Smart, B. A.; Tang, C. Y. *J. Chem. Soc. Dalton. Trans.* **1998**, 3685.

(53) Fraser, G. W.; Greenwood, N. N.; Straughan, P. B. *J. Chem. Soc.* **1963**, 3742.

(54) Heitsch, C. W.; Nordman, C. E.; Parry, R. W. *Inorg. Chem.* **1963**, *2*, 508.

(55) (a) Lobkovskii, E. B.; Korobov, I. I.; Semenenko, K. N. *Zh. Strukt. Khim.* **1978**, *19*, 1063. (b) Heitsch, C. W.; Nordman, C. E.; Parry, R. W. *Inorg. Chem.* **1963**, *2*, 508. (c) Gelbrich, T.; Dümichen, U.; Sieler, J. *Acta Crystallogr. Sect. C* **1999**, *55*, 1797.

(56) (a) Ecker, A.; Schnöckel, H. *Z. Anorg. Allg. Chem.* **1998**, *624*, 813. (b) Brown, M. A.; Tuck, D. G.; Wells, E. J. *Can. J. Chem.* **1996**, *74*, 1535. (c) Degnan, I. A.; Alcock, N. W.; Roe, S. M.; Wallbridge, M. G. H. *Acta Crystallogr. Sect. C* **1992**, *48*, 995. (d) Clegg, W.; Norman, N. C.; Pickett, N. L. *Acta Crystallogr. Sect. C* **1994**, *50*, 36.

(57) Baker, L.-J.; Kloo, L. A.; Rickard, C. E. F.; Taylor, M. J. *J. Organomet. Chem.* **1997**, *545*, 249.

(58) Schulz, S. Unpublished results.

(59) Reactions of hydrazine $N_2H_4$ and their organic derivatives with metalorganic group 13 compounds have been the object of several studies in recent years. Both adduct formation and elimination reactions have been reported. See the following and references cited therein: (a) Nöth H.; Seifert T. *Eur. J. Inorg. Chem.* **2002**, 602. (b) Uhl, W.; Molter, J.;

Neumüller, B. *Inorg. Chem.* **2001**, *40*, 2011. (c) Uhl, W.; Molter, J.; Neumüller, B. *Chem. Eur. J.* **2001**, *7*, 1510. (d) Uhl, W.; Molter, J.; Neumüller, B. *Organometallics* **2000**, *19*, 4422. (e) Uhl, W.; Molter, J.; Saak, W. *Z. Anorg. Allg. Chem.* **1999**, *625*, 321. (f) Kim, Y.; Kim, J. H.; Park, J. E. *J. Organomet. Chem.* **1997**, *546*, 99.

(60) See for a historical review: Seyferth, D. *Organometallics* **2001**, *20*, 1488.

(61) See the following and references cited therein: (a) Fritz, G.; Scheer, P. *Chem. Rev.* **2000**, *100*, 3341. (b) Krautscheid, H.; Matern, E.; Olkowska-Oetzel, J.; Pikies, J.; Fritz, G. *Z. Anorg. Allg. Chem.* **2001**, *627*, 1505. (c) Balazs, G.; Breunig, H. J.; Lork, E. *Z. Anorg. Allg. Chem.* **2001**, *627*, 1855. (d) Balazs, G.; Breunig, H. J. *Phosphorous Sulfur* **2001**, *168*, 625. (e) Breunig, H. J.; Rösler, R. *Chem. Soc. Rev.* **2000**, *29*, 403.

(62) (a) Breunig, H. J.; Rösler, R. *Coord. Chem. Rev.* **1997**, *163*, 33. (b) Silvestru, C.; Breunig, H. J.; Althaus, H. *Chem. Rev.* **1999**, *99*, 3277. (c) Sharma, P.; Cabrera, A.; Jha, N. K.; Rosas, N.; LeLagadec, R.; Sharma, M.; Arias, J. L. *Main Group Metal Chem.* **1997**, *20*, 697. (d) Sharma, P.; Rosas, N.; Hernandez, S.; Cabrera, A. *J. Chem. Soc. Chem. Commun.* **1995**, 1325. (e) Breunig, H. J.; Pawlik, J. *Z. Anorg. Allg. Chem.* **1995**, *621*, 817. (f) von Seyerl, J.; Huttner, G. *Cryst. Struct. Commun.* **1980**, *9*, 1099. (g) Dickson, R. S.; Heazle, K. D.; Pain, G. N.; Deacon, G. B.; West, B. O.; Fallon, G. D.; Rowe, R. S.; Leech, P.; Faith, M. *J. Organomet. Chem.* **1993**, *449*, 131. (h) Breunig, H. J.; Fichtner, W. *Z. Anorg. Allg. Chem.* **1981**, *477*, 119.

(63) Very recently, Breunig *et al.* reported on the synthesis of a dibismuthene–tungsten complex. Balázs, L.; Breunig, H. J.; Lork, E. *Angew. Chem. Int. Ed. Engl.* **2002**, *41*, 2309.

(64) Samaan, S. Houben Weyl: Methoden der Organischen Chemie, Metallorganische Verbindungen des Arsens, Antimons und Bismuts, Thieme Verlag (Stuttgart), 1978, 4. Aufl.

(65) (a) Nöth, H. *Z. Naturforsch. B* **1960**, *15*, 327. (b) Burg, A. B.; Wagner, R. I. *J. Am. Chem. Soc.* **1953**, *75*, 3872. (c) Burg, A. B.; Brendel, J. *J. Am. Chem. Soc.* **1958**, *80*, 3198. (d) Burg, A. B. *J. Am. Chem. Soc.* **1961**, *83*, 2226. (e) Carrell, H. L.; Donohue, J. *Acta Crystallogr. Sect. B* **1968**, *24*, 699. (f) Schmidbaur, H.; Wimmer, T.; Grohmann, A.; Steigelmann, O.; Müller, G. *Chem. Ber.* **1989**, *122*, 1607.

(66) Kuczkowski, A.; Schulz, S.; Nieger, M.; Saarenketo, P. *Organometallics* **2001**, *20*, 2000.

(67) Kuczkowski, A.; Schulz, S.; Nieger, M. *Angew. Chem. Int. Ed. Engl.* **2001**, *40*, 4222.

(68) Nieger, M. Private communication 2000, CCDC 139656.

(69) (a) Garton, G.; Powell, H. M. *J. Inorg. Nucl. Chem.* **1957**, *4*, 84. (b) Akitt, J. W.; Greenwood, N. N.; Storr, A. *J. Chem. Soc.* **1965**, 4410.

(70) Beamish, J. C.; Small, R. W. H.; Worrall, I. J. *Inorg. Chem.* **1979**, *18*, 220.

(71) Ecker, A.; Baum, E.; Friesen, M. A.; Junker, M. A.; Uffing, C.; Koppe, R.; Schnöckel, H. *Z. Anorg. Allg. Chem.* **1998**, *624*, 513.

(72) (a) Small, R. W. H.; Worrall, I. J. *Acta Crystallogr. Sect. B* **1982**, *38*, 86. (b) Beamish, J. C.; Boardman, A.; Small, R. W. H.; Worrall, I. J. *Polyhedron* **1985**, *4*, 983. (c) Gordon, E. M.; Hepp, A. F.; Duraj, S. A.; Habash, T. S.; Fanwick, P. E.; Schupp, J. D.; Eckles, W. E.; Long, S. *Inorg. Chim. Acta* **1997**, *257*, 247. (d) Pashkov, A. Y.; Bel'sky, V. K.; Bulychev, B. M.; Zvukova, T. M. *Izv. Akad. Nauk SSSR Ser. Khim.* **1996**, 2078. (e) Doriat, C. U.; Friesen, M.; Baum, E.; Ecker, A.; Schnöckel, H. *Angew. Chem. Int. Ed. Engl.* **1997**, *36*, 1969. (f) Schnepf, A.; Doriat, C.; Mollhausen, E.; Schnöckel, H. *J. Chem. Soc. Chem. Commun.* **1997**, 2111. (g) Beagley, B.; Godfrey, S. M.; Kelly, K. J.; Kungwankunakorn, S.; McAuliffe, C. A.; Pritchard, R. G. *J. Chem. Soc. Chem. Commun.* **1996**, 2179.

(73) Godfrey, S. M.; Kelly, K. J.; Kramkowski, P.; McAuliffe, C. A.; Pritchard, R. G. *J. Chem. Soc. Chem. Commun.* **1997**, 1001.

(74) Mocker, M.; Robl, C.; Schnöckel, H. *Angew. Chem. Int. Ed. Engl.* **1994**, *33*, 1754.

(75) Ecker, A.; Schnöckel, H. *Z. Anorg. Allg. Chem.* **1998**, *624*, 813.

(76) (a) Chang, C.-C.; Ameerunisha, M. S. *Coord. Chem. Rev.* **1999**, *189*, 199. (b) Carmalt, C. J. *Coord. Chem. Rev.* **2001**, *223*, 217. (c) Wells, R. L. *Coord. Chem. Rev.* **1992**, *112*, 273. (d) Cowley, A. H.; Jones, R. A. *Angew. Chem. Int. Ed. Engl.* **1989**, *28*, 1208.

(77) (a) Cowley, A. H.; Jones, R. A.; Kidd, K. B.; Nunn, C. M.; Westmoreland, D. L. *J. Organomet. Chem.* **1988**, *341*, C1. (b) Barron, A. R.; Cowley, A. H.; Jones, R. A.; Nunn, C. M.; Westmoreland, D. L. *Polyhedron* **1988**, *7*, 77. (c) Cowley, A. H.; Jones, R. A.; Nunn, C. M.; Westmoreland, D. L. *Chem. Mater.* **1990**, *2*, 221.

(78) Schulz, S.; Schoop, T.; Roesky, H. W.; Häming, L.; Steiner, A.; Herbst-Irmer, R. *Angew. Chem. Int. Ed. Engl.* **1995**, *34*, 919.

(79) Pitt, C. G.; Purdy, A. P.; Higa, K. T.; Wells, R. L. *Organometallics* **1986**, *5*, 1266.

(80) (a) Baldwin, R. A.; Foos, E. E.; Wells, R. L.; White, P. S.; Rheingold, A. L.; Yap, G. P. A. *Organometallics* **1996**, *15*, 5035. (b) Wells, R. L.; Foos, E. E.; White, P. S.; Rheingold, A. L.; Liable-Sands, L. M. *Organometallics* **1997**, *16*, 4771. (c) Schulz, S.; Nieger, M. *J. Organomet. Chem.* **1998**, *570*, 275. (d) Foos, E. E.; Wells, R. L.; Rheingold, A. L. *J. Cluster Sci.* **1999**, *10*, 121. (e) Foos, E. E.; Jouet, R. J.; Wells, R. L.; Rheingold, A. L.; Liable-Sands, L. M. *J. Organomet. Chem.* **1999**, *582*, 45. (f) Foos, E. E.; Jouet, R. J.; Wells, R. L.; White, P. S. *J. Organomet. Chem.* **2000**, *598*, 182.

(81) Lide D.R. CRC Handbook of Chemistry and Physics, 78th ed., CRC Press (New York), 1997–1998, 9–51.

(82) (a) Wells, R. L.; McPhail, A. T.; Self, M. F.; Laske, J. A. *Organometallics* **1993**, *12*, 3333. (b) Wells, R. L.; McPhail, A. T.; Speer, T. M. *Eur. J. Solid State Inorg. Chem.* **1992**, *29*, 63.

(83) (a) Jouet, R. J.; Wells, R. L.; Rheingold, A. L.; Incarvito, C. D. *J. Organomet. Chem.* **2000**, *601*, 191. (b) Wells, R. L.; Baldwin, R. A.; White, P. S.; Pennington, W. T.; Rheingold, A. L.; Yap, G. P. A. *Organometallics* **1996**, *15*, 91. (c) Barry, S. T.; Belhumeur, S.; Richeson, D. S. *Organometallics* **1997**, *16*, 3588. (d) Aubuchon, S. R.; McPhail, A. T.; Wells, R. L.; Giambra, J. A.; Bowser, J. R. *Chem. Mater.* **1994**, *6*, 82. (e) Douglas, T.; Theopold, K. H. *Inorg. Chem.* **1991**, *30*, 596. (f) Laske Cooke, J. A.; Wells, R. L.; White, P. S. *Organometallics* **1995**, *14*, 3562. (g) Foos, E. E.; Jouet, R. J.; Wells, R. L.; Rheingold, A. L.; Liable-Sands, L. M. *J. Organomet. Chem.* **1999**, *582*, 45.

(84) Wells, R. L.; McPhail, A. T.; Speer, T. M. *Eur. Organomet.* **1992**, *11*, 960.

(85) (a) Nöth, H.; Schrägle, W. *Z. Naturforsch. B* **1961**, *16*, 473. (b) Nöth, H.; Schrägle, W. *Chem. Ber.* **1965**, *98*, 473. (c) Wood, G. L.; Dou, D.; Narula, C. K.; Duesler, E. N.; Paine, R. T.; Nöth, H. *Chem. Ber.* **1990**, *123*, 1455.

(86) (a) Fritz, G.; Emül, R. *Z. Anorg. Allg. Chem.* **1975**, *416*, 19. (b) Krannich, L. K.; Watkins, C. L.; Schauer, S. J. *Organometallics* **1995**, *14*, 3094.

(87) Krannich, L. K.; Watkins, C. L.; Schauer, S. J.; Lake, C. H. *Organometallics* **1996**, *15*, 3980.

(88) (a) Schulz, S.; Nieger, M. *Organometallics* **1998**, *17*, 3398. (b) Schulz, S.; Kuczkowski, A.; Nieger, M. *Organometallics* **2000**, *19*, 699.

(89) Thomas, F.; Schulz, S.; Nieger, M. *Z. Anorg. Allg. Chem.* **2002**, *628*, 235.

(90) Schulz, S.; Nieger, M. *Angew. Chem. Int. Ed.* **1998**, *38*, 967.

(91) Thomas, F.; Schulz, S.; Nieger, M. *Organometallics* **2002**, *21*, 2793.

(92) Breunig, H. J.; Fichtner, W. *Z. Anorg. Allg. Chem.* **1981**, *477*, 119.

(93) Breunig, H. J.; Stanciu, M.; Rösler, R.; Lork, E. *Z. Anorg. Allg. Chem.* **1998**, *624*, 1965.

(94) CSD database, August 2002.

(95) Charton, M. *Top. Curr. Chem.* **1982**, *114*, 57.

(96) (a) Laske Cook, J. A.; Purdy, A. P.; Wells, R. L.; White, P. S. *Organometallics* **1996**, *15*, 84. (b) Atwood, J. L.; Stucky, G. D. *J. Am. Chem. Soc.* **1970**, *92*, 185.

(97) Beachley, O. T., Jr.; Racette, K. C. *Inorg. Chem.* **1976**, *15*, 2110.

(98) (a)Boron and Nitrogen. Gmelin Handbook of Inorganic and Organometallic Chemistry, 4th Supplement, Springer-Verlag (Berlin), 1991 (Vol. 3a) and 1992 (Vol. 3b). (b) Paine, R. T.; Nöth, H. *Chem. Rev.* **1995**, *95*, 343.

(99) (a) Xie, Y.; Grev, R. S.; Gu, J.; Schaefer, H. F., III; Schleyer, P.v.R.; Su, J.; Li, X. W.; Robinson, G. H. *J. Am. Chem. Soc.* **1998**, *120*, 3773. (b) Cotton, F. A.; Cowley, A. H.; Feng, X. *J. Am. Chem. Soc.* **1998**, *120*, 1795. (c) Müller, J. *J. Am. Chem. Soc.* **1996**, *118*, 6370. (d) Fink, W. H.; Power, P. P.; Allen, T. L. *Inorg. Chem.* **1997**, *36*, 1431. (e) Allen, T. L.; Schreiner, A. C.; Schaefer, H. F., III *Inorg. Chem.* **1990**, *29*, 1930.

(100) For excellent overviews see: (a) Power P. P.; Brothers P. J. *Adv. Organomet. Chem.* **1996**, *39*, 1. (b) Power P. P. *Chem. Rev.* **1999**, *99*, 3463.

(101) (a) Atwood, J. L.; Koutsantonis, G. A.; Lee, F.-C.; Raston, C. L. *J. Chem. Soc. Chem. Commun.* **1994**, 91. (b) Gardiner, M. G.; Koutsantonis, G. A.; Lawrence, S. M.; Lee, F.-C.; Raston, C. L. *Chem. Ber.* **1996**, *129*, 545. (c) Janik, J. F.; Duesler, E. N.; Paine, R. T. *J. Organomet. Chem.* **1997**, *539*, 19. (d) Janik, J. F.; Wells, R. L.; White, P. S. *Inorg. Chem.* **1998**, *37*, 3561.

(102) Atwood, D. A.; Contreras, L.; Cowley, A. H.; Jones, R. A.; Mardones, M. A. *Organometallics* **1993**, *12*, 17.

(103) Janik, J. F.; Wells, R. L.; Young, V. G., Jr.; Rheingold, A. L.; Guzei, I. A. *J. Am. Chem. Soc.* **1998**, *120*, 532.

(104) (a) Schulz, S.; Nieger, M. *Organometallics* **2000**, *19*, 2640. (b) Thomas, F.; Schulz, S.; Nieger, M. *Eur. J. Inorg. Chem.* **2001**, 161. (c) Thomas, F.; Schulz, S.; Nieger, M.; Nättinen, K. *Chem. Eur. J.* **2002**, *8*, 1915.

(105) Lesley, G. M. J.; Woodward, A.; Taylor, N. J.; Marder, T. B. *Chem. Mater.* **1998**, *10*, 1355.

(106) Wehmschulte, R. J.; Ruhlandt-Senge, K.; Power, P. P. *Inorg. Chem.* **1994**, *33*, 3205.

(107) Knabel, K.; Krossing, I.; Nöth, H.; Schwenk-Kirchner, H.; Schmidt-Amelunxen, M.; Seifert, T. *Eur. J. Inorg. Chem.* **1998**, 1095.

(108) Driess, M.; Kunz, S.; Merz, K.; Pritzkow, H. *Chem. Eur. J.* **1998**, *4*, 1628.

(109) Koide, Y.; Francis, J. A.; Bott, S. G.; Barron, A. R. *Polyhedron* **1998**, *17*, 983.

(110) Brain, P. T.; Brown, H. E.; Downs, A. J.; Greene, T. M.; Johnson, E.; Parsons, S.; Rankin, D. W. H.; Smart, B. A.; Tang, C. Y. *J. Chem. Soc. Dalton Trans.* **1998**, 3685.

(111) Rahbarnoohi, H.; Wells, R. L.; Liable-Sands, L. M.; Yap, G. P. A.; Rheingold, A. L. *Organometallics* **1997**, *16*, 3959.

(112) Luo, B.; Young, V. C., Jr.; Gladfelter, W. L. *Inorg. Chem.* **2000**, *39*, 1705.

(113) For a recent review see: Fischer, R. A.; Weiß, J. *Angew. Chem. Int. Ed. Engl.* **1999**, *38*, 2830.

(114) Several Lewis acid–base adducts of group 13 trialkyls $R_3M$ or trihalides $Cl_3M$ and transition metal complexes of the type $L_nFe=E=CR_2$ (E = P, As; Weber L.; Scheffer, M. H.; Stammler, H. G.; Stammler, A. *Eur. J. Inorg. Chem.* **1999**, 1607) and $L_nW\equiv P$ (Scheer, M.; Müller, J.; Baum, G.; Häser, M. *J. Chem. Soc. Chem. Commun.* **1998**, 1051) have been structurally characterized.

(115) Tessier-Youngs, C.; Bueno, C.; Beachley, O. T. Jr.; Churchill, M. R. *Inorg. Chem.* **1983**, *22*, 1054. Nöth et al. synthesized a comparable compound, $(Me_2N)_2BPPh_2$–$Cr(CO)_5$, but its solid state structure could not be verified by single crystal X-ray diffraction (Nöth, H.; Sze, S. N. *Z. Naturforsch. B: Anorg. Chem. Org. Chem.* **1978**, *33*, 1313).

(116) Beachley, O. T., Jr.; Banks, M. A.; Kopasz, J. P.; Rogers, R. D. *Organometallics* **1996**, *15*, 5170.

(117) Vogel, U.; Timoshkin, A. Y.; Scheer, M. *Angew. Chem. Int. Ed. Engl.* **2001**, *40*, 4409.

(118) (a) Thomas, F.; Schulz, S.; Nieger, M. *Organometallics* **2001**, *20*, 2405. (b) Thomas, F.; Schulz, S.; Nieger, M.; Nättinen, K. *Chem. Eur. J.* **2002**, *8*, 1915.

(119) The electronic Tolman parameter $\chi$ for a given ligand L is defined as the difference of the wavenumbers of the $A_1$ vibrations for a nickel carbonyl complex $[(L)Ni(CO)_3]$ compared to $t$-$Bu_3P$–$Ni(CO)_3$ containing the sterically hindered and weak $\pi$-accepting $t$-$Bu_3P$ ligand. Tolman, C. A. *Chem. Rev.* **1977**, *77*, 313. Examples of $\chi$ values are 13.25 for $PPh_3$ and 0.8 for the weak $\pi$-acceptor $P(SiMe_3)_3$. $\chi = \nu(A_1, [(L)Ni(CO)_3)] - \nu(A_1, [(tBu_3P) Ni(CO)_3)]$. Bartik, T.; Himmler, T.; Schulte, H.-G.; Seevogel, K. *J. Organomet. Chem.* **1985**, *293*, 343.

(120) Bodner, G. M.; May, M. P.; McKinney, L. E. *Inorg. Chem.* **1980**, *19*, 1951.

(121) Braga, D.; Grepioni, F.; Orpen, G. *Organometallics* **1993**, *12*, 1481.

(122) Chen, Y.; Hartmann, M.; Frenking, G. *Z. Anorg. Allg. Chem.* **2001**, *627*, 1985 and references cited herein.

(123) Cambridge Structural Database (ver. 5.21, April 2001). Allen F. H.; Kennard O. *Chem. Des. Autom. News* **1993**, *8*, 31.

(124) Weber, L.; Kirchhoff, R.; Boese, R. *J. Chem. Soc. Chem. Commun.* **1992**, 1182.

(125) Pickardt, J.; Rösch, L.; Schumann, H. *Z. Anorg. Allg. Chem.* **1976**, *426*, 66.

(126) (a) Carty, A. J.; Taylor, N. J.; Coleman, A. W.; Lappert, M. F. *J. Chem. Soc. Chem. Commun.* **1979**, *639*. (b) Plastas, H. J.; Steward, J. M.; Grim, S. O. *Inorg. Chem.* **1973**, *12*, 265.

(127) Weber, L.; Scheffer, M. H.; Stammler, H.-G.; Neumann, B. *Eur. J. Inorg. Chem.* **1998**, 55.

(128) See for example: van Wüllen, C. *J. Comput. Chem.* **1997**, *18*, 1985 and references cited herein.

(129) (a) Anton, K.; Fusstetter, H.; Nöth, H. *Chem. Ber.* **1981**, *114*, 2723. (b) Anton, K.; Nöth, H. *Chem. Ber.* **1982**, *115*, 2668. (c) Anton, K.; Euringer, C.; Nöth, H. *Chem. Ber.* **1984**, *117*, 1222. (d) Anton, K.; Nöth, H.; Pommerening, H. *Chem. Ber.* **1984**, *117*, 2495. (e) Hellmann, K. W.; Bergner, A.; Gade, L. H.; Scowen, I. J.; McPartlin, M. *J. Organomet. Chem.* **1999**, *573*, 156. (f) Robinson, G. H.; Pennington, W. T.; Lee, B.; Self, M.; Hrncir, D. C. *Inorg. Chem.* **1991**, *30*, 809.

(130) Schulz, S.; Thomas F.; Nieger M. *Organometallics*, submitted.

(131) Mitzel, N. W.; Lustig, C.; Berger, R. J. F.; Runeberg, N. *Angew. Chem. Int. Ed. Engl.* **2002**, *41*, 2519. According to a single crystal X-ray analysis, $GaMe_3$ forms a tetramer in the solid state. Each Ga center is coordinated by three Me-groups and by a Me-group of an adjacent $GaMe_3$ molecule. Ga$\cdots$H distance range from 295 to 296 pm and the Ga–C bond length was determined to 313.4 pm.

(132) (a) Jegier, J. A.; McKernan, S.; Gladfelter, W. L. *Inorg. Chem.* **1999**, *38*, 2726. (b) Jegier, J. A.; McKernan, S.; Purdy, A. P.; Gladfelter, W. L. *Chem. Mater.* **2000**, *12*, 1003.

(133) (a) Fisher, J. D.; Shapiro, P. J.; Yap, G. P. A.; Rheingold, A. L. *Inorg. Chem.* **1996**, *35*, 271. (b) Wehmschulte, R. J.; Power, P. P. *J. Am. Chem. Soc.* **1996**, *118*, 791. Very recently, we synthesized Lewis base-stabilized dimeric iminoalanes (Bauer, T.; Schulz, S.; Hupfer, H.; Nieger, M. *Organometallics* **2002**, *21*, 2931) and von Hänisch synthesized comparable Ga–P and Ga–As compounds (Hänisch, C. *Z. Anorg. Allg. Chem.* **2001**, *627*, 68).

(134) Waggoner, K. M.; Hope, H.; Power, P. P. *Angew. Chem. Int. Ed. Engl.* **1988**, *27*, 1699. Six-membered heterocycles $[RMER']_3$ have been subject to detailed preparative and theoretical studies since they are isoelectronic to borazine $B_3N_3H_6$. However, it was demonstrated that only $B_3P_3$-heterocycles exhibit a considerable delocalization of the 6 $\pi$-electrons, whereas the delocalization in other six-membered heterocycles of the desired type contributes very little to their stabilization. Power P. P. *J. Organomet. Chem.* **1990**, *400*, 49.

(135) See the following and references cited: (a) Schnitter, C.; Waezsada, S. D.; Roesky, H. W.; Teichert, M.; Usón, I.; Parisini, E. *Organometallics* **1997**, *16*, 1197. (b) Waggoner, K. M.; Power, P. P. *J. Am. Chem. Soc.* **1991**, *113*, 3385. (c) Belgardt, T.; Roesky, H. W.;

Noltemeyer, M.; Schmidt, H.-G. *Angew. Chem. Int. Ed. Engl.* **1993**, *32*, 1056. (d) Kühner, S.; Kuhnle, R.; Hausen, H.-D.; Weidlein, J. *Z. Anorg. Allg. Chem.* **1997**, *623*, 25.

(136) Veith, M. *Chem. Rev.* **1990**, *90*, 3.

(137) Cowley, A. H.; Jones, R. A.; Mardones, M. A.; Atwood, J. L.; Bott, S. G. *Angew. Chem. Int. Ed. Engl.* **1990**, *29*, 1409.

(138) (a) Hope, H.; Pestana, D. C.; Power, P. P. *Angew. Chem. Int. Ed. Engl.* **1991**, *30*, 691. (b) Niediek, K.; Neumüller, B. *Chem. Ber.* **1994**, *127*, 67. (c) Cowley, A. H.; Jones, R. A.; Mardones, M. A.; Ruiz, J.; Atwood, J. L.; Bott, S. G. *Angew. Chem. Int. Ed. Engl.* **1990**, *29*, 1150. (d) Atwood, D. A.; Cowley, A. H.; Jones, R. A.; Mardones, M. A. *J. Am. Chem. Soc.* **1991**, *113*, 7050.

(139) (a) Cucinella, S.; Salvatori, T.; Busetto, C.; Perego, G.; Mazzei, A. *J. Organomet. Chem.* **1974**, *78*, 185. (b) Nöth, H.; Wolfgardt, P. *Z. Naturforsch. B* **1976**, *31*, 697.

(140) See also: (a) Cesari M.; Cuccinella S. The Chemistry of Inorganic Homo- and Heterocycles, Eds. Haiduc I.; Sowerby D. B., Academic Press (London), 1987, Vol. 1, Chapter 6. (b) Robinson G. H. Coordination Chemistry of Aluminum, Ed. Robinson G. H., VCH (New York), 1993, Chapter 2.

(141) Perego, G.; Del Piero, G.; Cesari, M.; Zazzetta, A.; Dozzi, G. *J. Organomet. Chem.* **1975**, *87*, 53.

(142) See the following and references cited therein: (a) Driess, M.; Kuntz, S.; Monsé, C.; Merz, K. *Chem. Eur. J.* **2000**, *6*, 4343. (b) Driess, M.; Kuntz, S.; Merz, K.; Pritzkow, H. *Chem. Eur. J.* **1998**, *4*, 1628. (c) Styron, E. K.; Lake, C. H.; Powell, D. H.; Krannich, L. K.; Watkins, C. L. *J. Organomet. Chem.* **2002**, *649*, 78. (d) Luo, B.; Gladfelter, W. L. *Inorg. Chem.* **2002**, *41*, 590. (e) Schmid, K.; Niemeyer, M.; Weidlein, J. *Z. Anorg. Allg. Chem.* **1999**, *625*, 186. (f) Park, J. E.; Bae, B.-J.; Kim, Y.; Park, J. T.; Suh, I.-H. *Organometallics* **1999**, *18*, 1059. (g) Cesari, M.; Perego, G.; Del Piero, G.; Cucinella, S.; Cernia, E. *J. Organomet. Chem.* **1974**, *78*, 203. (h) Del Piero, G.; Perego, G.; Cucinella, S.; Cesari, M.; Mazzei, A. *J. Organomet. Chem.* **1977**, *136*, 13.

(143) Hitchcock, P. B.; Smith, J. D.; Thomas, K. M. *J. Chem. Soc. Dalton Trans.* **1976**, 1433.

(144) (a) Del Piero, G.; Cesari, M.; Perego, G.; Cucinella, S.; Cernia, E. *J. Organomet. Chem.* **1977**, *129*, 289. (b) Amirkhalili, S.; Hitchcock, P. B.; Smith, J. D. *J. Chem. Soc. Dalton Trans.* **1979**, 1206.

(145) (a) Schulz, S.; Häming, L.; Herbst-Irmer, R.; Roesky, H. W.; Sheldrick, G. M. *Angew. Chem. Int. Ed. Engl.* **1994**, *33*, 969. (b) Schulz, S.; Voigt, A.; Roesky, H. W.; Häming, L.; Herbst-Irmer, R. *Organometallics* **1996**, *15*, 5252.

(146) von Hänisch, C.; Scheer, P.; Rolli, B. *Eur. J. Inorg. Chem.* **2002**, 3268.

(147) It should be stated that reactions of $GaCl_3$ with $Sb(SiMe_3)_3$ resulted in the formation of nanocrystalline GaSb. (a) Schulz, S.; Martinez, L.; Ross J. L. *Adv. Mater. Optics Electron.* **1996**, *6*, 185. (b) Baldwin, R. A.; Foos, E. E.; Wells, R. L. *Mater. Res. Bull.* **1997**, *32*, 159. (c) Schulz, S.; Assenmacher, W. *Mater. Res. Bull.* **1999**, *34*, 2053.

(148) von Hänisch, C. *Z. Anorg. Allg. Chem.* **2001**, *627*, 68.

(149) von Hänisch, C.; Weigend, F. *Z. Anorg. Allg. Chem.* **2002**, *628*, 389.

(150) Bauer, T.; Schulz, S.; Hupfer, H.; Nieger, M. *Organometallics* **2002**, *21*, 2931.

(151) (a) Hannay, N. B. Semiconductors, Reingold Publishing Corporation, New York, 1959. (b) Madelung, O. Semiconductors, Group IV Elements and III–V Compounds, Springer-Verlag, Berlin, 1991. (c) Jones, A. C.; Whitehouse, C. R.; Roberts, J. S. *Chem. Vap. Deposition* **1995**, *1*, 65.

(152) Manasevit, H. *Appl. Phys. Lett.* **1968**, *12*, 156.

(153) (a) Polyakov, A. Y.; Stam, M.; Milnes, A. G.; Wilson, R. G.; Fang, Z. Q.; Rai-Choudhury, P.; Hillard, R. J. *J. Appl. Phys.* **1992**, *72*, 131. (b) Bolognesi, C. R.; Caine, E. J.; Kroemer, H. *IEEE Electron Device Lett.* EDL-15 **1994**, 16. (c) Samoska, L.; Brar,

B.; Kroemer, H. *Appl. Phys. Lett.* **1993**, *62*, 2539. (d) Zhang, Y.; Baruch, N.; Wang, W. I. *Appl. Phys. Lett.* **1993**, *63*, 1068. (e) Choi, H. K.; Turner, G. W. *Appl. Phys. Lett.* **1995**, *67*, 332. (f) Zhang, Y.-H. *Appl. Phys. Lett.* **1995**, *66*, 118. (g) Levi, A. F. J.; Chiu, T. H. *Appl. Phys. Lett.* **1987**, *51*, 984. (h) Lee, H.; York, P. K.; Menna, R. J.; Martinelli, R. U.; Garbuzov, D. Z.; Narayan, S. Y.; Connolly, J. C. *Appl. Phys. Lett.* **1995**, *66*, 1942.

(154) For an excellent review see the following and the references cited therein: Aardvark, A.; Mason, N. J.; Walker, P. J. *Prog. Cryst. Growth Charact.* **1997**, *35*, 207.

(155) See for example: (a) Subekti, A.; Goldys, E. M.; Paterson, M. J.; Drozdowics-Tomsia, K.; Tansley, T. L. *J. Mater. Res.* **1999**, *14*, 1238. (b) Alphandéry, E.; Nicholas, R. J.; Mason, N. J.; Zhang, B.; Möck, P.; Booker, G. R. *Appl. Phys. Lett.* **1999**, *74*, 2041. (c) Yi, S. S.; Hansen, D. M.; Inoki, C. K.; Harris, D. L.; Kuan, T. S.; Kuech, T. F. *Appl. Phys. Lett.* **2000**, *77*, 842. (d) Müller-Kirsch, L.; Pohl, U. W.; Heitz, R.; Kirmse, H.; Neumann, W.; Bimberg, D. *J. Cryst. Growth* **2000**, *221*, 611. (e) Yi, S. S.; Moran, P. D.; Zhang, X.; Cerrina, F.; Carter, J.; Smith, H. I.; Kuech, T. F. *Appl. Phys. Lett.* **2001**, *78*, 1358. (f) Agert C.; Lanyi P.; Bett A. W. *J. Cryst. Growth* **2001**, *225*, 426.

(156) (a) Stauff, G. T.; Gaskill, D. K.; Bottka, N.; Gedridge, R. W., Jr. *Appl. Phys. Lett.* **1991**, *58*, 1311. (b) Chen, C. H.; Stringfellow, G. B.; Gordon, D. C.; Brown, D. W.; Vaartstra, B. A. *Appl. Phys. Lett.* **1992**, *61*, 204. (c) Grahem, R. M.; Jones, A. C.; Mason, N. J.; Rushworth, S. A.; Smith, L.; Walker, P. J. *J. Cryst. Growth* **1994**, *145*, 363. (d) Chen, C. H.; Chiu, C. T.; Stringfellow, G. B.; Gedridge, R. W., Jr. *J. Cryst. Growth* **1992**, *124*, 88. (e) Biefeld, R. M.; Allerman, A. A.; Kurtz, S. R. *J. Cryst. Growth* **1997**, *174*, 593. (f) Cao, D. S.; Fang, Z. M.; Stringfellow, G. B. *J. Cryst. Growth* **1991**, *113*, 441. (g) Wang, C. W.; Finn, M. C.; Salim, S.; Jensen, K. F.; Jones, A. C. *Appl. Phys. Lett.* **1995**, *67*, 1384. (h) Wang, C. W.; Jensen, K. F.; Jones, A. C.; Choi, H. K. *Appl. Phys. Lett.* **1996**, *68*, 400. (i) Biefeld, R. M.; Allerman, A. A.; Pelczynski, M. W. *Appl. Phys. Lett.* **1996**, *68*, 932.

(157) (a) Shin, J.; Verma, A.; Stringfellow, G. B.; Gedridge, R. W. *J. Cryst. Growth* **1994**, *143*, 15. (b) Shin, J.; Hsu, Y.; Stringfellow, G. B.; Gedridge, R. W., Jr. *J. Electron. Mater.* **1995**, *24*, 1563. (c) Chun, Y. S.; Stringfellow, G. B.; Gedridge, R. W., Jr. *J. Electron. Mater.* **1996**, *25*, 1539.

(158) See the following and the references cited therein: (a) Agert, C.; Lanyi, P.; Bett, A. W. *J. Cryst. Growth* **2001**, *225*, 426. (b) Giesen, C.; Szymakowski A.; Rushworth, S.; Heuken, M.; Heime, K. *J. Cryst. Growth* **2000**, *221*, 450.

(159) (a) Leroux, M.; Tromson-Carli, A.; Gibart, P.; Vérié, C.; Bernard, C.; Schouler, M. C. *J. Cryst. Growth* **1980**, *48*, 367. (b) Cao, D. S.; Fang, Z. M.; Stringfellow, G. B. *J. Cryst. Growth* **1991**, *113*, 441. (c) Jaw, D. H.; Cao, D. S.; Stringfellow, G. B. *J. Appl. Phys.* **1991**, *69*, 2552. (d) Chen, W.-K.; Ou, J.; Lee, W.-I. *Jpn. J. Appl. Phys.* **1994**, *33*, L402.

(160) (a) Biefeld, R. M.; Allerman, A. A.; Kurtz, S. R. *J. Cryst. Growth* **1997**, *174*, 593. (b) Wang, C. W.; Finn, M. C.; Salim, S.; Jensen, K. F.; Jones, A. C. *Appl. Phys. Lett.* **1995**, *67*, 1384. (c) Wang, C. W.; Jensen, K. F.; Jones, A. C.; Choi, H. K. *Appl. Phys. Lett.* **1996**, *68*, 400.

(161) Hwang, J.-W.; Hanson, S. A.; Britton, D.; Evans, J. F.; Jensen, K. F.; Gladfelter, W. L. *Chem. Mater.* **1990**, *2*, 342.

(162) (a) MacInnes, A. N.; Power, M. B.; Barron, A. R. *Chem. Mater.* **1992**, *4*, 11. (b) MacInnes, A. N.; Power, M. B.; Barron, A. R. *Chem. Mater.* **1993**, *5*, 1344.

(163) Gysling, H. J.; Wernberg, A. A.; Blanton, T. N. *Chem. Mater.* **1992**, *4*, 900.

(164) Staring, E. G. J.; Meekes, G. J. B. M. *J. Am. Chem. Soc.* **1989**, *111*, 7648.

(165) Sauls, F. C.; Interrante, L. V.; Jiang, Z. P. *Inorg. Chem.* **1990**, *29*, 2989.

(166) (a) Jones, A. C.; Rushworth, S. A.; Houlton, D. J.; Roberts, J. S.; Roberts, V.; Whitehouse, C. R.; Critchlow, G. W. *Chem. Vapor Deposition* **1996**, *2*, 5. (b) Interrante, L. V.; Lee, W.; McConnell, M.; Lewis, N.; Hall, E. *J. Electrochem. Soc.* **1989**, *136*, 472.

(167) Park, H. S.; Schulz, S.; Wessel, H.; Roesky, H. W. *Chem. Vap. Deposition* **1999**, *5*, 179.

(168) Kuczkowski, A.; Schulz, S.; Assenmacher, W. *J. Mater. Chem.* **2001**, *11*, 3241.

(169) Jones, A. C.; O'Brien, P. CVD of Compound Semiconductors **1997**, p. 146, VCH, Weinheim.

(170) GaSb, space group *F*-43*m*, *a* = 609.5 pm; Natl. Bur. Stand. (U.S.), Circ. 539 6 30 (1956) JCPDF-Card P070215).

(171) The VLS mechanism was confirmed at the nanometer scale by direct, *in situ* observation of nano-wire growth in a transmission electron microscope at high temperatures. Wu, Y.; Yang, P. *J. Am. Chem. Soc.* **2001**, *123*, 3165.

# Index

319

# Cumulative List of Contributors for Volumes 1–36

Abel, E. W., **5**, 1; **8**, 117
Aguiló. A., **5**,321
Akkerman, O. S., **32**, 147
Albano, V. G., **14**, 285
Alper, H., **19**, 183
Anderson, G. K., **20**, 39; **35**, 1
Angelici, R. J., **27**, 51
Aradi, A. A., **30**,189
Armitage, D. A., **5**, 1
Armor, J. N., **19**, 1
Ash, C. E., **27**, 1
Ashe, A. J., III. **30**, 77
Atwell, W. H., **4**, 1
Baines. K. M., **25**, 1
Barone, R., **26**, 165
Bassner, S. L., **28**, 1
Behrens, H., **18**, 1
Bennett, M. A., **4**, 353
Bickelhaupt, F., **32**, 147
Binningham, J., **2**, 365
Blinka, T. A., **23**, 193
Bockman, T. M., **33**, 51
Bogdanović, B., **17**, 105
Bottomley, F., **28**, 339
Bradley, J. S., **22**, 1
Brew, S. A., **35**, 135
Brinckman, F. E., **20**, 313
Brook. A. G.. 7, 95; **25**, 1
Bowser, J. R., **36**, 57
Brown, H. C., **11**, 1
Brmcn, T. L., **3**, 365
Bruce, M. 1., **6**, 273, **10**, 273; **11**, 447: **12**, 379: **22**, 59
Brunner, H., **18**, 151
Buhro, W. E., **27**, 311
Byers, P. K., **34**, 1
Cais, M., **8**, 211
Calderon, N., **17**, 449
Callahan, K. P., **14**, 145
Canty, A. J., **34**, 1
Cartledge, F. K., **4**, 1
Chalk, A. J., **6**, 119
Chanon, M., **26**, 165

Chatt, J., **12**, 1
Chini, P., **14**,285
Chisholm, M. H., **26**, 97; **27**,311
Chiusoli, G. P., **17**, 195
Chojinowski, J., **30**, 243
Churchill, M. R., **5**, 93
Coates, G. E., **9**, 195
Collman, J. P., **7**,53
Compton, N. A., **31**, 91
Connelly, N. G., **23**, 1; **24**, 87
Connolly, J. W., **19**, 123
Corey, J. Y., **13**, 139
Corriu, R. J. P., **20**, *265*
Courtney, A., **16**, 241
Coutts, R. S. P., **9**, 135
Coville, N. J., **36**, 95
Coyle. T. D., **10**, 237
Crabtree, R. H., **28**, 299
Craig, P J., **11**, 331
Csuk, R., **28**, 85
Cullen, W R., **4**, 145
Cundy, C. S., **11**, 253
Curtis, M. D., **19**, 213
Darensbourg, D. J., **21**, 113; **22**, 129
Darensbourg, M. Y., **27**, 1
Davies, S. G., **30**, 1
Deacon, G. B., **25**, 337
de Boer, E., **2**, 115
Deeming, A. J., **26**, 1
Dessy, R. E., **4**, 267
Dickson, R. S., **12**, 323
Dixneuf, P H., **29**, 163
Eisch, J. J., **16**, 67
Ellis, J. E., **31**, 1
Emerson, G. F., **1**, 1
Epstein, P. S., **19**, 213
Erker. G., **24**, 1
Ernst. C. R., **10**, 79
Errington, R. J., **31**, 91
Evans, J., **16**, 319
Evan, W. J., **24**, 131
Faller, J. W., **16**, 211
Farrugia, L. J., **31**, 301

# Cumulative Index
## for Volumes 37–49